COMPUTATIONAL AND NUMERICAL CHALLENGES IN ENVIRONMENTAL MODELLING

STUDIES IN COMPUTATIONAL MATHEMATICS 13

Editors:

C.K. CHUI
Stanford University
Stanford, CA, USA

L. WUYTACK
University of Antwerp
Antwerp, Belgium

ELSEVIER

Amsterdam – Boston – Heidelberg – London – New York – Oxford – Paris
San Diego – San Francisco – Singapore – Sydney – Tokyo

COMPUTATIONAL AND NUMERICAL CHALLENGES IN ENVIRONMENTAL MODELLING

Zahari ZLATEV
National Environmental Research Institute
Frederiksborgvej 399, PO Box 358
DK-4000 Roskilde, Denmark

Ivan DIMOV
ACET Centre, University of Reading
Whiteknights, PO Box 217
Reading RG6 6AH, U.K.

on leave from
Institute of Parallel Processing
Bulgarian Academy of Sciences, Acad. G. Bonchev 25A
1113 Sofia, Bulgaria

ELSEVIER

Amsterdam – Boston – Heidelberg – London – New York – Oxford – Paris
San Diego – San Francisco – Singapore – Sydney – Tokyo

Elsevier
Radarweg 29, PO Box 211, 1000 AE Amsterdam, The Netherlands
The Boulevard, Langford Lane, Kidlington, Oxford OX5 1GB, UK

First edition 2006

Library of Congress Cataloging-in-Publication Data
A catalog record for this book is available from the Library of Congress

British Library Cataloguing in Publication Data
A catalogue record for this book is available from the British Library

ISBN-13: 978-0-444-52209-2
ISBN-10: 0-444-52209-3
ISSN: 1570-579X

Printed and bound in The Netherlands

06 07 08 09 10 10 9 8 7 6 5 4 3 2 1

To our wives and children

Preface

The protection of our environment is one of the most important problems facing modern society. The importance of this problem has been steadily increasing during the last three to four decades, and protecting the environment is becoming even more important now, in the 21st century. Reliable and robust control strategies for keeping the pollution caused by harmful chemical compounds below certain safe levels have to be developed and used in a routine way. Large mathematical models, in which all of the important physical and chemical processes are adequately described, can successfully be used to support this task. However, the use of large-scale mathematical models in which all of the important physical and chemical processes are adequately described leads, after the application of appropriate discretization and splitting procedures, to the treatment of huge computational tasks. In a typical simulation one has to perform several hundred runs. In each of these runs one has to carry out several thousand time-steps and at each time-step one has to solve numerically systems of coupled ordinary differential equations containing up to several million equations. Therefore, it is difficult to treat such large mathematical models numerically even when fast modern computers are available. Combined research by specialists from the fields of

- environmental and ecological modelling,
- numerical analysis, and
- scientific computing

must be carried out in an attempt to resolve successfully the challenging computational problems that appear when comprehensive environmental studies are to be carried out.

In some areas of Europe, as well as in some other parts of the world, the pollution levels are so high that preventive actions are urgently needed. Therefore, robust and reliable control strategies must be developed in order to find out where and by how much the emissions of harmful pollutants should be reduced. The solutions found must be optimal (or, at least, close to optimal), because the reduction of the emissions is as a rule an expensive process. This means that great economical problems may appear when the task of optimal reduction of the emissions

- The systems of ordinary differential equations (ODEs), which arise during the treatment the chemical part of the air pollution models, can efficiently be handled by using partitioning procedures. The use of partitioning procedures (and mainly the theoretical justification of using partitioning procedures during the treatment of large-scale applications, which are not necessarily air pollution models) is described in detail in the **fifth chapter**.

- The application of numerical methods in the treatment of comprehensive air pollution models on computers leads to different kinds of matrix computations. It is very important to carry out all these computations in an efficient way. The efficient performance of the matrix computations during the treatment of the air pollution models is described in the **sixth chapter**. Efficient treatment of *very large* sparse matrices (in this book *"very large"* means that the matrices are of order $N > 10^6$) is a great computational challenge. The efficient treatment of very large sparse matrices is also discussed in the sixth chapter.

- While it is clear that the use of parallel computers can improve significantly the performance of the code, it is well known that the efficient use of a parallel computer is not an easy task when large-scale applications are to be run in parallel. Again, this statement holds not only for large-scale air pollution models, but also for large-scale models arising in other fields of science and engineering. The use of standard tools for achieving both efficiency and portability when large-scale air pollution models are run on parallel computers is discussed in the **seventh chapter**. The ideas are fairly general and, therefore, can also be applied when large-scale models arising in some other areas are to be treated on parallel computers.

- Some typical applications related to different environmental studies (including here the important study of the influence of the biogenic emissions on the high ozone levels that can cause damages on plants, animals and/or human health) are shortly described in the **eighth chapter**.

- The future climate changes due to the green house effect have been the major topic of many studies in the last two-three decades. In the **ninth chapter** we study the impact of future climate changes on pollution levels.

- Data assimilation is becoming more and more popular when large-scale mathematical models arising in different fields of science and engineering are handled. Different problems related to the implementation of data assimilation algorithms are discussed in the **tenth chapter**.

- There are still many open problems related to environmental and ecological modelling. Some of these open problems are discussed in the **eleventh chapter** of the book.

Preface

The protection of our environment is one of the most important problems facing modern society. The importance of this problem has been steadily increasing during the last three to four decades, and protecting the environment is becoming even more important now, in the 21st century. Reliable and robust control strategies for keeping the pollution caused by harmful chemical compounds below certain safe levels have to be developed and used in a routine way. Large mathematical models, in which all of the important physical and chemical processes are adequately described, can successfully be used to support this task. However, the use of large-scale mathematical models in which all of the important physical and chemical processes are adequately described leads, after the application of appropriate discretization and splitting procedures, to the treatment of huge computational tasks. In a typical simulation one has to perform several hundred runs. In each of these runs one has to carry out several thousand time-steps and at each time-step one has to solve numerically systems of coupled ordinary differential equations containing up to several million equations. Therefore, it is difficult to treat such large mathematical models numerically even when fast modern computers are available. Combined research by specialists from the fields of

- environmental and ecological modelling,
- numerical analysis, and
- scientific computing

must be carried out in an attempt to resolve successfully the challenging computational problems that appear when comprehensive environmental studies are to be carried out.

In some areas of Europe, as well as in some other parts of the world, the pollution levels are so high that preventive actions are urgently needed. Therefore, robust and reliable control strategies must be developed in order to find out where and by how much the emissions of harmful pollutants should be reduced. The solutions found must be optimal (or, at least, close to optimal), because the reduction of the emissions is as a rule an expensive process. This means that great economical problems may appear when the task of optimal reduction of the emissions is not correctly solved. Optimal (or nearly optimal) solutions can successfully be found only by performing long series of simulation experiments consisting of many hundreds of runs of several comprehensive mathematical models. Then the results obtained by different models must be carefully compared in order to answer the following questions:

- *are there any discrepancies, and*
- *what are the reasons for discrepancies?*

In the cases where the answer to the first question is positive, the needed corrections of some of the models have to be made and the simulation experiments must be repeated for the corrected models. This shows that the process of finding an optimal solution will in general be very long and, thus, efficiency of the codes is highly desirable. Achieving efficiency of the codes by selecting faster (but still sufficiently accurate) numerical algorithms and by improving the performance on different high speed computers will be the major topics of this book.

Running mathematical models in real time can be of essential and even life saving importance in the case of accidental, hazardous releases (as, for example, the accident at Chernobyl in 1986). Real time calculations are also needed in connection with short periods (episodes) with very high pollution levels, which might have damaging effects on human health. A special abatement strategy in such periods can be based on temporal reductions of emissions from specific sources; e.g. traffic regulations, the energy sector, large industries (such as petrochemical plants). Reliable mathematical models coupled with weather forecasting models are run operationally in order

- to predict the appearance of harmful pollution levels, and
- to make the right decision in such situations.

Running **comprehensive mathematical** models (in which all of the important physical and chemical processes are adequately described) in real time is another very difficult and very challenging problem that must urgently be solved.

Computer architectures are becoming faster and faster. Supercomputers that have top performance of several Tflops will become easily available in the near future. New and efficient numerical methods, by which the great potential power of the parallel computers can be better exploited, are also gradually becoming available. However, many unresolved problems are still remaining. One of the most important problems is **the great gap** between the top performance of a modern supercomputer and the speeds actually achieved when large application codes are run. The task of achieving high computational speeds close to the top performance of the computers available, is very difficult and, at the same time, very challenging for the comprehensive mathematical models (not only for the large-scale environmental models). This is also true for the heterogeneous computations, where it is possible to achieve very high computational speed when the computational process is properly organized.

Some non-realistic simplifying assumptions are always made, in all of the existing large-scale ecological and environmental models, in order to be able to treat them numerically on the computers that are available. Many such assumptions are no longer absolutely necessary, because both the computers and the numerical algorithms are much faster than before, and will become even faster in the near future. Therefore, it becomes more and more essential to describe all of the important physical and chemical processes in an adequate way (according to the present knowledge) and to try to solve the problems by exploiting extensively the modern computational and numerical tools.

This short analysis of the major tasks, which must be solved in the field of environmental and ecological modelling, shows clearly that there are great computational and numerical challenges in the attempts to resolve efficiently these tasks. The computational and numerical challenges related to the treatment of large-scale air pollution models will be the main topic of this book. However, the mathematical problems that have to be handled in the treatment of the large-scale air pollution models are rather general (systems of partial differential equations, systems of ordinary differential equations, systems of linear algebraic equations, etc.). Such mathematical problems appear also during the treatment of models arising in other fields in science and engineering. Therefore, the approaches discussed in this book might also be useful when different large-scale scientific models are treated on computers.

This book is divided into eleven chapters. The contents of the different chapters can shortly be described as follows.

- A general discussion of the systems of partial differential equations (PDEs), which arise in air pollution modelling and in several other fields of science and engineering, is presented in the **first chapter**.

- It is very difficult to handle directly the systems of PDEs, by which large-scale environmental and ecological models are described. Therefore, different splitting techniques are commonly used in the computer treatment of such models. The use of splitting techniques is discussed in the **second chapter**.

- One of the most important, and also one of the most difficult, processes involved in an air pollution model is the horizontal transport. The description of the horizontal transport by mathematical terms and some numerical algorithms that are used to treat these terms are discussed in the **third chapter**.

- General ideas about the treatment of the chemical part of an air pollution model are presented in the **fourth chapter**. The major numerical methods, which are used (can be used) in this part of the models, are also discussed in the **fourth chapter**. It is explained that the same technique can be used in connection with many other large-scale models.

- The systems of ordinary differential equations (ODEs), which arise during the treatment the chemical part of the air pollution models, can efficiently be handled by using partitioning procedures. The use of partitioning procedures (and mainly the theoretical justification of using partitioning procedures during the treatment of large-scale applications, which are not necessarily air pollution models) is described in detail in the **fifth chapter**.

- The application of numerical methods in the treatment of comprehensive air pollution models on computers leads to different kinds of matrix computations. It is very important to carry out all these computations in an efficient way. The efficient performance of the matrix computations during the treatment of the air pollution models is described in the **sixth chapter**. Efficient treatment of *very large* sparse matrices (in this book *"very large"* means that the matrices are of order $N > 10^6$) is a great computational challenge. The efficient treatment of very large sparse matrices is also discussed in the sixth chapter.

- While it is clear that the use of parallel computers can improve significantly the performance of the code, it is well known that the efficient use of a parallel computer is not an easy task when large-scale applications are to be run in parallel. Again, this statement holds not only for large-scale air pollution models, but also for large-scale models arising in other fields of science and engineering. The use of standard tools for achieving both efficiency and portability when large-scale air pollution models are run on parallel computers is discussed in the **seventh chapter**. The ideas are fairly general and, therefore, can also be applied when large-scale models arising in some other areas are to be treated on parallel computers.

- Some typical applications related to different environmental studies (including here the important study of the influence of the biogenic emissions on the high ozone levels that can cause damages on plants, animals and/or human health) are shortly described in the **eighth chapter**.

- The future climate changes due to the green house effect have been the major topic of many studies in the last two-three decades. In the **ninth chapter** we study the impact of future climate changes on pollution levels.

- Data assimilation is becoming more and more popular when large-scale mathematical models arising in different fields of science and engineering are handled. Different problems related to the implementation of data assimilation algorithms are discussed in the **tenth chapter**.

- There are still many open problems related to environmental and ecological modelling. Some of these open problems are discussed in the **eleventh chapter** of the book.

Difficult computational problems, which arise when large-scale scientific and engineering models have to be efficiently treated on computers, are discussed in this book. Efficiency is essential, because the computational tasks are normally huge. We are mainly discussing the basic ideas, which can be applied in the efforts to deal with the challenges when large-scale models are to be treated on modern computers, without going into details that are related to the numerical methods used. Including many details will make the book unnecessarily long and, what is even more important, the main topic will be diffused in long explanations. However, adequate references to all appropriate numerical methods are given and/or commented in the book. Moreover, some new methods, which have not yet been treated in detail elsewhere, are fully described. This is done, for example, when

- the stability control of the solution of the semi-discretized transport equations,
- the analysis of the error of the partitioned systems of ordinary differential equations arising from the chemical schemes used in air pollution models,
- some problems related to very large sparse matrices, and
- the problems arising in the implementation of data assimilation algorithms in large-scale mathematical problems

are studied. We are sure that the approach, which was sketched in this paragraph and which was used systematically in the preparation of all eleven chapters, will make our book easily understandable for specialists working in the following four fields:

- environmental modelling,
- scientific computing,
- applied mathematics, and
- numerical analysis.

It is important to illustrate the fact that the ideas described in the book can successfully be applied to develop different air pollution models which can be used in the solution of a large class of important applications, such as studying effects of climate changes on pollution levels in Europe, investigating the influence of natural emissions on high ozone levels, etc. The success of the chosen algorithms and devices in treating efficiently such applications is demonstrated in Chapter 8 and Chapter 9 of the book.

Finally, it should be emphasized, once again, that many of the particular algorithms and devices that are discussed in this book are applicable not only to large-scale air pollution models, but also for large scale problems arising in other fields of science and engineering.

Acknowledgements

Most of the numerical methods and results discussed in this book have been implemented after many discussions with specialists from

- the National Environmental Research Institute (Ruwim Bertkowicz, Jesper Christensen, Jørgen Brandt, Lise Frohn and Carsten Ambelas Skjøth),
- the Bulgarian Academy of Sciences (Krassimir Georgiev and Tzvetan Ostromsky),
- the Department of Applied Analysis of the University of Budapest (István Faragó and Ágnes Havasi),
- the Institute of Informatics and Mathematical Modelling at the Technical University of Denmark (Per Grove Thomsen, Per Christian Hansen and Vincent Alan Barker),
- the Computer Science Department of the University of Copenhagen (Stig Skelboe),
- the Computer Science Department of the Purdue University, Indiana, USA (Ahmed Sameh).

The authors should like to thank very much all of them.

All parallel methods were implemented at the computers of UNI-C (the Danish Center for Research and Education) and DCSC (the Danish Center for Scientific Computing). The specialists from these two institutions helped us to run the models or some modules of the models on different vector and parallel architectures. We should like to thank very much all of these specialists for their great help in different runs of the models on high performance computers; especially Bjarne S. Andersen, Jørgen Moth, Carl Niels Hansen, Bernd Dammann, Wojciech Owczarz and Jerzy Wasniewski.

The different mechanisms implemented in the physical and chemical parts of the model where developed in cooperation with many scientists from different European countries (the cooperation being documented in many scientific publications):

- Lars P. Prahm (director of the Danish Meteorological Institute),

- Dimiter Syrakov (from the Bulgarian Academy of Sciences),
- Adolf Ebel and his co-workers (from the University of Cologne, Germany),
- Heinz Hass and his co-workers (from the Ford Research Centre in Aachen, Germany),
- Anton Eliassen and Øystein Hov (from the Norwegian Meteorological Institute).

The authors should like to thank very much all of these scientists for the helpful discussions related to different issues from the field of large-scale air pollution modelling.

We are also very much obligated to Simon Branford from the University of Reading, UK, who read the manuscript and made a number of suggestions that improved the manuscript.

The research on many of the topics discussed in this book was a part of many different projects funded by

- the European Union (EU),
- the North Atlantic Treaty Organization (NATO),
- the Nordic Council of Ministers (NMR),
- the Danish Research Council,
- the National Science Foundation of Bulgaria,
- the Bulgarian Academy of Sciences,
- the Danish Centre for Scientific Computing (DCSC),
- the Danish Environmental Protection Agency.

The authors should like to thank very much all of these institutions for the financial support of our research.

Zahari Zlatev was invited to give a course on "Numerical and Computational Challenges in Environmental Modelling" at the Fields Institute for Research in Mathematical Sciences (University of Toronto, Canada) in 2002. Zahari Zlatev had many useful discussions there with Kenneth R. Jackson, Wayne Enright and other participants in the course. This book is based on the lectures given in Toronto.

Contents

Chapter 1

PDE systems arising in air pollution modelling and justification of the need for high speed computers

It is well-known that mathematics bridges gaps in culture, science and technology. This statement will be illustrated in this book

- by studying a large class of mathematical models arising in the area of air pollution modelling (as well as in mathematical models that arise in many other areas of science and engineering), and
- by discussing how long sequences of challenging problems from different scientific and engineering areas can be resolved by using powerful tools from the numerical mathematics and scientific computing during the treatment of advanced models in which all underlying physical and chemical processes are adequately described.

Air pollution, especially the reduction of the air pollution to some acceptable levels, is a highly relevant environmental problem, which is becoming more and more important. This problem can successfully be studied only when high-resolution comprehensive mathematical models, described by systems of partial differential equations (PDEs), are developed and used on a routine basis. However, such models are very time-consuming, even when modern high-speed computers are available. Indeed, if an air pollution model is to be applied on a large space domain by using fine grids, then its discretization will always lead to huge computational problems, even in the case when two-dimensional versions of the models are used. Assume, for example, that the two-dimensional space domain on which the model under consideration is defined is discretized by using a (480×480) grid and that the number of chemical species studied by the model is 35. Then several systems of ordinary differential equations (ODEs) containing 8064000 equations have to be treated at every time-step (the number of time-steps typically being several hundred thousand). If a three-dimensional version of the air pollution model is to be used, then the above quantity must be multiplied by the number of layers. Moreover, hundreds and even thousands of simulation runs have to be carried out in most of the studies related to policy making. Therefore, it is extremely difficult to treat such large computational problems. This is true even when the fastest computers that are available at present are used.

The computational time needed to run such a model causes, of course, the major difficulty. However, there is another difficulty which is at least as important as the problem with the computational time. The models need a great amount of input data (meteorological, chemical and emission data). Furthermore, the model produces huge files of output data, which have to be stored for future uses (for visualization and animation of the results). Finally, huge sets of measurement data (normally taken at many stations located in different countries) have to be used in the efforts to validate the model results.

The short discussion of the difficulties, which arise when large-scale mathematical models are treated on high-speed computers, shows clearly that both

- the computational time requirements, and
- the storage requirements

impose many challenging problems when large-scale air pollution models (or other large-scale models) are to be developed and used in different studies. A general discussion of the major numerical and computational challenges in air pollution modelling will be presented in this chapter.

After that, specific numerical and computational challenges will be studied in detail in the following chapters.

It must be emphasized here that many results from this book, results that are related to the numerical solution of systems of PDEs and/or ODEs, might be used in some other fields of science and engineering (several examples arising in other areas of environmental modelling are given in Section 7 of this chapter).

1.1 Need for large-scale air pollution modelling

The control of the pollution levels in different highly developed regions of Europe and North America (and also in other highly industrialized regions of the world) is an important task for modern society. Its relevance has been steadily increasing during the last two to three decades. The need to establish reliable control strategies for the air pollution levels will become even more important in the future. Large-scale air pollution models can successfully be used to design reliable control strategies. Many different tasks must be solved before starting to run operationally an air pollution model. The following tasks are most important:

1. Describe in an adequate way all important physical and chemical processes.
2. Apply fast and sufficiently accurate numerical methods in the different parts of the model.
3. Ensure that the model runs efficiently on modern high-speed computers (and, first and foremost, on different types of parallel computers).
4. Use high quality input data (both meteorological data and emission data) in the runs.
5. Verify the model results by comparing them with reliable measurements taken in different parts of the space domain of the model.
6. Carry out some sensitivity experiments to check the response of the model to changes of different key parameters.
7. Visualize and animate the output results to make them easily understandable even to non-specialists.

It is absolutely necessary to solve efficiently all these tasks. However, the treatment of such large-scale air pollution models, in which all relevant physical and chemical processes are adequately described, leads to great computational difficulties related both to the CPU time needed to run the models and to the input and output data used during the runs and saved for future applications. Therefore, it is necessary to give an answer to the question:

Why are large-scale comprehensive mathematical models needed (and also used) in many environmental studies?

A short, but also adequate, answer to this question can be given in the following way. It is well-known, by the broad public in Europe, North America and all other parts of the world, that the air pollution levels in a given region depend not only on the emission sources located in it, but also on emission sources located outside the region under consideration, and even on sources that are far away from the studied region. This is due to the transboundary transport of air pollutants. The atmosphere is the major medium where pollutants can quickly be transported over long distances. Harmful effects on plants, animals and humans can also occur in areas which are far away from big emission sources. Chemical reactions take place during the transport. This leads to the creation of secondary pollutants, which can also be harmful. The air pollution levels in densely populated and highly developed regions of the world, such as Europe and North America (but this will soon become true also for many other regions), must be studied carefully to find out how the air pollution can be reduced to safe levels and, moreover, to develop reliable control strategies by which air pollution can be kept under certain prescribed critical levels. The reduction of the air pollution is an expensive process. This is why one must try to optimize this process by solving two important tasks:

- the air pollution levels must be reduced to reliably determined critical levels (but no more if costs are high), and

- since the air pollution levels in different parts of a big region (as, for example, Europe or North America) are varying in a quite wide range, the optimal solution (a solution which is as cheap as possible) will require reducing the emissions by different amounts in different parts of a big region.

Reliable mathematical models, in which all relevant physical and chemical processes are adequately described (according to the available at present knowledge about these processes) are needed in the efforts to solve successfully these two tasks. Large comprehensive models for studying air pollution phenomena were first developed in North America: see, for example, Binkowski and Shankar [21], Carmichael and Peters [38], Carmichael et al. [39], [40], [41], Chang et al. [42], Dabdub and Seinfeld [51], Peters et. al. [194], Stockwell et al. [234] and Venkatram et al. [257]. There are several such large models in Europe: see Ackermann et al. [2], Aloyan [7], Builtjes [34], [35], D'Ambra et al. [57], Elbern et al. [91], Hongisto, [138], Hongisto et al. [139], Langner et al. [156], Jonson et al. [149], San Jose et al. [214], Simpson [222], [223], Tomlin et al. [250] and Zlatev [282].

The recent development of many of the existing large-scale air pollution models is described in several papers from the proceedings of two NATO Advanced Research Workshops:

- *"Large Scale Computations in Air Pollution Modelling"*, see Zlatev et al. [291], and

- *"Advances in Air Pollution Modelling for Environmental Security"*, see Faragó et al. [99].

A comprehensive review of latest results achieved in the efforts to improve the performance of the large-scale environmental models is given in a very well-written paper of Ebel, see [84].

The use of the Danish Eulerian Model (DEM), which is fully described in Zlatev [282] (see also the web-site of the model [284]), for the solution of the tasks listed above will be discussed in this book. However, **the particular choice of this model is made only in order to facilitate the understanding of the statements.** The ideas used in the development of all comprehensive mathematical models for studying air pollution phenomena are rather similar. Therefore, the consideration of any other large-scale air pollution model will in general lead to similar results and conclusions. It should be emphasized, once again, that some other applications from different fields of science and engineering are described mathematically by similar systems of PDEs and, thus, the methods and techniques used in the following chapters in relation to DEM can also be applied to such applications (this is demonstrated by discussing several examples, which are taken from other areas of environmental modelling, in Section 7 of this chapter).

1.2 The Danish Eulerian Model (DEM)

Some information about the last versions of the Danish Eulerian Model (DEM), which was developed at the National Environmental Research Institute (Roskilde, Denmark) and used in many environmental studies, will be presented in this and the following sections.

1.2.1 Some historical information

The work on the development of DEM has been initiated in the beginning of 80's. Only two species, sulphur dioxide and sulphate, were studied in the first version of the model. The space domain was rather large (the whole of Europe), but the discretization was very crude; the space domain was discretized by using of a (32×32) grid, which means, roughly speaking, that $(150\ km \times 150\ km)$ grid-squares were used. Both a two-dimensional version and a three-dimensional version of this model were developed. In both versions the chemical scheme was very simple. There was only one, **linear**, chemical reaction: it was assumed that a fixed part of the sulphur dioxide concentration is transformed, at each time step, into sulphate concentration. The two-dimensional version of this model is fully described in Zlatev [277], while the three-dimensional version was discussed in Zlatev et al. [288].

This first version was gradually upgraded to much more advanced versions by carrying out successively the following actions:

1. Introducing physical mechanisms in a more adequate way.

2. Increasing the number of chemical compounds which can be studied by the model (chemical schemes with 35, 56 and 168 compounds were implemented).

3. Improving the spatial resolution of the model by transition from the original (32×32) grid to a (96×96) grid, a (288×288) grid and a (480×480) grid, i.e. by reducing the $(150\ km \times 150\ km)$ grid-squares used in the original versions to $(50\ km \times 50\ km)$, $(16.667\ km \times 16.667\ km)$ and $(10\ km \times 10\ km)$ grid-squares.

4. Developing more advanced three-dimensional versions of the model with ten vertical layers.

5. Attaching new and faster numerical algorithms to the different parts of the model.

6. Preparing the model for efficient runs on different high-speed computers (both vector processors and parallel computer architectures).

DEM is well-structured and the above changes have been made by replacing only one module at a time. In this way the response of the model to different changes was also studied. Different versions of the model obtained during the upgrading procedure were described and discussed in many publications; see, for example, Zlatev [282], Zlatev et al. [294] and [295].

Results, obtained in comparisons of the concentrations and depositions calculated by different versions of the model with measurements taken at stations located in different European countries, were presented in Zlatev [282] and Zlatev et al. [292], [293]. Some comparisons of model results with measurements taken over sea were also carried out; see Harrison et al. [128].

Different versions of the model were also used in long simulations connected with studies of the relationships between high ozone concentrations in Europe and related emissions (NO_x emissions and VOC emissions). Many results, which were obtained in these simulations, were discussed in Ambelas Skjøth et al. [9], Bastrup-Birk et al. [18] and Zlatev et al. [297].

The improvement of the physical and chemical mechanisms used in the model demanded an improvement also of the numerical algorithms used in the computer treatment of the model. Furthermore, high-speed computers became necessary. It is normally very difficult to achieve a good performance on the new modern computer architectures. A long and careful work, which is still carried out, is necessary in order to achieve a high computational speed when the model is run on such architectures. On the other hand, it will be impossible to carry out several hundreds (and in many cases even several thousands) of runs with DEM in the long simulation process when its code does not run very efficiently on the computers available. Different procedures used in the attempts to achieve high computational speeds on different computers are presented in Brown et al. [32], Georgiev and Zlatev [116], [117], Owczarz and Zlatev [188], [189], [190] and Zlatev [278], [280], [282].

Some more details about the last versions of DEM will be given in the following sub-sections. Recently, the different versions of DEM were united in a flexible and powerful model, the Unified Danish Eulerian Model (UNI-DEM). UNI-DEM will be discussed in Chapter 7.

1.2.2 Chemical species treated in the model

The European pollution levels of all important chemical pollutants can be studied by using DEM. The chemical species involved in the model are:

- sulphur pollutants,
- nitrogen pollutants,
- ozone,
- ammonia-ammonium,
- several radicals, and
- a large number of relevant hydrocarbons.

1.2.3 Mathematical formulation of the model

DEM, as many other large-scale air pollution models, is described mathematically by a system of partial differential equations (**PDEs**):

$$
\begin{aligned}
\frac{\partial c_s}{\partial t} &= -\frac{\partial (u c_s)}{\partial x} - \frac{\partial (v c_s)}{\partial y} - \frac{\partial (w c_s)}{\partial z} \\
&\quad + \frac{\partial}{\partial x}\left(K_x \frac{\partial c_s}{\partial x}\right) + \frac{\partial}{\partial y}\left(K_y \frac{\partial c_s}{\partial y}\right) + \frac{\partial}{\partial z}\left(K_z \frac{\partial c_s}{\partial z}\right) \\
&\quad + E_s + Q_s(c_1, c_2, \ldots, c_{N_s}) \\
&\quad - (\kappa_{1s} + \kappa_{2s}) c_s, \\
&\quad s = 1, 2, \ldots, N_s.
\end{aligned}
\tag{1.1}
$$

The different quantities that are involved in the mathematical model have the following meaning:

- the concentrations are denoted by c_s,
- u, v and w are wind velocities,
- K_x, K_y and K_z are diffusion coefficients,
- the emission sources in the space domain are described by the functions E_s,
- κ_{1s} and κ_{2s} are deposition coefficients,
- the chemical reactions used in the model are described by the non-linear functions $Q_s(c_1, c_2, \ldots, c_{N_s})$.

The number of equations N_s is equal to the number of chemical species that are involved in the model and varies in different studies. Until now, the model has mainly been used with a chemical scheme containing 35 species (it may be necessary to involve more species in the future; experiments with chemical schemes containing 56 and 168 species have recently been carried out). The chemical scheme with 35 species is the well-known **CBM IV** scheme (described in Gery et al. [118]) with a few enhancements which have been introduced in order to make it possible to use the model in studies concerning the distribution of ammonia-ammonium concentrations in Europe as well as the transport of ammonia-ammonium concentrations from Central and Western Europe to Denmark. Recently, some other chemical schemes were developed and successfully tested in different models; see, for example, Stockwell et al. [233]. There are plans to implement some of these new chemical mechanisms in DEM in the near future.

The number of chemical species used in different large air pollution models varies from 20 to about 200 (see, for example, Borrell et al. [22], Carmichael and Peters [38], Carmichael et al. [39], Chang et al. [42], Ebel et al. [85], Hass et al. [129], [130], Peters et. al. [194], Stockwell et al. [234] and Venkatram et al. [257]). The use of less than 20 chemical species will require crude parameterization of some chemical processes and, thus, the use of such chemical schemes is in general not advisable, when the accuracy requirements that are to be satisfied are not very crude. On the other hand, the use of more than 200 chemical species is connected with huge computational tasks that cannot be handled, at least when long-term simulations are to be carried out, on the computers available at present without imposing a sequence of simplifying assumptions, which will lead to a crude parameterization of the other physical processes involved in the air pollution models (advection, diffusion, deposition and emission) and/or to discretizations of the models on coarse grids. Thus, the chemical scheme containing 35 species seems to be a good choice. Nevertheless, as mentioned above, experiments with two other chemical schemes, containing 56 and 168 chemical species, are at present carried out.

It should be pointed out that although the use of more than 100 chemical species in the chemical scheme (but less than 200) leads at present, as mentioned above, to the necessity both

- to apply certain simplifying assumptions in the description of the physical and chemical mechanisms, and/or
- to apply coarse grids in the dsicretization of the space domain of the model under consideration,

the computers are becoming faster and faster and, therefore, such actions will probably not be necessary in the near future.

The non-linear functions Q_s from (1.1) can as a rule be rewritten in the following form:

$$Q_s(c_1, c_2, \ldots, c_{N_s}) = -\sum_{i=1}^{N_s} \alpha_{si} c_i + \sum_{i=1}^{N_s} \sum_{j=1}^{N_s} \beta_{sij} c_i c_j, \qquad (1.2)$$

where $s = 1, 2, ..., N_s$.

This is a special kind of non-linearity, but it is not obvious how to exploit this fact during the numerical treatment of the model.

It follows from the above description of the quantities involved in the mathematical model that all five physical processes (advection, diffusion, emission, deposition and chemical reactions) can be studied by using the system of **PDEs** described by (1.1). The most important processes, when the computations are considered, are the horizontal advection (the horizontal transport) and the chemical reactions. Kernels for these two parts of the model must be treated numerically by using both fast and sufficiently accurate algorithms.

1.2.4 Space domain

The space domain of the model (it contains the whole of Europe together with parts of Asia, Africa and the Atlantic Ocean) is discretized by using a (96×96) grid in the version of DEM which is operationally used at present. This means that

- Europe is divided into 9216 grid-squares, and
- the grid resolution is approximately $(50\ km \times 50\ km)$.

It has been mentioned above that some work is carried out at present to run DEM by using much higher resolution applying either a (288×288) grid or a (480×480) grid. If these two grids are used, then

- Europe is divided into 82944 grid-squares for the (288×288) grid and 230400 grid-squares for the (480×480) grid, and
- the grid resolution is approximately $(16.67\ km \times 16.67\ km)$ for the (288×288) grid and $(10\ km \times 10\ km)$ for the (480×480) grid.

It will be shown in the next sections that the high resolution versions of DEM impose severe storage requirements; both the input files and the output files that are needed in the runs become much larger than those used in the operational version. Also, of course, the computational time needed to run this version is increased very considerably. Nevertheless, some rather comprehensive studies were recently performed by using fine resolution options of DEM (see Zlatev and Syrakov [299], [300]).

1.2.5 Initial and boundary conditions

Appropriate initial and boundary conditions are needed. If initial conditions are available (for example from a previous run of DEM and/or another model), then these are read from the files where they are stored. If initial conditions are not available, then a five day start-up period is used to obtain initial conditions (i.e. the computations are started five days before the desired starting date with some background concentrations and the concentrations found at the end of the fifth day are actually used as starting concentrations).

The choice of lateral boundary conditions is in general very important. This issue has been discussed in Brost [31]. However, if the space domain is very large, then the choice of lateral boundary conditions becomes less important; which is stated on p. 2386 in Brost [31]: *"For large domains the importance of the boundary conditions may decline"*. The lateral boundary conditions are represented in DEM with typical background concentrations which are varied, both seasonally and diurnally. The use of background concentrations is justified by the facts that:

- the space domain is very large (approximately $4800\ km \times 4800\ km$), and

- the boundaries are located far away from the highly polluted regions (in the Atlantic Ocean, North Africa, Asia and the Arctic areas).

Nevertheless, it would perhaps be better to use values of the concentrations at the lateral boundaries that are calculated by a hemispheric or global model.

For some chemical species, as for example ozone, it is necessary to introduce some exchange with the free troposphere. Three such rules have been tested in Zlatev et al. [292]. The third rule described in Zlatev et al. [292] is at present used in DEM, because the experiments, which were carried out there, indicated that this rule performs in general better than the other two rules.

1.2.6 Need for splitting procedures

It is difficult to treat the system of **PDEs** (1.1) directly. This is the reason for using different kinds of splitting.

The mathematical model is **divided into several simpler sub-models** when a splitting procedure is applied. Consider any of these sub-models. Assume that the space domain is a parallelepiped which is discretized by using a grid with $N_x \times N_y \times N_z$ grid-points, where N_x, N_y and N_z are the numbers of the grid-points along the grid-lines parallel to the Ox, Oy and Oz axes. Assume further that the number of chemical species involved in the model is N_s. Finally, assume that the spatial derivatives in (1.1) are discretized by some numerical algorithm. Then the system of PDEs, by which the sub-model under consideration is represented, will be transformed into a system of ODEs (ordinary differential equations)

$$\frac{dg}{dt} = f(t, g), \tag{1.3}$$

where $g(t)$ is a vector-function with $N_x \times N_y \times N_z \times N_s$ components. Moreover, the components of function $g(t)$ are the concentrations (at time t) at all grid-points and for all species. The right-hand side $f(t, g)$ of (1.3) is also a vector function with $N_x \times N_y \times N_z \times N_s$ components which depends both on the particular discretization method used and on the concentrations of the different chemical species at the grid-points. If the space discretization method is fixed and if the concentrations are calculated (at all grid-points and for all species), then the right-hand side vector in (1.3) can also be calculated.

The number of systems of ODEs of type (1.3) is equal to the number of sub-models obtained when the selected splitting procedure is applied. If no splitting is used and the spatial derivatives in (1.1) are discretized, then only one system of type (1.3) will appear. The latter system and the systems of type (1.3) obtained by applying some splitting procedure are of the same order. Therefore, the first impression is that the computational work will be increased when splitting is used, because one has to handle several systems of ODEs instead of only one for the case where (1.1) is treated without splitting. However, this is in general not true, because the systems (1.3) induced by the selected splitting procedure are much simpler. In fact, these systems are normally formed of many independent relatively small sub-systems (see also Chapter 7). The advantage of solving several simpler and relatively small systems of type (1.3) when splitting is used has much greater effect on the efficiency of the solution process than the disadvantage that several such systems are to be handled instead of only one system in the case where (1.1) is solved without splitting. This is why splitting is used in **all** operational large-scale air pollution models.

Splitting procedures will be studied in the next chapter. The space discretization of the spatial derivatives in the different sub-models, which are obtained by using the selected splitting procedure, will be studied in detail in Chapter 3. The time discretization of the systems of ODEs which arise in the different sub-models will be studied in Chapter 3 to Chapter 6. The exploitation of the properties of different splitting procedures in the efforts to obtain efficient performance on parallel computers will be discussed in Chapter 7.

1.2.7 Need for high speed computers

The size of the systems, which arise after the space discretization and the splitting procedures used to treat (1.1) numerically, is enormous. Consider the case where the model is two-dimensional. Let us assume that

- the model is discretized on a (96×96) grid (such a grid has been used in DEM after 1993), and
- $N_s = 35$.

As mentioned in Sub-section 1.2.6, the model is split to several sub-models by the selected splitting procedure. After the application of a semi-discretization, each sub-model is reduced to a system of ODEs. The number of equations in each of these systems of ODEs is $96 \times 96 \times 35 = 322560$. The time-step size used in the advection step is 15 min The chemical sub-model cannot be treated with such a large time-step size (because it is very stiff; especially when photochemical reactions are involved). Therefore six small chemical time-steps are carried out for each advection time-step (this means that the chemical time-step size is 2.5 min).

Assume, furthermore, that a one-year run (and, more precisely, 365 days + 5 days to start-up the computations) is to be carried out. This will result in 35 520 advection time-steps and in 213 120 chemical time-steps.

Consider now the case where the model is three-dimensional. Assume that ten layers are used in the vertical direction. Then the number of equations in every system of ODEs is 3 225 600 (i.e. ten times greater than in the previous case). The numbers of advection and chemical time-steps remain the same, 35 520 and 213 120 respectively.

More general, the number of equations in each ODE system is equal to the product of the number of the grid-points and the number of chemical species. Therefore,

Table 1.1: The numbers of equations per system of ODEs that are to be treated at every time-step when different discretizations and different chemical schemes (number of species) are used.

Number of species	$(32 \times 32 \times 10)$	$(96 \times 96 \times 10)$	$(288 \times 288 \times 10)$	$(480 \times 480 \times 10)$
1	10240	92160	829440	2304000
2	20480	184320	1658880	4608000
10	102400	921600	8294400	23040000
35	358400	3225600	29030400	80640000
56	573440	5160960	46448640	129024000
168	1720320	15482880	139345920	387072000

this number grows very quickly when the number of grid-points and/or the number of chemical species is increased. The numbers of equations treated at every time-step, which are obtained for different numbers of grid-points and for different numbers of chemical species, are given in Table 1.1. It should be mentioned that the coarsest grid, i.e. the $32 \times 32 \times 10$ grid is no longer used.

It is clear (and this has already been mentioned) that such large problems can be solved **only** if new and modern high speed computers are used. Moreover, it is necessary to select the right numerical algorithms (which are most suitable for the high speed computers available) and to perform the programming work very carefully in order to exploit fully the great potential power of the vector and/or parallel computers. Different versions of DEM have already been successfully run on several high-speed computers (see Brown et al. [32], Georgiev and Zlatev [116], [117], Owczarz and Zlatev [188], [189], [190], Zlatev [278], [280], [282] and Zlatev et al. [294]).

Standard tools for achieving parallelism, such as

- OpenMP ([265]),
- the Parallel Virtual Machine (PVM, Geist et al. [114]) and
- the Message Passing Interface (MPI, Gropp et al. [124]),

have been used to facilitate the transition of the code from one high-speed computer to another and, hopefully, to facilitate the implementation of the code on computer architectures which will appear in the near future.

However, we are still not able to solve all the problems listed in Table 1.1. Some of the problems are only treated as two-dimensional models. This means that it is still necessary to improve the performance of the different algorithms on different high speed computers. Therefore, many more experiments are needed (and will be carried out in the future). High-quality algorithms and software for solving some standard mathematical problems are now available (see, for example, Anderson et al. [11], Barrett et al. [15], Demmel [60], Dongarra et al. [80] and Trefethen and Bau III [252]) and one should try to use them extensively in the attempts to achieve high efficiency when large air pollution models are to be treated numerically on high performance computers.

Finally, it should be emphasized here that there are many other activities in the efforts to utilize in a better way the great potential power of the modern high-speed computers; see, for example, Bruegge et al. [33], D'Ambra et al. [57], Dabdub and Mahonar [50], Dabdub and Seinfeld [52], Elbern [87], Elbern et al. [91], Jacobsen and Turco [143].

1.3 Input data

It is necessary to prepare large files of input data which are to be used when an air pollution model is run on the available computers. Of course, this is also true when large-scale models arising in other fields are to be handled. Some of the problems connected with the necessity to handle huge input data sets will be discussed in this section.

1.3.1 General information about the input data

Two types of input data are needed when a large-scale air pollution model is to be handled numerically on a computer:

- meteorological data, and
- emission data.

The input data needed in DEM (both the meteorological data and the emission data) have been prepared within EMEP (the European Monitoring and Evaluation Programme). EMEP is a common European project in which nearly all European countries participate. The project was initiated in the 70s and is supported financially by the United Nations Economic Commission for Europe, UN-ECE. In the past we used the data prepared by DNMI (the Norwegian Meteorological Institute). Now a lot of data can be obtained from specialized Internet sites; see, for example, EMEP Home Page [94].

Input data, which have been received from other sources, have occasionally been used.

Measurement data are also to be used for different purposes (as, for example, in the validation of the reliability of the model results). Measurement data from the network of the EMEP measurement stations are also available on the Internet (see again EMEP Home Page [94]), and have been regularly used in

- order to verify the model results, and

Table 1.2: Meteorological fields used at present in the two-dimensional versions of DEM

Meteorological input data	Level
Horizontal wind velocities	$\sigma = 0.925, z = 750m$
Vertical wind velocities	$\sigma = 0.850, z = 1500m$
Mixing height	Above ground
Average precipitation	Ground
Cloud covers	Free troposphere
Temperature in the mixing layer	$\sigma = 0.925$
Surface temperature	$z = 2m$
Relative humidity	$\sigma = 0.925$

- an attempt to improve the quality of some fields used in the model (initial concentrations, emissions, etc.) by applying different data assimilation procedures.

1.3.2 Meteorological data

The meteorological data for the **two-dimensional** versions of DEM contain horizontal wind velocity fields, vertical wind velocity fields (at the top of the boundary layer), temperatures (both surface temperatures and temperatures of the boundary layer), precipitation fields, mixing heights, cloud covers and pressures. The resolution of the meteorological fields is normally coarser than the resolution used in the model: the time resolution for all fields except the mixing heights fields is at present six hours (the resolution for the mixing heights fields is 12 hours), the spatial resolution of the available now data is 150 $km \times 150$ km. Simple linear interpolation rules are used both

- in time (because the time-steps used are 15 min for the 50 $km \times 50$ km grid and 2.5 min for 10 $km \times 10$ km grid), and

- in space (because we are not running the model with a grid resolution 150 $km\times$ 150 km anymore).

The main meteorological fields which are used at present in the two-dimensional versions of DEM are given in Table 1.2. For the **three-dimensional versions** most of these data are given in several vertical layers, which as a rule do not correspond to the vertical layers that are actually used in DEM (i.e. interpolation is needed when this is the case).

The total amount of data stored in the meteorological fields, which are needed to run the model over a time period of one month, is about 8 MBytes. It is clear that this figure is increased by a factor of 12 if a one-year run is to be carried out (which is the operational mode when the code is run on the 50 $km \times 50$ km grid) and by a factor of 120 when the code is run over a time period of ten years. Such runs were carried out in the study described in Zlatev et al. [296]. Some results will be presented in Chapter 9.

In the near future it is expected that the resolution of the meteorological data will be improved (in both space and time). This process has already been started. The meteorological data used by us after 1999 is prepared on a 50 $km \times 50$ km grid and the time resolution is one hour (instead of six hours). If all the data are prepared in this way, then the amount of meteorological data needed for a ten year run will be about 432 MBytes.

Let us consider the three-dimensional case now. The meteorological centers which prepare the data are normally calculating the data on layers which are not quite compatible with the layers used in the air pollution model. Therefore, every time new data are received one should either adapt the model to the data used or interpolate the data (adapt the data to the model). Anyway, normally the amount of data is proportional to the number of layers. This means that the three-dimensional runs have severe storage requirements. The problems are tackled either by using some kind of secondary storage or by preparing a long series of runs (instead of running one job over a very long time-interval).

1.3.3 Emission data

Five fields containing human-made (anthropogenic) emissions are currently used in DEM:

- sulphur dioxide, SO_2, emissions,
- nitrogen oxides, NO_x, emissions,
- volatile organic compounds, VOC, emissions,
- carbon monoxide, CO, emissions, and
- ammonia-ammonium, $NH_3 + NH_4$, emissions.

The emission data in all of these five fields are available on a 50 $km \times 50$ km grid. However, **only** the annual means of the human-made emissions are given (i.e. only one number per grid-square per year). Simple rules must be used to get seasonal variations for the SO_2 and the ammonia-ammonium emissions. Both seasonal variations and diurnal variations are simulated for the NO_x emissions, for the human-made VOC emissions and for the CO emissions. Some more details about the mechanisms used to obtain an appropriate temporal resolution can be found in Havasi and Zlatev [131].

While interpolation in time is needed when the air pollution model is discretized on a 50 $km \times 50$ km grid, interpolation in both space and time is needed when the fine resolution version, discretized on a 10 $km \times 10$ km grid, is run.

The natural (biogenic) emissions should also be taken into account. Only natural VOC emissions are at present used in DEM. These emissions are calculated on an hourly basis by using a mechanism based on ideas proposed by Lübkert and Schöpp [169] and Simpson et al. [225]. Information about the forest distribution in Europe and about the surface temperatures is needed in this algorithm.

The amount of emission data is much lower than the amount of meteorological data, because the emissions are at present given on annual basis. The total amount of data needed to run the two-dimensional version of the model on a time-period of ten years (1989-1998) is more than 4 MBytes. However, improvement of the temporal resolution of the emissions is highly desirable. If the emission inventories are prepared on a monthly basis (it should be noted that even this is a rather crude approximation), then the amount of the emission data will be increased by a factor of twelve. If the emission inventories are prepared on a 10 $km \times 10$ km grid, then the amount of the emission data will be further increased by a factor of 25.

The amount of emission data does not grow too much in the transition from two-dimensional computations to three-dimensional computations, because most of the emission sources are located on the surface. However, it may be necessary to distinguish between low sources and higher sources. The latter sources must be put in higher layers (in most of the cases, in the second layer).

Emissions from the aircraft traffic are normally not used in the large-scale air pollution models, in spite of the fact that such emissions have been the central topic in several studies. It should be emphasized here that the emissions from the aircraft traffic have been steadily growing during the last decade (and they will continue to grow in the near future). Therefore, the modellers must start to include such emissions in their models. Of course, it is necessary, first and foremost, to prepare high-quality emission inventories for the emissions from the aircraft traffic and to make them easily available for the modellers. Some more details on this issue can be found in Forster et al. [109].

Emission inventories on a (16.667 $km \times 16.667$ km) grid were recently prepared at NERI (the National Environmental Research Institute), see Hertel et el. [133]. Again, only annual values of the emissions are available in these inventories.

It should be mentioned that some emission fields with very high resolution (both a very high time resolution and a very high spatial resolution) are available for rather short time periods (from several days to several weeks); see, for example, Friedrich [108]. These fields are mainly used to study episodes (short periods with very high concentrations of certain pollutants).

1.3.4 Input data for air pollution forecasts

The prediction of high air pollution levels (which may be harmful for plants, animals and humans) is becoming more and more important. In some extreme cases, when certain pollution levels are likely to become too high in the next two-three days, the population must be warned (see, for example, EU Ozone Directive [96] and the Ozone Position Paper [95]). This can be done if air pollution forecasts for the next two-three days become available on a routinely basis. If such an air pollution forecast is to be prepared one needs the following data:

- data from some weather forecast for the same period (the period for which an air pollution prediction is required),
- an initial field of the concentrations for the starting point of the air pollution forecast, and
- reliable emissions for the period for which air pollution prediction is required.

Some meteorological input data are to be received in the first stage. Such data are normally calculated at meteorological offices and have to be received by using some fast and reliable tools (such as ftp, Internet protocols or others). Some limited area model (LAM) for weather forecasts might be selected and used in parallel with the model for preparing the air pollution forecast. The major advantage of such an approach is the possibility to prepare the needed meteorological data on precisely the same grid as that used in the air pollution model. It should be emphasized here, however, that the necessity of receiving data is not completely avoided when some limited area model for weather forecasts is run in parallel with the air pollution model. The problem is that one still needs reliable starting meteorological data which can only be prepared at a meteorological office where one has access to many meteorological measurements (which are necessary for obtaining, by using data assimilation procedures, reliable initial fields for the meteorological input data). Furthermore, some meteorological data are needed at the boundaries of the space domain used in the limited area weather forecast models. These data have also to be transmitted from some meteorological office. The amount of meteorological data which has to be received is reduced considerably when some limited area model for weather forecasts is run in parallel with the air pollution model. Interpolation rules have to be applied in order to attach the received data to the limited area model for weather forecasts. It is important to reiterate here that the limited area meteorological weather forecast model and the air pollution prediction model must be run on the same grid; if this is not done, then some interpolation of the meteorological data is necessary (at every time step) and this may seriously affect the air pollution forecast. The best solution is to totally merge the two models (i.e. at every time step to prepare first the meteorological data and then the air pollution forecast).

Some measurements in the beginning phase of the computations, if such measurements are available, and data assimilation techniques (similar to the data assimilation technique used for the chemical scheme in air pollution models by Daescu and Carmichael [53], Daescu and Navon [54], Daescu et al. [55], Elbern and Schmidt [88] [89] and Elbern et al. [90], [91], [92]) can in principle be applied to prepare the initial concentration fields for the air pollution forecast model. It is very difficult to provide the measurement data needed in the beginning of the computations. However, the situation is gradually improved. Some measurement data are already available on the Internet. The use of data assimilation techniques in this field is still quite new, but some progress has been made during the last decade; see again Daescu and Carmichael [53], Daescu and Navon [54], Daescu et al. [55], Elbern and Schmidt [88] [89] and Elbern et al. [90], [91], [92].

The third task (i.e. the task of obtaining reliable emissions) is still not sufficiently well resolved. More efforts are needed here to prepare reliable and robust algorithms by which the needed emissions can be prepared by using some given (and not very accurate) emission inventories. The major problem here is that the time resolution of the existing emission inventories is very poor. Therefore, the attempt to improve the available emission inventories by using appropriate data assimilation techniques is crucial when air pollution forecasts are to be prepared. Some progress in this direction was recently reported in Elbern et al. [90], [91], [92].

Some work related to the implementation of variational data assimilation procedures in air pollution models is carried out at present at the National Environmental Research Institute, Roskilde, Denmark. When such procedures are prepared, they will be attached to the existing software for air pollution forecasts (this software is described, for example, in Brandt et al. [27]). The models for air pollution forecasts can also be modified for usage in case of nuclear accidents, see Brandt et al. [28]. Data assimilation procedures will be very useful also when nuclear accidents are studied (assuming that the needed measurement data are available).

1.4 Output data

When a run with the model is completed, several very big output data files containing digital information are prepared and moved from the high speed computers at the Danish Centre for Scientific Computing (DCSC) to several powerful work-stations at the National Environmental Research Institute (NERI). Thus, the output data at present are processed using these work-stations. The main principles that are applied in the treatment of the output data will be discussed in this section.

1.4.1 Visualization of the output data

The digital data cannot directly be used in the efforts to understand better the relationships between different quantities. These data must be visualized by using powerful graphic tools. Four types of visualizations are currently used (according to the processes that are studied and to the requirements for presentation of the relationships of interest):

1. scatter plots where the mean values of the calculated concentrations and depositions are compared with corresponding measurements (mainly on monthly, seasonal and yearly basis),

2. time-series plots where temporal variations of the calculated concentrations and depositions at a given measurement station are compared with the corresponding observations,

3. colour plots presenting concentration and deposition levels of the calculated quantities in a given area (the whole of Europe or a prescribed part of Europe),

4. animations of the concentration and deposition levels in Europe or in a part of Europe in a given period of time (say, the ozone concentration levels during an episode with high ozone concentrations).

Several examples, which illustrate the application of different visualization tools in several air pollution studies, will be presented in this chapter, in Chapter 8 and in Chapter 9.

1.4.2 Types of output data

The output quantities, which are used in the visualization procedures, are given in Table 1.3. Average values of the concentrations and depositions are normally used during the visualization procedures. The different averaging lengths are given in Table 1.4.

Table 1.3: Output quantities used in the visualization programs.

Output quantities
Air concentrations
Concentrations in precipitation
Wet depositions
Dry depositions
Total depositions

1.4.3 Amount of the output data

The information given above shows clearly that the visualization of the results obtained by a large-scale air pollution model is by no means an easy task. A large environmental model, in which all physical and chemical processes are adequately described, produces huge output files. The output data produced after a long simulation (as the simulations carried out in Bastrup-Birk et al., [18] and Zlatev et al., [297]) contain normally several hundred MBytes, or even several GBytes, of

Table 1.4: Time averages used in the visualization programs.

Averaging periods	Remark
Hourly mean values	Only for ozone at present
Daily mean values	
Monthly mean values	
Seasonal mean values	For pollutants with a strong seasonal variation
Yearly mean values	Mainly in connection with critical loads

digital information. Fast and efficient visualization routines have to be used, as stated in the previous paragraph, in order to represent the great amount of digital information in a form from which the main trends and relationships, which are hidden behind millions and millions of numbers, can be easily seen and understood (also by non-specialists).

The evaluations of the amounts of data, which are given above, are for the two-dimensional case. In the three dimensional case the amount of the required data may become very large. However, one is very often interested only in the situation on the surface layer. If this is the case, then nearly the same amount of output data as in the two-dimensional case is required.

The amount of output data depends very essentially on the particular requirements. If, for example, the variation of the daily means of the concentrations has to be studied over a long time-period, then the amount of output data can become very large. However, this is true when we do not know in advance the points where data are needed (in such a case we have to store information at all grid-points in the space domain). Normally, it is known where such information is needed (for example, at some measurement stations). The amount of output data can be reduced very considerably in such a case.

The main conclusion is that the amount of output data can vary, from one run with the model to another run, depending of the particular study which is carried out. Therefore, one should be prepared to take into account the specific requirements for output data for the run under consideration by doing the necessary changes in the program in order to meet these requirements.

1.4.4 Need for fast graphical tools

Modern graphical tools, both hardware and software, are needed in the solution of this task. Therefore, a powerful computer, an **ONYX2** from Silicon Graphics, was bought jointly by NERI (the National Environmental Research Institute) and the Danish Computing Centre (UNI-C). This computer gave us the possibility to solve efficiently the tasks connected with the visualization and animation of the processes described in the output files. It should be emphasized, however, that much more powerful graphical tools are necessary for the new more advanced versions of DEM (these new versions are united in a new model, UNI-DEM, see Chapter 7).

It is not sufficient to have a fast and big computer in order to visualize and animate the output information. It is also necessary to use efficient software. A powerful program has been developed. This program is based on new and modern ideas used at present in the visualization techniques. Some more efforts are needed to make this program both faster and more user friendly.

1.4.5 Requirement for flexible visualization programs

Sometimes it is required to investigate the variation of the pollution levels in the whole space domain (in the whole of Europe). In other cases one needs more detailed information for a part of the space domain (for example, in a part containing Denmark and some surroundings of Denmark). Therefore, the visualization program must allow the user to get such more detailed information when this is required by applying the following devices:

- using different zooming procedures,
- representing the grid-lines on the plots,
- giving the numerical values of the changes in each grid-square,
- plotting the wind fields or the temperatures when the changes of the pollution levels are animated.

The visualization program developed at the National Environmental Research Institute can be used to get more detailed and more representative information about the studied processes by applying the above devices.

1.5 Measurement data

The reliability of the measurement data is an important issue. This is true both in the case where the measurement data are used to verify the reliability of the model results (see the results presented in Chapter 9) and in the case where the measurement data are applied in data assimilation algorithms (such algorithms are discussed in Daescu and Carmichael [53], Daescu and Navon [54], Daescu et al. [55], Elbern and Schmidt [88] [89] and Elbern et al. [90], [91], [92]). The incorporation of measurements in large-scale air pollution models by using variational data assimilation techniques will be discussed in Chapter 10.

1.5.1 Measurements used in the verification procedures

Mainly measurements collected at the EMEP network of measurement stations have been used to verify the reliability of the model output results. There are more than

200 such stations located in different European countries. The measurement data that are taken at the EMEP network of stations are collected at the Norwegian Air Research Institute (NILU, Oslo, Norway), where these are analyzed and then many modellers in Europe can obtain these data and use them in their studies. There are many NILU publications about the measurements collected there; see, for example, Hjellbrekke [136].

Many chemical species are measured at the stations of the EMEP network of measurement stations. The measurements, which are relevant for the validation of the results from DEM, can be divided into two groups:

- measurements of concentrations in air, and
- measurements of concentrations in precipitation.

The measurements from the **first group**, which have mainly been used until now in comparisons with results obtained by using DEM, are:

- nitrogen dioxide,
- nitrate,
- sulphur dioxide,
- sulphate,
- ozone, and
- ammonia + ammonium.

The measurements of the **second group**, which have mainly been used until now in comparisons with results obtained by using DEM, are:

- nitrate,
- sulphate, and
- ammonium.

It should be mentioned here that not all measurement stations measure all of the concentrations listed above. Moreover, there are periods (some days, some months and even some years) in which some measurements at some stations are missing.

Recently, concentrations of several hydrocarbons have been measured at some of the measurement stations in the EMEP network. A few comparisons of model results with the corresponding measurements of hydrocarbons have been carried out. This activity is still in a very preliminary stage.

Finally, it should be mentioned that not only measurements from the EMEP network have been used in the efforts to verify the reliability of the results obtained by DEM. Occasionally, measurements from some other sources have also been used in some studies.

1.5.2 Time and space resolution of the measurement data

Most of the measurement data are available as daily mean values of the concentrations over a long time-period (the measurement data used in DEM are from the period 1989-2003).

The ozone concentrations can also be obtained as hourly mean values (the ozone concentrations vary on a diurnal basis; the availability of hourly mean values of the ozone concentrations allows the modellers to study the diurnal variations of the ozone concentrations in different places in Europe in different seasons).

The space distribution of the available measurements depends very much on the measured compound and on the year of consideration. For some compounds (as, for example ozone and sulphate) the number of stations that measure regularly (at least 15 measurements per every month of the year under consideration) is about 60. For the remaining compounds the number of measurement stations is considerably smaller. The number of stations which measure regularly (at least 5 measurements per every month of the year under consideration) concentrations in precipitation is also rather small.

In the ten-year period, from 1989 to 1998, which has been used in the study related to the impact of climate changes on the pollution levels in Europe (some of the results, which were obtained in this study will be presented and discussed in detail in Chapter 9), it has been discovered that the number of measurement stations in the EMEP network tends to grow during the latter years. However, most of the measurement stations are still mainly located in the countries of Western Europe and Scandinavia.

1.5.3 Amount of the measurement data

The files containing measurement data are, at present, smaller than the files containing input data and output data. Anyway, if the model results are to be compared with measurements over a long time-period, then the files of the measurement data needed are quite large; several hundreds of MBytes. The amount of these data tends to grow, because more and more stations have been opened during the last years and more and more of the existing stations are starting to measure more compounds.

It should be mentioned here that the EMEP measurement data are available on the Internet. Therefore, one can download the needed data for the particular study under consideration and delete the downloaded files when the study is completed. Moreover, if the study is very large (over a time period of many years), then it can be performed in parts (say, studies over one year at a time). If the work is organized in such a way, then one can

- download only the data that are needed in the current part of the study,
- delete the downloaded files after completing the current part of the study, and

- down-load data for the next part of the study and carry out the same procedure until all parts of the study are finished.

The storage requirements can be reduced considerably when this approach is used.

1.5.4 Measurement data for real time computations

It has already been stated that measurement data are needed for preparation of high-quality initial fields of concentrations (by using data assimilation techniques) when air pollution forecasts are to be prepared. It is important to have these data as quickly as possible (say, one or two hours after performing the measurements). This was, of course, a very big problem only two to three years ago. Now the situation is rapidly changing due to the possibilities to use the Internet and to carry out fast transition of large amounts of data. The needed data can be taken from many sites where such data are available. It should be mentioned here that also the National Environmental Research Institute is regularly exposing measurement data on the Internet.

1.6 Some results obtained by DEM

DEM, together with the graphical tools associated by this model, can be considered as a powerful tool for

- treating very large sets of input data,
- preparing very large sets of output data,
- verifying the sets of output data by using very large sets of measurement data, and
- applying the verified sets of output data in different studies by also adding results from long sequences of scenarios performed by the model.

A few results will be presented in this section as well as in Chapter 8 and Chapter 9. Many more results obtained by this model have been described and discussed in Ambelas Skjøth et al. [9], Bastrup-Birk et al. [18], [19] Harrison et al. [128], Zlatev [282], Zlatev and Ambelas Skjøth [284] and Zlatev et al. [292], [293], [294], [295], [297].

Only three features of DEM, but these are very important issues, will be discussed here and/or illustrated by some results, which will be presented in the next subsections:

- the ability of the model to calculate output results by using very high resolution and to zoom at sub-domains that are of particular interest,

- the ability of the model to produce reliable results, and
- the use of the model with different scenarios.

1.6.1 Results obtained with the high resolution version

In the two-dimensional high-resolution version of DEM the space domain (containing the whole of Europe together with parts of Asia, Africa and the Atlantic Ocean) is discretized by using a (480×480) grid of $(10\ km \times 10\ km)$ grid-squares. It is still not possible, as mentioned in Section 1.2, to run operationally this version of the model on long time-intervals (up to ten years). Therefore, only some interesting situations, which occur on shorter time-intervals (the typical length of such a time-interval being a month or a year), were studied by this version until now. Some results obtained when the model was run with meteorological data for 1997 are shown in Fig. 1.1 and Fig. 1.2. More results can be found in Zlatev and Syrakov [299], [300].

The distribution of the nitrogen dioxide pollution in Europe is shown in Fig. 1.1. It is seen that the most polluted areas in Europe are

- parts of United Kingdom,
- the Netherlands,
- Belgium,
- Germany
- and parts of France,
- the Check Republic, and
- Poland

as well as

- the Northern part of Italy.

Moreover, in the parts of Europe which are not very polluted one can locate large cities, such as Madrid, Rome, Oslo, Stockholm, Helsinki, St. Petersburg and Moscow, which are large sources of nitrogen pollution (the most important reason being that pollution from the traffic is one of the major sources for nitrogen emissions).

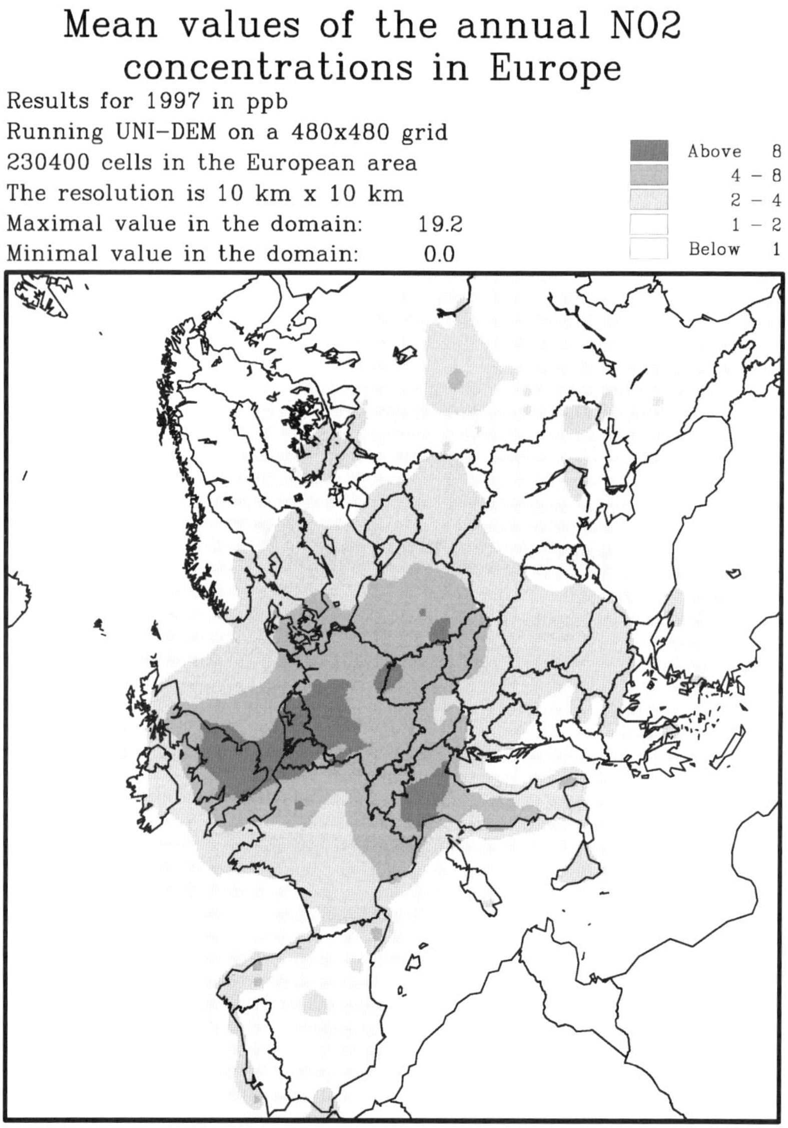

Figure 1.1: The distribution of the nitrogen dioxide concentrations in different parts of Europe and its surroundings.

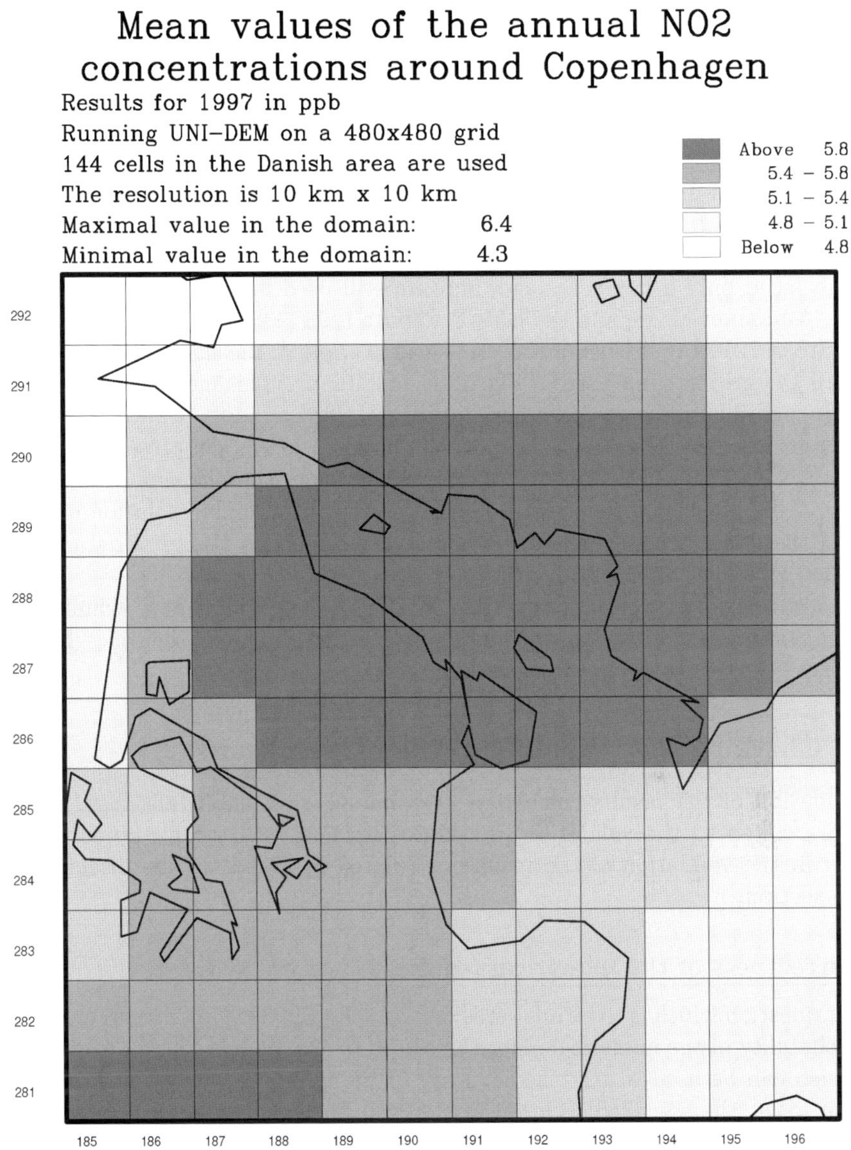

Figure 1.2: The nitrogen dioxide concentrations in area around the Danish capital Copenhagen and the Swedish city Malmö (the Øresund region).

The same output data as used to draw Fig. 1.1, are also used to draw Fig. 1.2. Indeed, Fig. 1.2 could be viewed as a result from zooming in Fig. 1.1 onto the area around Copenhagen, the capital of Denmark, and the Swedish city Malmö.

The grid-lines are drawn in Fig. 1.2. This is impossible when the whole space domain is used. Indeed, Fig. 1.1 will become completely black if the option for drawing the grid-lines is switched on when results on the whole space domain, containing 230 400 grid-squares, are plotted.

The number of grid-squares used in Fig. 1.2 is 144, obtained by using only a tiny part, a (12×12) sub-grid, of the whole (480×480) grid. At the same time, this area is nearly twice smaller than one grid-square of the (32×32) grids which were commonly used only a few years ago, and are still used in some models (see, for example, Amann et al. [8] and the EMEP Report 1/98 [93]). While the size of one grid-square is 22500 km^2 when a (32×32) grid is applied, the size of the combined area of all 144 grid-squares used in Fig.1.2 is 14400 km^2. This means that, while it will not be possible to see any difference in the area shown in Fig. 1.2 when a coarse (32×32) grid is used, the results from the high resolution model presented in Fig. 1.2 show clearly that there are several different levels of the nitrogen pollution in this area. This illustrates the great potential power of the high resolution models. However, there is a price that is to be paid: very large sets of digital data are to be handled when such models are run.

It is also necessary to emphasize here that Fig. 1.2 is only one example for zooming. Zooming could also be done for any other sub-domain of the space domain of DEM.

1.6.2 Checking the reliability of the model results

The validation of the results calculated by a model is an important task. Many questions related to the validation procedures used are still open. Some results obtained in the validation of the results calculated by DEM will be discussed in this sub-section.

(A) Correctness of the numerical results.

The first question which is to be answered during the validation process is connected to the reliability of the numerical algorithms used in the model. There are different tests which can be used to test the accuracy of the numerical algorithms. For the advection part, the so-called rotation test is well known. It has been introduced simultaneously by Crowley [48] and Molenkampf [183] in 1968. Both the classical rotation test and many modifications of this test have been used in several hundred publications in the field of air pollution modelling; see, for example, Chock [43], [44], Chock and Dunker [45], Fornberg [103] and [104], Pepper and Baker [192] and Pepper et al. [193].

A set of tests concerning both the advection and the diffusion part is discussed in Zlatev [282] and Zlatev et al. [287].

An extension of the rotation test for the combined advection-chemistry module of an air pollution model has been proposed and tested in Hov et al. [141]; see also Zlatev [282].

It should be emphasized that successful tests of the performance of the numerical method **do not** ensure the reliability of the results when the full model is run. The success of the numerical tests is only indicating that if there are errors, then these errors are not caused by the numerical methods and, thus, the source for the errors should be searched in other places (as, for example, uncertainties in the emission inventories, uncertainties in the meteorological data, uncertainties in parameters used in the description of some physical and/or chemical mechanisms, etc.). The conclusion is that one should not overestimate the importance of the numerical tests. On the other hand, one should not underestimate the importance of the numerical tests either. Successful numerical tests indicate that

- one of the major sources for errors in the model results has been removed (it should be mentioned here that this is crucial when **Eulerian models** are used in different environmental studies),

and, more importantly,

- there will be no interference of numerical errors with other errors when the sensitivity of the model results to variation of some key parameters (as, for example, emissions and meteorological conditions) is studied.

(B) Qualitative validation.

Comparison of the distribution of the concentrations in the different parts of the space domain with the distribution of related emissions can give us some qualitative estimation of the reliability of the results. Comparing the results in Fig. 1.1 with the distribution of the NO_x emissions in Europe (the distribution of the NO_x emissions in Europe can be seen in upper right-hand-side plot in Fig. 8.2; see also Bastrup-Birk et al. [18], [19]) we can conclude that

- DEM is producing highest levels of the nitrogen dioxide pollution precisely in the parts of the space domain where the NO_x emissions are highest, and
- in the large European cities (which are in general also big sources of NO_x emissions), the levels of the nitrogen dioxide pollution is also high.

The qualitative correctness of the model results allows us to draw two conclusions. The first conclusion is that there are no crude errors in the description of the physical and chemical mechanisms. The second conclusion is that there are not too big uncertainties in the meteorological data used in the model. It should be noted here that both conclusions are drawn under an assumption that the numerical methods work sufficiently well (which emphasizes once again the importance of checking the accuracy of the numerical algorithms that are implemented in the model).

(C) Quantitative validation.

Of course, quantitative correctness of the model results is the most desirable property of an air pollution model. Most of the unresolved problems become immediately visible when one attempts to validate quantitatively the model results. The best that we can do at present (in the attempts to validate quantitatively the model results) is to compare model results with measurements taken at representative measurement stations located in different parts of the space domain.

There are two major ways to present the results which are obtained when calculated by the model concentrations are compared with corresponding measurements. The results from such comparisons can be represented:

- by scatter plots where the mean values of the calculated concentrations and depositions are compared with corresponding measurements (mainly on monthly, seasonal and yearly basis), and
- by time-series plots where temporal variations of the calculated concentrations and depositions at a given measurement station are compared with the corresponding observations.

Both approaches have extensively been used when different versions of DEM have been developed. Results from validation procedures applied to different versions of DEM can be found in many publications; see, for example, Ambelas Skjøth et al. [9], Bastrup-Birk et al. [18], [19], Harrison et al. [128], Zlatev [277] [282], Zlatev and Berkowicz [286], Zlatev and Ambelas Skjøth [284] and Zlatev et al. [292], [293], [294], [295], [297]. Some results are also presented and discussed in Chapter 9. Nevertheless, many more comparisons are necessary.

1.6.3 Running the model with different scenarios

Results calculated by using comprehensive mathematical models can successfully be used in different studies (assuming here that some validation process based on comparison of model results and measurements has successfully been completed). An example for such a study will be given here as an illustration. Other examples can be found in Ambelas Skjøth et al. [9], Bastrup-Birk et al. [18], [19], Zlatev [282], Zlatev and Ambelas Skjøth [284] and Zlatev et al. [297] as well as in Chapter 8 and Chapter 9.

Different critical levels of the concentrations of certain pollutants have been defined and used in many countries; especially in the European Union (EU), see, for example, Amann et al. [8]. Concerning ozone, one of the critical levels, which is under discussion now in the European Union, is the number of days in which the ozone concentrations were higher than 120 ppb for at least one hour (this critical level is introduced in the EU directive for ozone and the "Ozone Position Paper" of the Commission of EU, [96], [95]). According to the EU directive for ozone and the "Ozone Position Paper" of the Commission of EU ([96], [95]), the population must be warned if this is likely to happen in the next two-three days (this is one of the reasons for calculating air pollution forecasts - see also Section 1.4).

There exist some predictions concerning the anthropogenic (human-made) emissions in Europe for year 2010. These emission inventories have been prepared by a large group of European specialists in this field. It is agreed that the European countries will work for achieving the predicted emission levels by the year 2010. These emission levels are well-known (in Europe, at least) as Scenario 2010.

In connection with this critical level (number of days in which the ozone concentrations were higher than 120 ppb for at least one hour) and with the emissions from Scenario 2010, it is important to have the answers to the following two important questions:

- How frequently is the ozone level of 120 ppb exceeded?
- What will be the effect of using the emissions from Scenario 2010 on the number of days in which the ozone level of 120 ppb is exceeded?

DEM, and more precisely the version of this model which is discretized by using ($50\ km \times 50\ km$) grid-squares, have been run twice over a time-interval of ten years (from 1989 to 1998). For each year N, where $N = 1989, 1990, \ldots, 1998$, the existing emission inventories for year N were used in the first run. The predicted emissions in Scenario 2010 were used for all of the ten years in the second run. The meteorological data used in both runs were the same; for each year N, where $N = 1989, 1990, \ldots, 1998$, the meteorological data for year N were used. The first run will be called *Basic Scenario*, while the name *Scenario 2010* will be used for the second run.

Results concerning Denmark and the region surrounding Denmark are given in Fig. 1.3 (the Basic Scenario) and in Fig. 1.4 (Scenario 2010). The total numbers of days (for the whole period of ten years) in which the ozone level of 120 ppb has been exceeded for at least one hour are given in these two figures.

The results given in Fig. 1.3 show that the total numbers of days (for the all ten years) in which the ozone level of 120 ppb has been exceeded in at least one hour vary between 3 and 14 in the different parts of Denmark when the Basic scenario is used. This means that the answer to the first question for the different parts of Denmark is the following:

- *the exceedance of this critical level does not occur very often in Denmark, but may take place sometimes.*

The results given in Fig. 1.4 are the same as those in Fig. 1.3 (numbers of days in which the critical level of 120 ppb is exceeded), but here the emissions from Scenario 2010 are used. The results indicate that practically

- *no exceedance of the critical level for ozone of 120 ppb takes place in Denmark when the emissions from Scenario 2010 are used (excluding the island of Bornholm).*

Thus, the reduction of the emissions to the levels prescribed in Scenario 2010 will be very beneficial for Denmark in connection with the critical level of 120 ppb. Let us emphasize here that it is assumed in Scenario 2010 that the emissions in all European countries will be reduced.

Only results related to Denmark were presented here. However, the huge output data sets can be used to zoom at any other sub-domain of the space domain of DEM. It should be mentioned here that the results indicate that, while it is true that the number of occurrences of days in which the critical level of 120 ppb is reduced considerably in the countries in Central and Western Europe when Scenario 2010 is used, it is also true that the reduction of the European emissions as planned in Scenario 2010 will not totally remove the exceedances of this critical level for ozone in these countries. This is why some scenarios for additional reductions of the anthropogenic emissions in the 15 EU countries are currently being discussed (see, for example, Amann et al. [8]).

Most of the experiments are at present carried out by varying in a suitable way the anthropogenic emissions. However, it is also necessary to study the sensitivity of the concentrations and depositions to changes of the biogenic (natural) emissions. This issue is becoming more and more important, because recent investigations indicate that the biogenic emissions are strongly underestimated (by up to ten times: see, for example, Bouchet et al., [25], [26]). Moreover, the relative part of the biogenic emissions in Europe has clearly been increased during the last two decades, because the anthropogenic emissions were considerably decreased (the reduction of the anthropogenic emissions in Europe is documented in the annual reports of EMEP; see, for example, Vestreng and Støren [261]). Some preliminary results about the sensitivity of the ozone concentrations to variations of biogenic emissions are presented in Geernaert and Zlatev [112], [113]. This study is very important and, therefore, it will be further discussed in Chapter 8.

1.6.4 Sensitivity tests

Systematic sensitivity tests of the variations of the model results caused by the variations of certain physical parameters can give very useful results. If the results are very sensitive to the variations of the selected parameter, then this is an indication that it is necessary to use a very accurate value of the parameter under consideration. If we are not able to provide a sufficiently accurate value of the critical parameter, then it may be more profitable to replace the physical mechanism in which the critical parameter occurs with another mechanism which is perhaps not so accurate, but more robust (in the sense that the results are not very sensitive to the variation of the physical parameters involved in the new mechanism).

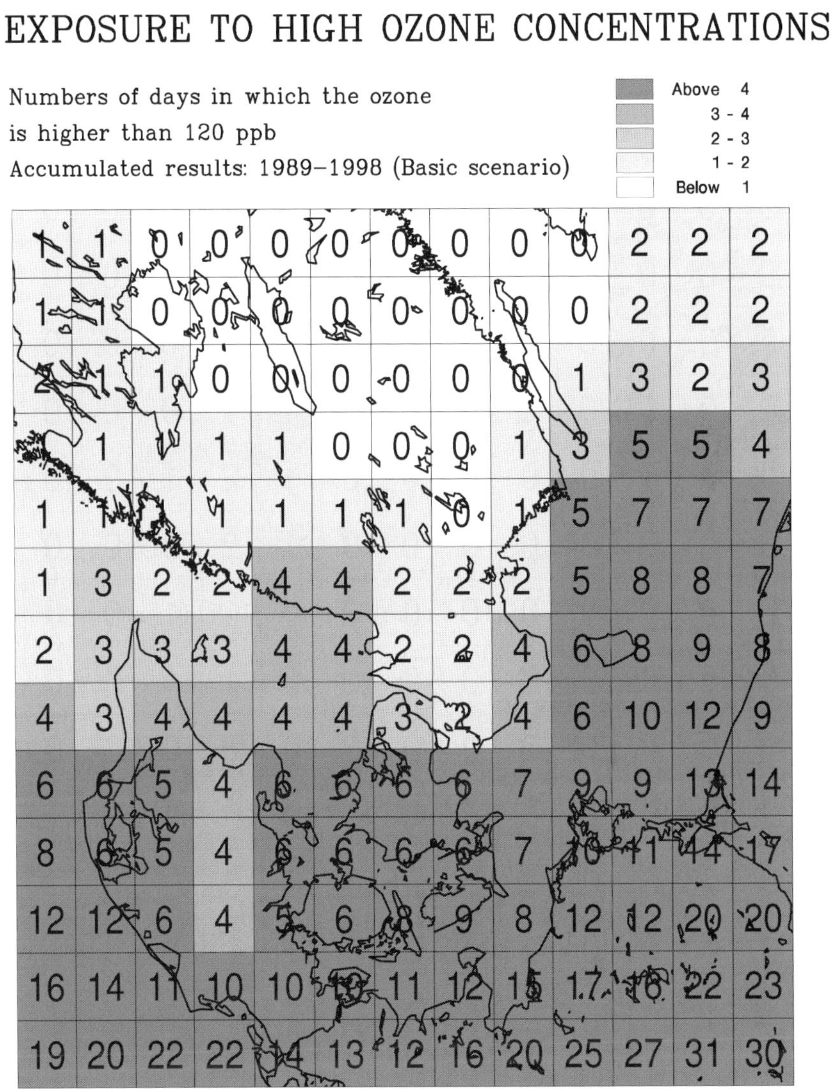

Figure 1.3: The total numbers of days in the period 1989-1998 in which the ozone concentrations in Denmark exceeded the critical level of 120 ppb - Basic Scenario.

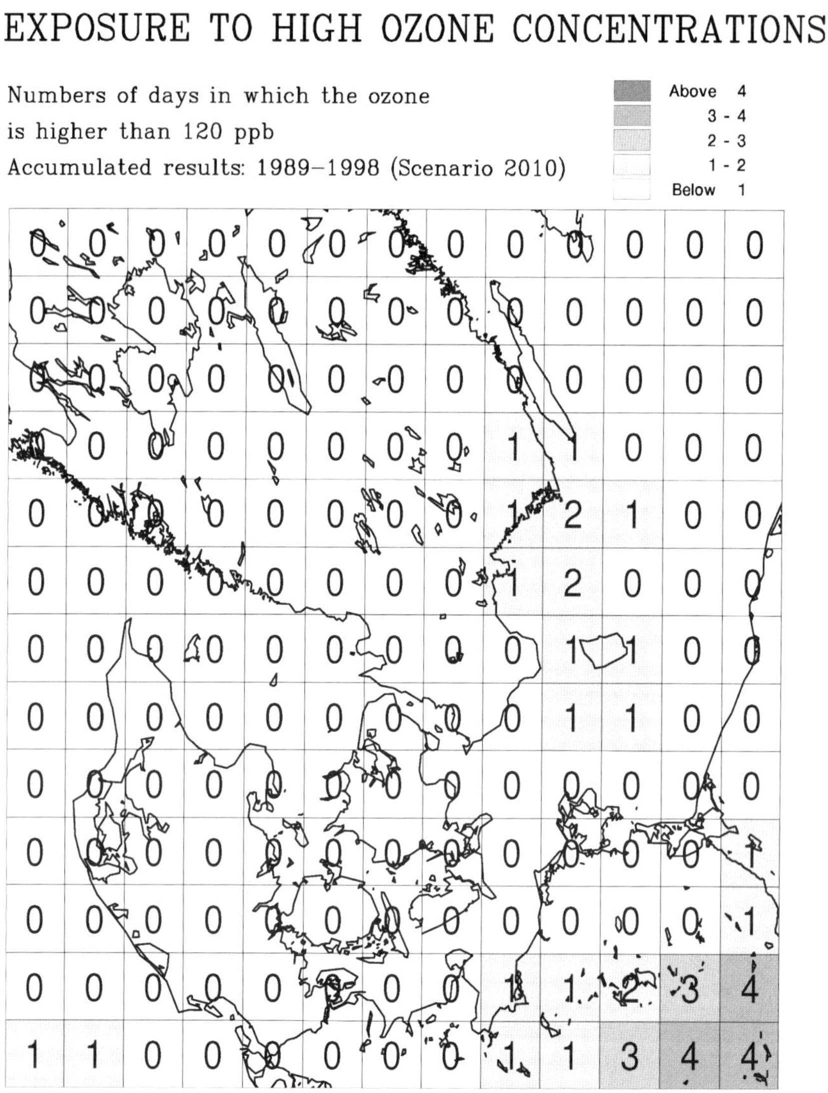

Figure 1.4: The total numbers of days in the period 1989-1998 in which the ozone concentrations in Denmark exceeded the critical level of 120 ppb - Scenario 2010.

Some carefully chosen tests, appropriate to study the sensitivity of the model results to variations of the rate coefficients of some of the chemical reactions, were carried out in Dimov and Zlatev [74]. Monte Carlo techniques were used in Dimov and Zlatev [74] in order

- to find the rate coefficients to the variation of which the model results are most sensitive, and
- to study in a more systematic way the variation of the model results caused by random variations in a prescribed range of the selected rate coefficients.

A simple box-model (one grid-point only and no other processes except the chemical reactions) is quite sufficient in the first part. The second part is much more complicated, because it requires a large number of runs of the whole air pollution model. It has been established in this study that the magnitude of the changes in the different parts of Europe was not the same. More details and other conclusions can be found in Dimov and Zlatev [74].

1.7 Applicability to PDEs arising in other areas

Air pollution models were discussed in the previous sections of this chapter. As mentioned several times in the previous sections of this chapter, this has been done only in order to facilitate the presentation of the results. All (or, at least, nearly all) methods and devices, which were discussed in the previous sections and will be discussed in the remaining part of the book, are also applicable when systems of PDEs arising in some other fields of science and engineering are to be handled on computers. Several examples will be given in this section to illustrate the correctness of this statement.

It should be mentioned here that the physical and chemical processes in the examples given in this section are slower than the corresponding processes taking place in the atmosphere. This means that the requirements that must be satisfied by the numerical methods used in the treatment of air pollution models are normally stronger than the requirements that are to be satisfied when similar systems of ODEs arising from other fields of science and engineering are handled. Two conclusions can be drawn by using this fact:

- The numerical methods used in an air pollution model can be applied in the treatment of models arising in many other areas (however, the opposite is not necessarily true), and
- it might be more efficient to select some simpler numerical methods when the physical and chemical processes studied by the model under consideration are not so fast as the corresponding processes in the atmosphere.

1.7.1 Models used to study multicomponent transport in ground-water

Ground-water resources are becoming more and more threatened by

- contaminants containing industrial and household waste,
- infiltration of pesticides and fertilizers from agricultural areas, and
- leakage of organic pollutants from petrol stations, refineries and pipelines.

Mathematical models of the following type can be used in ground-water studies (see Prommer et al. [197] and Mao et al. [171]):

$$\begin{aligned}\frac{\partial c_s}{\partial t} &= -\frac{\partial (u c_s)}{\partial x} - \frac{\partial (v c_s)}{\partial y} - \frac{\partial (w c_s)}{\partial z} \\ &+\frac{\partial}{\partial x}\left(K_x \frac{\partial c_s}{\partial x}\right) + \frac{\partial}{\partial y}\left(K_y \frac{\partial c_s}{\partial y}\right) + \frac{\partial}{\partial z}\left(K_z \frac{\partial c_s}{\partial z}\right) \\ &+Q_s(c_1, c_2, \ldots, c_{N_s}) \\ &-\frac{q_k}{\theta}\, c_{ks}, \\ &s = 1, 2, \ldots, N_s.\end{aligned} \tag{1.4}$$

The different quantities that are involved in the mathematical model have the following meaning:

- the concentrations are denoted by c_s,
- u, v and w are fluid velocities,
- K_x, K_y and K_z are hydrodynamic dispersion coefficients,
- c_{ks} are the total aqueous concentrations in the water coming from sinks and sources,
- q_k is the volumetric flow rate per unit volume of the aquifer representing sources and sinks,
- θ is the volumetric porosity,
- the chemical reactions used in the model are described by the non-linear functions $Q_s(c_1, c_2, \ldots, c_{N_s})$.

It is seen that (1.4) is nearly the same as (1.1). Of course, the physical interpretation of the involved quantities is different. However, since the mathematical description is quite similar, the same methods and devices, which are used in treatment of the PDE system (1.1), can also be applied in the treatment of (1.4).

More details about mathematical models used in ground-water studies can be found, for example, in Prommer et al. [197] and Mao et al. [171].

1.7.2 PDEs arising in oil spill studies

Oil spill spreading and drifting can cause serious environmental problems. Mathematical models, which are developed to study these problems, lead to systems of PDEs similar to (1.1). An example, which is taken from Tkalich [249], is given in this subsection to illustrate the similarity. The model consists of two parts. The first part is describing the thickness of the oil slick

$$\begin{aligned}\frac{\partial h}{\partial t} &= -\frac{\partial(u_h h)}{\partial x} - \frac{\partial(v_h h)}{\partial y} \\ &\quad + \frac{\partial}{\partial x}\left(D_x \frac{\partial h}{\partial x}\right) + \frac{\partial}{\partial y}\left(D_y \frac{\partial h}{\partial y}\right) \\ &\quad + Q_h(h)\end{aligned} \tag{1.5}$$

while the second one is describing the processes in the water column:

$$\begin{aligned}\frac{\partial c_s}{\partial t} &= -\frac{\partial(u c_s)}{\partial x} - \frac{\partial(v c_s)}{\partial y} - \frac{\partial(w c_s)}{\partial z} \\ &\quad + \frac{\partial}{\partial x}\left(K_x \frac{\partial c_s}{\partial x}\right) + \frac{\partial}{\partial y}\left(K_y \frac{\partial c_s}{\partial y}\right) + \frac{\partial}{\partial z}\left(K_z \frac{\partial c_s}{\partial z}\right) \\ &\quad + Q_s(c_1, c_2, c_3) \\ &\quad s = 1, 2, 3.\end{aligned} \tag{1.6}$$

The different quantities that are involved in the mathematical model described by (1.5) and (1.6) have the following meaning:

- h is the oil slick thickness, while the oil concentrations in droplet, dissolved and particulate phases in the water column are denoted by c_s,
- u_h and v_h are representing the slick drifting velocities, while u, v and w are representing the fluid velocity,

- D_x and D_y are diffusion coefficient in the oil slick,s
- K_x, K_y and K_z are diffusion coefficients in the water column,
- the chemical kinetic terms used in the model are described by the non-linear functions $Q_h(h)$ and $Q_s(c_1, c_2, c_3)$.

It is again seen that the system of PDEs defined by (1.5) and (1.6) is very similar to (1.1), which indicates that the numerical algorithms and the computational devices discussed in the remaining part of this book can successfully be used in the computer treatment of problems connected to oil spill.

More details about this class of environmental problems can be found in Fingas [101], Reed et al. [200], Spaulding [231] and Tkalich [249].

1.7.3 Contaminated disposal facilities

It is important to study the environmental significance of contaminated losses from in-lake disposal facilities. The mathematical modelling of this process leads (under an assumption for incompressible flow and homogeneous medium) to two equations: a flow equation

$$S\frac{\partial h}{\partial t} = \frac{\partial}{\partial x}\left(K_x\frac{\partial h}{\partial x}\right) + \frac{\partial}{\partial y}\left(K_y\frac{\partial h}{\partial y}\right) \tag{1.7}$$

and an equation describing the contaminant transport

$$\begin{aligned} S_0\frac{\partial c}{\partial t} &= -\frac{\partial(uc)}{\partial x} - \frac{\partial(vc)}{\partial y} \\ &+ \frac{\partial}{\partial x}\left[D_{xx}\frac{\partial c}{\partial x} + D_{xy}\frac{\partial c}{\partial y}\right] + \frac{\partial}{\partial z}\left[D_{yx}\frac{\partial c}{\partial x} + D_{yy}\frac{\partial c}{\partial y}.\right] \end{aligned} \tag{1.8}$$

The different quantities that are involved in the mathematical model described by (1.7) and (1.8) have the following meaning:

- h is representing the potential function of the flows,
- c is the contaminant concentration,
- S is representing the specific storage coefficient,
- S_0 is representing the porosity of the aquifer,
- u and v are representing the fluid velocities,
- K_x and K_y are hydraulic conductivity components,
- D_{xx}, D_{xy}, D_{yx} and D_{yy} are diffusion coefficients.

This example is taken from Priklonsky et al. [196]. Some other examples can be found in Su [238], Su et al. [239].

Note that there are no chemical reactions in this example. However, the methods used in the advection-diffusion part of (1.1) can also be used in the computer treatment of (1.7) and (1.8).

1.7.4 Other applications

It is difficult to prepare a full list of applications which lead to systems of PDEs that are similar to (1.1), because such a list will be very long. While it is true that it is not necessary to present in this book such a long list, we do believe that some readers might be interested in recent references where models which lead to systems of PDEs similar to (1.1) are discussed. Different such applications can be found in Jennings, [147], Gupta et al., [125], Umgiesser et al., [253], Cichota et al., [47] and Lee et al., [161].

1.8 Concluding remarks

Remark 1.1. Different problems connected with the needs of using large data sets in air pollution modelling have been discussed by using a particular model, DEM. It has already been emphasized that this approach has been chosen only in order to simplify the exposition of the results. The same conclusions could also be drawn if any other large-scale air pollution model is used. Moreover, similar problems also occur when models arising in other fields of science and engineering are to be handled (several examples were given in the previous section).

The development of DEM is similar to the development of many other models. The development of a more advanced air pollution model is normally carried out in two stages:

- *starting with models defined on coarse grids with a few chemical compounds, and*
- *arriving gradually to very advanced mathematical models utilizing fine grids and involving many more chemical compounds.*

Similar stages are also used in the development of advanced mathematical models in other areas of science and engineering.

Remark 1.2. The question:

> *Are the mathematical models used at present sufficient for studying the most important environmental problems?*

is relevant. The answer is, of course, no. While many problems can successfully be studied and **are** successfully studied by the models that are available at present, it is also true that many other problems cannot be handled with the models which exist at present. The question of having models discretized on finer grids (say, by using 2 $km \times 2$ km grid-squares) is already a topic of discussion. Also the high-resolution three-dimensional models are still not tractable even on the largest computers existing at present (at least in the case where calculations over a long time-interval with many scenarios are required). This indicates clearly that

- *there is a lot of work to be done in this field, and*
- *the need of developing good tools for handling efficiently very large data sets (input data sets, output data sets and data sets containing measurements) is steadily increasing in the environmental studies which are carried out by using large-scale mathematical models.*

Remark 1.3. The performance of the computers that are available at present is a central problem in the field of the large-scale environmental modelling (as well as in many other fields of science and engineering where large-scale mathematical models are to be treated on computers). We are going to handle larger and larger environmental models on computers (as well as larger and larger models that appear in other fields of science and engineering). Therefore, the following two questions, which arise in connection with this development, should be adequately answered:

- Are the computers that are available at present sufficiently fast and is it possible to resolve by using them all the comprehensive scientific problems that are to be treated now?
- Do we need bigger and faster computers for the treatment of the environmental problems that have to be resolved in the near future?

These two questions can best be answered by using a quotation taken from a paper *"Ordering the universe: The role of mathematics"* written by Arthur Jaffe ([144]):

> *"Although the fastest computers can execute millions of operations in one second they are always too slow. This may seem a paradox, but the heart of the matter is: the bigger and better computers become, the larger are the problems scientists and engineers want to solve."*

This paper of Jaffe was published in 1984. At that time it was difficult to handle the version of DEM containing, after the discretization, 2048 equations. One of the biggest problems solved, which can be treated successfully at present by DEM, contains, again after the discretization, more than 8000000 equations. One of the major reasons for being able to handle such huge problems, about 4000 times larger than the problems that were solved in 1984, is the availability of much bigger and much faster computers. There is no reason to assume that the development of bigger and faster computers will not continue. Therefore, the scientists could continue to plan the formulation and the treatment of bigger and bigger tasks that impose strong requirements for handling larger and larger data sets, because the solutions obtained in this way

- will be *closer* to the reality,
- will contain *more* details and *more useful* details, and
- will give *reliable* answers to questions which are at present open.

Remark 1.4. The conclusions that were made in the previous three remarks illustrate clearly the fact that the introduction of new and more advanced modules for the mathematical description of the physical and chemical processes that are involved in the large-scale models leads necessarily to bigger computational tasks. In other words, both

- the computational time, and
- the storage requirements

are normally increased very considerably when more advanced physical and chemical modules are developed and implemented in a large air pollution model. The efficient treatment of the more difficult tasks that appear when more adequate modules are incorporated in the models will be the main topic in most of the chapters in the remaining part of this book (and, more precisely, the discussion in Chapter 2 to Chapter 7 and Chapter 10 is completely related to the computational issues). However, we shall also illustrate the usefulness of the efficient treatment of the computational tasks by showing some results from comprehensive environmental studies in Chapter 8 and Chapter 9. It was possible to complete successfully these studies only because the computational tasks are efficiently carried out. Only a few examples for comprehensive studies, which were performed by the our air pollution model, are discussed in Chapter 8 and Chapter 9. It should be mentioned here that many other comprehensive studies were also successfully completed. However, it should also be mentioned that further improvements of the numerical methods and the computational devices is necessary, because there are still problems, which cannot be solved at present. Several examples for such difficult and/or open problems will be given in Chapter 11.

Chapter 2

Using splitting techniques in the treatment of air pollution models

The application of splitting procedures in the treatment of large scientific and engineering problems is an excellent tool (and, very often, the only tool) by which huge computational tasks can be made tractable on the available computers by dividing them to a sequence of smaller and simpler tasks.

A simple splitting procedure for partial differential equations was proposed, as an example, in 1957 by Bagrinovskii and Godunov in [14]. This was probably the first attempt to apply a splitting procedure in the numerical solution of partial differential equations. Different splitting procedures have been developed and/or studied in many scientific papers; see, for example, Csomos et al. [49], Dyakonov, [83], Lanser and Verwer, [157], Marchuk, [172], [173], [175], [177], Penenko and Obraztsov, [191], Strang, [235], Tikhonov and Samarski, [248], Yanenko, [269].

A detailed theoretical study and analysis of some splitting procedures can be found in Marchuk, [173], [176], Faragó and Havasi, [97] and Faragó et al., [100]. Some convergence properties of splitting procedures were recently discussed in Faragó and Havasi [98].

Results that are directly related to the use of splitting procedures in the field of air pollution modelling can be found in Bartholy et al. [16], Marchuk [174] and McRae et al. [181]. Splitting procedures for air pollution models are also discussed in detail in Hundsdorfer and Verwer [142].

Splitting techniques are, to our knowledge, used in **all** large-scale air pollution models, which are

- discretized using fine resolution grids,
- run over long time-intervals (time-intervals of ten or more years),

and/or

- run operationally with many scenarios (emission scenarios, meteorological scenarios, climatic scenarios, etc.).

Several different ways of implementing splitting procedures in large-scale models (not necessarily large-scale air pollution models) as well as the advantages and disadvantages of using splitting procedures in such models will be discussed in this chapter.

The discussion in this chapter is fairly general and, therefore, the basic ideas used in the different splitting techniques that are discussed in this chapter are applicable to many large-scale mathematical models arising in different fields of science and engineering.

2.1 Four types of splitting procedures

Let us reiterate here that splitting techniques are used not only in problems arising in air pollution modelling, but also in many other scientific and engineering problems. This is why it is worthwhile to give a general presentation of some of the basic ideas used when splitting techniques are to be designed.

We shall assume here that the following rather general mathematical problem is to be solved:

$$\frac{\partial \mathbf{c}(\mathbf{x},t)}{\partial t} = A(\mathbf{c}(\mathbf{x},t)), \tag{2.1}$$

where

- $t \in [0,T]$,
- $\mathbf{x} \in D$ (D being some space domain), and
- $A(\mathbf{c}(\mathbf{x},t))$ is a differential operator.

The abbreviation A will be used instead of A($\mathbf{c}$($\mathbf{x}$, t)) when there is no danger for misunderstanding.

We shall assume that operator A from (2.1) can be represented as a sum of m operators, which are in some sense simpler than the original operator A:

$$A = \sum_{i=1}^{m} A^{[i]}. \tag{2.2}$$

It should be noted here that the division of the operator A from (2.1) to a sum of simpler operators is in general not unique (this will be illustrated by several examples in the next sections of this chapter).

It is as a rule worthwhile to replace the numerical treatment of the original problem (2.1) with a numerical treatment of a sequence of smaller problems. Each of the smaller problems can formally be obtained from the original problem (2.1) by exchanging A with $A^{[i]}$. This means that the original problem (2.1) can be split into the following m sub-problems by replacing successively A with the simpler operators $A^{[i]}$ from (2.2):

$$\frac{\partial \mathbf{c}^{[i]}(\mathbf{x},t)}{\partial t} = A^{[i]}(\mathbf{c}^{[i]}(\mathbf{x},t)), \quad i = 1,2,\ldots,m. \tag{2.3}$$

Moreover, the time-interval $[0,T]$ is divided to $N = T/\tau$ small sub-intervals, where τ is called the splitting time-step size (it is assumed here that the splitting time-step size is constant, equal to τ, but this is done only in order to facilitate the explanation of the results; such an assumption is in fact not needed). The smaller problems from (2.3) are treated (in some chosen by the user ordering) at every splitting time-step (i.e., for $n = 1, 2, ..., N$).

It is said that a **splitting procedure** is used when an approximation of the solution of (2.1) is obtained by treating (in the manner sketched above) a sequence of smaller problems (2.3) involving the simpler operators $A^{[i]}$ from (2.2). The particular splitting procedure depends on

- the way in which the operator A from (2.1) is represented as a sum of simpler operators $A^{[i]}$, and
- the order in which the smaller and simpler problems that were defined by (2.3) are treated.

Assume that the representation (2.2) of the original operator A as a sum of simpler operators $A^{[i]}$ is fixed. Then four types of splitting procedures can be constructed by ordering the simpler operators from (2.2) in different ways:

- sequential splitting procedures,
- symmetric splitting procedures,
- weighted sequential splitting procedures,
- weighted symmetric splitting procedures.

The main ideas, on which these four types of splitting procedures are based, will be discussed in the following four sections. Some particular examples of splitting procedures used in large-scale air pollution models will also be presented in the next four sections.

2.2 Sequential splitting procedures

If the sub-problems from (2.3) are treated one after the other, then the resulting algorithm is called a sequential splitting procedure. Different sequential splitting procedures can be obtained if the ordering, in which the simpler problems (2.3) are treated, is varied. There are two requirements:

- each of the simpler problems must be treated only once per a splitting step, and
- each of the simpler problems must be treated with the same splitting time-step size τ.

Three examples for sequential splitting procedures used in air pollution models are given below. The description of the first example is given with some more details in the next sub-section (it is also explained there how the simpler models that are used in the splitting procedures are coupled).

It should be mentioned here that similar examples can be constructed in relation with many environmental models (as those described in Section 7 of Chapter 1) as well as in connection with many large-scale mathematical models arising in other fields of science and engineering.

2.2.1 The old splitting procedure used in DEM

It is sometimes more convenient to rewrite in a vector form the system of partial differential equations (**PDEs**), which is given in Chapter 1 and by which DEM is described mathematically:

$$
\begin{aligned}
\frac{\partial \mathbf{c}(\mathbf{x},t)}{\partial t} &= -\frac{\partial (u\mathbf{c})}{\partial x} - \frac{\partial (v\mathbf{c})}{\partial y} - \frac{\partial (w\mathbf{c})}{\partial z} \\
&\quad + \frac{\partial}{\partial x}\left(K_x \frac{\partial \mathbf{c}}{\partial x}\right) + \frac{\partial}{\partial y}\left(K_y \frac{\partial \mathbf{c}}{\partial y}\right) + \frac{\partial}{\partial z}\left(K_z \frac{\partial \mathbf{c}}{\partial z}\right) \\
&\quad + E + Q(\mathbf{c}) \\
&\quad - \kappa\ \mathbf{c},
\end{aligned}
\tag{2.4}
$$

where N_s is the number of chemical species and

- $\mathbf{x} = (x, y, z)$,
- $\mathbf{c} = (c_1, c_2, \ldots, c_{N_s})^T$,
- $E = (E_1, E_2, \ldots, E_{N_s})^T$,
- $Q = (Q_1, Q_2, \ldots, Q_{N_s})^T$, and
- κ is a diagonal matrix, the diagonal elements of which are $\kappa_{11} + \kappa_{21}, \kappa_{12} + \kappa_{22}, \ldots, \kappa_{1N_s} + \kappa_{2N_s}$.

By using this notation, the sequential splitting, which has been used in DEM until about two years ago, can be defined by the following five simpler sub-models:

$$\frac{\partial \mathbf{c}^{[1]}(\mathbf{x},t)}{\partial t} = -\frac{\partial(u\mathbf{c}^{[1]})}{\partial x} - \frac{\partial(v\mathbf{c}^{[1]})}{\partial y} \tag{2.5}$$

$$\frac{\partial \mathbf{c}^{[2]}(\mathbf{x},t)}{\partial t} = \frac{\partial}{\partial x}\left(K_x \frac{\partial \mathbf{c}^{[2]}}{\partial x}\right) + \frac{\partial}{\partial y}\left(K_y \frac{\partial \mathbf{c}^{[2]}}{\partial y}\right) \tag{2.6}$$

$$\frac{\partial \mathbf{c}^{[3]}(\mathbf{x},t)}{\partial t} = E + Q(\mathbf{c}^{[3]}) \tag{2.7}$$

$$\frac{\partial \mathbf{c}^{[4]}(\mathbf{x},t)}{\partial t} = -\kappa\, \mathbf{c}^{[4]} \tag{2.8}$$

$$\frac{\partial \mathbf{c}^{[5]}(\mathbf{x},t)}{\partial t} = -\frac{\partial(w\mathbf{c}^{[5]})}{\partial z} + \frac{\partial}{\partial z}\left(K_z \frac{\partial \mathbf{c}^{[5]}}{\partial z}\right). \tag{2.9}$$

Five physical and chemical processes:

- the horizontal advection,
- the horizontal diffusion,
- the chemistry,
- the deposition, and
- the vertical exchange

are described mathematically with the terms in the right-hand-side of the systems (2.5)-(2.9).

These five systems are solved successively at every time-step (i.e., for $t = t_1, t_2, \ldots, t_N$ with $t_0 = 0$ and $t_n - t_{n-1} = \tau$, where $n = 1, 2, \ldots, N$). The order of computations at time-step $n+1$ can be explained in the following way. Assume that the first n splitting steps are successfully completed. In other words, assume that some approximations of the values of the concentrations $\mathbf{c}$ are known for $t = t_n$, where $n \in \{0, 1, \ldots, N-1\}$. Then these approximations are used as initial values when solving system (2.5) at time-step $n+1$ (i.e., when an approximation to $\mathbf{c}^{[1]}$ is computed at $t = t_{n+1}$). After that the following steps are carried out:

- the solution $\mathbf{c}^{[1]}$ at $t = t_{n+1}$ is used as an initial value when (2.6) is solved at time-step $n+1$ (i.e., when an approximation to $\mathbf{c}^{[2]}$ is computed at time $t = t_{n+1}$),

- the solution $\mathbf{c}^{[2]}$ at $t = t_{n+1}$ is used as an initial value when (2.7) is solved at time-step $n+1$ (i.e., when an approximation to $\mathbf{c}^{[3]}$ is computed at time $t = t_{n+1}$),
- the solution $\mathbf{c}^{[3]}$ at $t = t_{n+1}$ is used as an initial value when (2.8) is solved at time-step $n+1$ (i.e., when an approximation to $\mathbf{c}^{[4]}$ is computed at time $t = t_{n+1}$),
- the solution $\mathbf{c}^{[4]}$ at $t = t_{n+1}$ is used as an initial value when (2.9) is solved at time-step $n+1$ (i.e., when an approximation to $\mathbf{c}^{[5]}$ is computed at time $t = t_{n+1}$).

The solution of the last sub-model, i.e., the approximation to $\mathbf{c}^{[5]}$ calculated at time $t = t_{n+1} = t_n + \tau$, is accepted as an approximation of the solution $\mathbf{c}$ of the original system (2.4) at splitting time-step $n+1$. This means that the splitting time-step $n+1$ is completed and everything is ready to carry out the same kind of computations at splitting time-step $n+2$.

The only question that remains to be answered is how to start the computations at the first splitting time-step (i.e., $n = 1$). This is not a problem, because the initial conditions can be used at the first splitting time-step.

It should be mentioned here that the same (or, at least, a very similar) principle can be used to couple the smaller problems in all other splitting procedures. Therefore, this issue will not be discussed in the description of the other splitting procedures in the following part of this chapter.

The relationship between the particular splitting procedure used in this sub-section and splitting procedures for the more general problem discussed in the previous section can be explained as follows. It is clear that if we denote the right-hand-side of (2.4) with A, then (2.1) will be obtained. In the same way, if we replace the right-hand-sides of (2.5) - (2.9) with $A^{[1]}$, $A^{[2]}$, $A^{[3]}$, $A^{[4]}$ and $A^{[5]}$ respectively, then simpler sub-models of type (2.3) will be obtained. Thus, (2.4) and (2.5) - (2.9) are the result of using the notation in the previous section for the particular case of problems arising in air pollution modelling with $m = 5$.

It is very often worthwhile to describe how the different splitting procedures are to be applied in practice by

- giving the order in which the the simple operators $A^{[i]}$, where $i = 1, 2, \ldots, m$, are applied, and
- indicating the splitting time-step size τ, which is actually used.

For the particular case discussed in this sub-section (i.e., for the sequential splitting procedure with $m = 5$ that is related to DEM), such a description is given by the following sequence:

$$(A^{[1]})_\tau,\ (A^{[2]})_\tau,\ (A^{[3]})_\tau,\ (A^{[4]})_\tau,\ (A^{[5]})_\tau. \tag{2.10}$$

2.2.2 The new splitting procedure used in DEM

There were good reasons for using the splitting procedure described in the previous sub-section when the pseudo-spectral method was applied in the numerical treatment of the horizontal transport (advection) and the horizontal diffusion. Truncated series of trigonometric functions are used in the pseudo-spectral method. Both sines and cosines are involved in the trigonometric series that are used in the discretization of the horizontal advection. It is more efficient to apply trigonometric series containing only cosines when the horizontal diffusion is treated. Thus, it is better to split the horizontal advection and the horizontal diffusion when the pseudo-spectral method is implemented in an air pollution model (see more details in Zlatev [276], [282]).

Finite elements are now used in the discretization of the horizontal transport and the horizontal diffusion. Therefore, there is no need to split these two processes and they are combined in the new splitting procedure used in DEM. Furthermore, the deposition process is added to the chemical part. In this way, the number of the sub-models, obtained when the new splitting procedure is implemented, is reduced from five to three.

The three sub-models that are obtained by applying the new DEM splitting procedure are given below:

$$\frac{\partial \mathbf{c}^{[1]}(\mathbf{x},t)}{\partial t} = -\frac{\partial (w\mathbf{c}^{[1]})}{\partial z} + \frac{\partial}{\partial z}\left(K_z \frac{\partial \mathbf{c}^{[1]}}{\partial z}\right) \tag{2.11}$$

$$\frac{\partial \mathbf{c}^{[2]}(\mathbf{x},t)}{\partial t} = -\frac{\partial (u\mathbf{c}^{[2]})}{\partial x} - \frac{\partial (v\mathbf{c}^{[2]})}{\partial y} + \frac{\partial}{\partial x}\left(K_x \frac{\partial \mathbf{c}^{[2]}}{\partial x}\right) + \frac{\partial}{\partial y}\left(K_y \frac{\partial \mathbf{c}^{[2]}}{\partial y}\right) \tag{2.12}$$

$$\frac{\partial \mathbf{c}^{[3]}(\mathbf{x},t)}{\partial t} = E + Q(\mathbf{c}^{[3]}) - \kappa\, \mathbf{c}^{[3]}. \tag{2.13}$$

The first of these sub-models, (2.11), describes the vertical exchange. The second sub-model, (2.12), describes the combined horizontal transport (the advection) and the horizontal diffusion. The last sub-model, (2.13), describes the chemical reactions together with emission sources and deposition terms.

Let us denote the operators in the right-hand-sides of (2.11), (2.12) and (2.13) by $A^{[1]}$, $A^{[2]}$ and $A^{[3]}$ respectively. Then the computations are to be carried out using the operators in the following order when the new splitting procedure is used with some splitting time-step size τ:

$$(A^{[1]})_\tau,\ (A^{[2]})_\tau,\ (A^{[3]})_\tau, \tag{2.14}$$

The correspondence between the operators used in the new splitting procedure and the operators used in the old one (see the previous sub-section) can be expressed by the following relationships:

- The operator $A^{[1]}$ in the new splitting procedure is equal to the operator $A^{[5]}$ in the old one.
- The operator $A^{[2]}$ in the new splitting procedure is equal to the sum of the operators $A^{[1]} + A^{[2]}$ in the old one.
- The operator $A^{[3]}$ in the new splitting procedure is equal to the sum of the operators $A^{[3]} + A^{[4]}$ in the old one.

The boundary conditions can be treated in a natural way when the splitting procedure described by (2.11) - (2.13) is used. The implementation of the boundary conditions is performed as follows:

- The boundary conditions on the top and the bottom of the space domain are treated in (2.11), where the computations are carried out along the vertical grid-lines.
- The lateral boundary conditions are handled in (2.12), where the computations are carried out in each of the horizontal grid-planes.
- The computations related to (2.13) are carried out by performing the chemical reactions at each grid-point. It is clear that the computations at any of the grid-points do not depend on the computations at the remaining grid-points. Therefore, no boundary conditions are needed when (2.13) is handled.

The fact that the boundary conditions can be introduced in a natural way when the splitting based on (2.11) - (2.13) is selected is a great advantage of this splitting procedure, because the boundary conditions are normally causing problems when splitting techniques are used.

2.2.3 A third sequential splitting procedure

The new DEM splitting procedure was obtained from the old one by reducing the number of operators (combining together some of them; see the previous section) as well as by changing the order of the computations. A new splitting procedure can also be obtained by keeping the number of simpler operators the same as in the old DEM splitting procedures (i.e., by keeping $m = 5$), but by taking another combination of the terms from the right-hand-side of (2.4) in the formation of the simpler operators. An example will be given in this sub-section as an illustration of this possibility.

Consider again (2.4). The operator in the right-hand-side can be represented as a sum of five simpler operators not only as shown in (2.5) - (2.9), but also in the following way:

$$\frac{\partial \mathbf{c}^{[1]}(\mathbf{x},t)}{\partial t} = -\frac{\partial (u\mathbf{c}^{[1]})}{\partial x} - \frac{\partial (v\mathbf{c}^{[1]})}{\partial y} - \frac{\partial (w\mathbf{c}^{[1]})}{\partial z} \tag{2.15}$$

$$\frac{\partial \mathbf{c}^{[2]}(\mathbf{x},t)}{\partial t} = \frac{\partial}{\partial x}\left(K_x \frac{\partial \mathbf{c}^{[2]}}{\partial x}\right) + \frac{\partial}{\partial y}\left(K_y \frac{\partial \mathbf{c}^{[2]}}{\partial y}\right) + \frac{\partial}{\partial z}\left(K_z \frac{\partial \mathbf{c}^{[2]}}{\partial z}\right) \tag{2.16}$$

$$\frac{\partial \mathbf{c}^{[3]}(\mathbf{x},t)}{\partial t} = -\kappa\, \mathbf{c}^{[3]} \tag{2.17}$$

$$\frac{\partial \mathbf{c}^{[4]}(\mathbf{x},t)}{\partial t} = E \tag{2.18}$$

$$\frac{\partial \mathbf{c}^{[5]}(\mathbf{x},t)}{\partial t} = Q(\mathbf{c}^{[5]}) \tag{2.19}$$

It is clear that if the operators in the right-hand-sides of (2.15) - (2.19) are denoted by $A^{[i]}$, $i = 1, 2, 3, 4, 5$, then the order in which the computations are carried out when this sequential splitting procedure is used is the same as in (2.10), i.e., as in the case where the old DEM sequential splitting procedure is used. However, note that the simpler operators used in splitting procedure described in this sub-section and in the old DEM splitting procedure are different; compare the right-hand-sides of (2.5) - (2.9) with the right-hand-sides of (2.15) - (2.19).

The sequential splitting procedure based on the formulae (2.15) - (2.19) has been used by Dimov et al. ([68], [69]) in the derivation of some important theoretical and practical results for splitting procedures that are applicable in air pollution modelling.

2.3 Symmetric splitting procedures

Consider (2.2) and (2.3). Assume that the first $m-1$ simpler sub-problems in (2.3) are solved using a splitting time-step size $\tau/2$. Assume that the last of the simpler sub-problems, the mth sub-problem, is after that treated using a splitting time-step size τ. Finally, assume that the computations at the splitting time-step under consideration are completed by performing (again, but in reverse order) the computations with the first $m-1$ simpler problems by using a splitting time-step size $\tau/2$. It is easy to express the computations using same representation as in (2.10) and (2.14). For the general case of m simpler operators, this can be done as follows:

$$(A^{[1]})_{\tau/2}, (A^{[2]})_{\tau/2}, \ldots, (A^{[m-1]})_{\tau/2}, (A^{[m]})_{\tau}, (A^{[m-1]})_{\tau/2}, \ldots, \tag{2.20}$$

$$\ldots, (A^{[2]})_{\tau/2}, (A^{[1]})_{\tau/2}.$$

The splitting procedure is called symmetric when it is defined by the equality (2.20). Different properties of symmetric splitting procedures are discussed in McRae et al. [181], Hundsdorfer and Verwer [142], Marchuk, [172], [173], [175], [177] and Strang, [235]. Splitting procedures of this type are rather popular in air pollution modelling.

2.4 Weighted sequential splitting procedures

Consider again (2.2) and (2.3). Assume that the computations at the splitting time-step under consideration are carried out by applying successively the sub-problems as shown by the following two sequences of operators:

$$(A^{[1]})_\tau, (A^{[2]})_\tau, \ldots, (A^{[m-1]})_\tau, (A^{[m]})_\tau \tag{2.21}$$

and

$$(A^{[m]})_\tau, (A^{[m-1]})_\tau, \ldots, (A^{[2]})_\tau, (A^{[1]})_\tau. \tag{2.22}$$

It is seen from (2.21) and (2.22) that only the order in which the simpler sub-problems are treated is different. In the first of these two formulae the computations are performed by taking sequentially the operators (the index numbers of the operators being increasing). The splitting procedure in the second formula (2.22) is also sequential, but now the index numbers of the operators are decreasing.

Denote by $\mathbf{c}_{forward}$ and $\mathbf{c}_{backward}$ the results obtained when (2.21) and (2.22) are applied at a given splitting time-step. Then an approximation:

$$\mathbf{c} = \theta \ \mathbf{c}_{forward} + (1 - \theta) \ \mathbf{c}_{backward} \tag{2.23}$$

of the solution of the original problem (2.1) at the splitting time-step under consideration is calculated using some weighting parameter θ and one can proceed with the calculations for the next splitting time-step.

The splitting procedures based on the performance of the calculations at each splitting time-step using the formulae (2.21), (2.22) and (2.23) are called weighted sequential splitting procedures.

The definition of weighted sequential splitting procedures given in this subsection can be slightly generalized. In (2.22) one can apply ordering with upper indices $\{r_1, r_2, \ldots, r_{m-1}, r_m\}$ instead of ordering with upper indices $\{m, m-1, \ldots, 2, 1\}$, where $\{r_1, r_2, \ldots, r_{m-1}, r_m\}$ is any permutation of $\{1, 2, \ldots, m-1, m\}$. Then the procedure defined by (2.21) and (2.22) is a special case of the generalized weighted splitting procedures. Probably, it is most natural to use the weighted splitting procedure defined by (2.21) and (2.22).

2.5 Weighted symmetric splitting procedures

Assume that the calculations at each splitting time-step are carried out in the orders that are shown in the two sequences given below; see (2.24) and (2.25):

$$(A^{[1]})_{\tau/2}, (A^{[2]})_{\tau/2}, \ldots, (A^{[m-1]})_{\tau/2}, (A^{[m]})_{\tau}, (A^{[m-1]})_{\tau/2}, \ldots, \\ \ldots, (A^{[2]})_{\tau/2}, (A^{[1]})_{\tau/2} \tag{2.24}$$

and

$$(A^{[m]})_{\tau/2}, (A^{[m-1]})_{\tau/2}, \ldots, (A^{[2]})_{\tau/2}, (A^{[1]})_{\tau}, (A^{[2]})_{\tau/2}, \ldots, \\ \ldots, (A^{[m-1]})_{\tau/2}, (A^{[m]})_{\tau/2}. \tag{2.25}$$

It is seen that the same principles as in the previous section are used. However, symmetric splitting procedures are used both in the forward mode and in the backward mode; in fact, formula (2.24) is precisely the same as (2.20).

Denote again (as in the previous section) by $\mathbf{c}_{forward}$ and $\mathbf{c}_{backward}$ the results obtained when (2.24) and (2.25) are applied at a given splitting time-step. Then, as in the previous section, the approximation:

$$\mathbf{c} = \theta\ \mathbf{c}_{forward} + (1-\theta)\ \mathbf{c}_{backward} \tag{2.26}$$

of the solution of the original problem (2.1) at the splitting time-step under consideration is calculated using some weighting parameter θ and one can proceed by the calculations for the next splitting time-step.

Some results showing that weighted splitting procedures have some advantages with regard to accuracy when compared with other splitting procedures can be found in Botchev et al. [23].

The splitting procedures based on the performance of the calculations at each splitting time-step using the formulae (2.24), (2.25) and (2.26) are called weighted symmetric splitting procedures.

It should be mentioned here that the definition of weighted symmetric splitting procedures can be generalized in a similar way as the generalization of the weighted sequential procedures, which was described at the end of the previous sub-section.

2.6 Advantages and disadvantages of the splitting procedures

The replacement of the original problem (2.1) by a series of simpler problems containing the operators given in (2.2) has several advantages:

- It is in general much easier to find optimal (or, at least, good) numerical methods for the simpler systems that are generated by the operators in (2.2) than for the big system arising after the direct discretization of operator A in the right-hand-side of (2.1).

- It is easy to change numerical algorithms and physical mechanisms used in the different sub-problems arising as a results of the implementation of the splitting procedure chosen. Thus, improvements of the code can easily be achieved.

- The application of splitting procedure facilitates the preparation of efficient codes for different parallel computers. In many cases, the splitting procedure is leading directly to parallel tasks. Consider, for example, the second of the sub-models defined by (2.12) in the new DEM splitting procedure. It is immediately seen that this sub-model consists of N_s parallel tasks (where N_s is the number of chemical species studied by the model under consideration). The ability of the splitting procedures to create directly parallel tasks will be studied in detail in Chapter 7.

It must be emphasized here that, while the use of splitting procedures facilitates the treatment of large-scale air pollution models (as well as the treatment of many large-scale models arising in different fields of science and engineering), the splitting procedures introduce splitting errors. It is difficult to control these errors. Therefore, it is desirable to avoid the use of splitting procedures when large-scale mathematical models are treated numerically on computers. Unfortunately, it is hard to see how this can be done at present. This is why, as mentioned in the beginning of this chapter, splitting procedures are still used in all operational large-scale air pollution models (and also in many other large-scale models).

More research results related to the splitting procedures are needed. It is especially important to develop reliable and robust tools for the automatically evaluating the error caused by the use of splitting procedures when these are applied in large scientific and engineering problems.

2.7 Comparison of different splitting procedures

It is desirable to define some clear rules for choosing the best possible splitting procedure. However, this is not an easy task (and, in fact, the solution of this task is a great challenge). The choice depends essentially on the following factors:

- on the particular application (i.e., on the particular model that is to be handled),

- on the numerical methods selected for the different sub-problems,

- on the grid resolution that is used in the numerical treatment of the model, and

- on several other issues.

An attempt to give some recommendations about the choice of a suitable splitting procedure according to two important criteria:

- the computational cost, and
- the accuracy of the results

will be made in this section. We shall also try, in the next section, to justify the conclusions by performing some numerical experiments. However, it is important to emphasize, as at the end of the previous section, that much more research effort is still needed in this area.

2.7.1 Comparing the computational cost

Assume that the number of simpler operators m used in (2.2) is fixed. Assume also that the same numerical methods are used in the treatment of the sub-problems containing the simpler operators $A^{[i]}$, $i = 1, 2, \ldots, m$, when different types of splitting procedures are used (more precisely, the numerical methods may be different for two operators $A^{[i]}$ and $A^{[j]}$, but the same numerical method for treatment of $A^{[i]}$ is to be used for the different types of splitting procedures). Some recommendations related to the choice of a splitting procedure can be given when such assumptions are made.

- The sequential splitting procedures have the cheapest computational cost (among the four types of splitting procedures) when the number of simple operators m is fixed.
- If a symmetric splitting procedure is to be selected, then it is best (in the efforts to reduce the computational work as much as possible) to put the most expensive operator at the end of (2.2) or, in other words, in the middle of (2.20). In this way the total computational cost of the symmetric splitting procedure is reduced, but this cost will always be higher than the cost of the corresponding sequential procedure.
- The computational cost of a weighted sequential splitting procedure is higher, by a factor approximately equal to two, than that of the corresponding sequential procedure. If two processors are available when the weighted sequential procedure is used and only one processor is applied with the sequential splitting procedure, then the computer time needed will be nearly the same for the two splitting procedures. This is due to the fact that

  (a) the two tasks (2.21) and (2.22) can be carried out in parallel, and

  (b) the parallelization will be efficient, because the computational cost of (2.21) and (2.22) is much greater than the computational cost of (2.23).

  However, it must also be emphasized here that if the code based on the use of a sequential splitting procedure is parallelized and if the fact that two processors are available is utilized also when the sequential splitting procedure is run, then this is no more true (it will be shown in Section 7 that it is possible to develop efficient parallel codes in the case when sequential splitting procedures are used). Of course, a similar conclusion can be drawn if $2p$ processors are available when the weighted sequential splitting procedure is used, while the number of processors available when the sequential splitting procedure is used is only p: if the parallel codes for the two splitting procedures are scalable (see Chapter 7), then approximately the same computational time will be used for the two splitting procedures. However, if $2p$ processors are also available when the sequential splitting procedure is used, then the above conclusion is in general no longer true (assuming again that the sequential splitting procedure is efficiently parallelized and that the parallel algorithm is scalable; see Chapter 7).

- Similar conclusions (as those for the relationship between weighted sequential splitting procedures and sequential splitting procedures) are valid for the relationship between weighted symmetric splitting procedures and symmetric splitting procedures.

It follows from the above discussion that the choice of a sequential splitting procedure is more preferable than the choice of a splitting procedure from the other three cases in the following two situations:

- when the accuracy requirements are relatively low

and/or

- when the computational time is the most important issue.

2.7.2 Comparing the accuracy properties

The accuracy of computations is another very important requirement. The accuracy of the different types of splitting procedures will be discussed in this subsection.

- All sequential splitting procedures are of first order, i.e., of order $O(\tau)$. This means that the errors due to these splitting procedures are proportional to the splitting time-step size τ. Thus, if the time-step size τ is sufficiently small and if the error due to the use of the splitting procedure chosen is dominant, then one should expect the errors caused by the splitting procedure to be reduced by a factor of two when the splitting time-step size is reduced from τ to $\tau/2$.

- Symmetric splitting procedures of second order, of order $O(\tau^2)$, can be constructed. This means that if the splitting time-step size τ is sufficiently small and if the error due to the use of the splitting procedure chosen is dominant, then one should expect the errors caused by the splitting procedure to be reduced by a factor of four when the splitting time-step size is reduced from τ to $\tau/2$.

- Weighted sequential and symmetric splitting procedure of order higher than two can be constructed.

The discussion in this subsection shows that if the accuracy is the most important issue, then symmetric splitting procedures and even weighted (either sequential or symmetric) splitting procedures are to be selected.

Combining the conclusions in this subsection with those in the previous subsection makes it clear that the selection of a splitting procedure is a trade off between two requirements (the requirement for reducing the computational time and the requirement for achieving sufficiently accurate results). Unfortunately, these two requirements are working in opposite directions. This is why in the general case it is very difficult to find a choice that satisfies sufficiently well both requirements.

Normally, the results related to the order of the local error are derived by investigating only the magnitude of the first non-vanishing term in the difference between the expansion of the unknown function in Taylor series and the splitting procedure. Some stronger results related to the convergence of the splitting procedures are even more desirable. Such results were recently reported by Faragó and Havasi [98].

If some assumptions are satisfied, then the errors due to the splitting procedures can vanish. Such results were proved by Lanser and Verwer ([157]) using the concept of "L-commutativity of two operators". It is not necessary to define formally the concept of "L-commutativity of two operators" in this book. It is sufficient to point out that

- if the operators are linear, then L-commutativity reduces to ordinary commutativity,
- if the representation from (2.2) contains only two operators (i.e., if $m = 2$) and if these operators L-commute then the errors due to the splitting procedure disappear, and
- if more than two operators are involved in the splitting procedure under consideration, then in the general case the errors due to splitting will vanish only if all pairs of operators L-commute.

Some of the results proved by Lanser and Verwer ([157]) were slightly improved by Dimov et al. ([68]). The results proved in Dimov et al. [68] are related to the third sequential splitting procedure defined by (2.15) - (2.19). The main result can be formulated by the following theorem.

Theorem 2.1. *Consider the splitting defined by (2.15 - 2.19). The following statements are proved in Dimov et al. ([68]).*

(A) If

$$\frac{\partial u}{\partial x} + \frac{\partial v}{\partial y} + \frac{\partial w}{\partial z} = 0, \tag{2.27}$$

$$\frac{\partial \kappa_s}{\partial x} + \frac{\partial \kappa_s}{\partial y} + \frac{\partial \kappa_s}{\partial z} = 0, \quad s = 1, 2, \ldots, N_s \tag{2.28}$$

$$\frac{\partial E_s}{\partial x} + \frac{\partial E_s}{\partial y} + \frac{\partial E_s}{\partial z} = 0, \quad s = 1, 2, \ldots, N_s, \tag{2.29}$$

then the pairs $(A^{[1]}, A^{[3]})$, $(A^{[1]}, A^{[4]})$, $(A^{[1]}, A^{[5]})$, $(A^{[2]}, A^{[3]})$ *and* $(A^{[1]}, A^{[5]})$ *L-commute.*

(B) If furthermore

$$K_x, K_y, K_z \quad \text{and} \quad \kappa \quad \text{are} \quad \text{constants}, \tag{2.30}$$

while u, v, w, Q *and* c *are linear functions, then the pairs* $(A^{[1]}, A^{[2]})$, $(A^{[2]}, A^{[5]})$ *and* $(A^{[3]}, A^{[5]})$ *L-commute.*

(C) Finally, if also E *lies in the null space of the Jacobian of* Q *(i.e., if* $E \in \ker(\partial Q/\partial c)$*), then the pair* $(A^{[4]}, A^{[5]})$ *L-commutes.*

The conditions of Theorem 2.1 are too restrictive and one should not expect these conditions to be satisfied in realistic cases related to large-scale air pollution models. However, similar results for splitting procedures applied to models arising in some other fields of science and engineering might be very useful.

2.8 Numerical experiments

Many numerical experiments were carried out in the efforts to justify the reliability of the splitting procedure in the solution of some classes of large-scale problems. The selected tests are rather artificial, but they allow us to study different properties of the splitting procedures and to draw some valuable conclusions.

Some of the numerical experiments will be presented in this section. Four important cases were studied by these experiments:

- The behaviour of the results in the case where only splitting errors appear in the computations.
- The behaviour of the results in the case where the splitting errors are appearing together with errors due to the spatial discretization and the errors due to the time integration.
- The behaviour of the results in the case where there are no splitting errors (i.e., all pairs of the involved operators L-commute).
- The behaviour of the results in the case where different splitting procedures and different numerical methods are used.

Some information about the way in which the numerical results will be presented will be given in the next subsection.

2.8.1 Information about the presentation of the results

The analytical solution of all examples presented in the following subsections is known and this fact is exploited to calculate an estimation of the numerical errors made in the computer treatment of these examples.

The magnitude of the errors is measured by the parameter $ERRMAX$, which is calculated in the following way. Assume that N_x and N_y are the numbers of grid-points in the Ox and Oy axes, respectively. At every time-step n, $n \in \{1, 2, \ldots, N\}$, approximations of the solution at the $N_x \times N_y$ grid-points are to be calculated. We shall denote these approximations by

$$c_{i,j,n} \approx c(x_i, y_j, t_n) \tag{2.31}$$

with

$$i \in \{1, 2, \ldots, N_x\}, \;\; j \in \{1, 2, \ldots, N_y\}, \;\; n \in \{1, 2, \ldots, N\}. \tag{2.32}$$

Then the error ERR_n at time-step n is evaluated by

$$ERR_n = \frac{DIFFERENCE_n}{MAXVALUE_n}, \tag{2.33}$$

where

$$DIFFERENCE_n = \max_{i=1,2,\ldots,N_x,\; j=1,2,\ldots,N_y} (\mid c_{i,j,n} - c(x_i, y_j, t_n) \mid) \tag{2.34}$$

$$MAXVALUE_n = \max_{i=1,2,\ldots,N_x,\; j=1,2,\ldots,N_y} (\mid c(x_i, y_j, t_n) \mid). \tag{2.35}$$

The global error ERRMAX (which is the maximal error over the whole time-integration interval) is evaluated by

$$ERRMAX = \max_{n=1,2,\ldots,N} (ERR_n). \tag{2.36}$$

The values of $ERRMAX$, rounded to the third significant digit, will be considered in the tables.

The two-dimensional case is discussed in this section. Three-dimensional examples will also be used in this chapter. It is clear that values of $ERRMAX$ for three-dimensional problems can be calculated in a similar manner.

It is necessary here to discuss shortly the choice of boundary conditions. Boundary conditions are, of course, very important. This is a problem, which is, in general, not fully resolved yet. The boundary conditions normally cause great difficulties in the large-scale models, even when no splitting is applied and sometimes even when perfect numerical methods are used. Therefore, the boundary condition should be carefully chosen. In all examples used in this chapter we applied exact Dirichlet boundary conditions for all modules (i.e., for all sub-models obtained when the selected splitting procedure was applied). This was possible, because we know the exact solutions of all these sub-models. Such an approach is fully justified, because we are only interested in the investigation of the relationship between splitting errors and errors caused by the numerical methods (and, therefore, we do not want the errors caused by the boundary conditions to interfere with the other errors).

The purpose with the numerical examples, which are given in this chapter, is not to test the performance of different splitting procedures. The main purpose is to show that the errors caused by any splitting procedure are only a part of the total error. Therefore, it is essential to know both when these errors become important and why they are becoming important. To answer these questions, we have designed examples that allow us to quantify the significance of the splitting errors in four carefully chosen and very important situations that were formulated in the beginning of this section. If we treat these examples with different splitting procedures, then we shall obtain results that are qualitatively the same. This is why the use of only one splitting procedure is quite sufficient when the relationship between splitting errors and other errors is studied. We have shown, among other things, that the behaviour of the total error reflects the order of the splitting procedure

only when the splitting errors are
dominat over the other errors.

This explains the importance of our analysis "accuracy versus computational cost" in a previous section of this chapter: there is no use to select a more accurate but also a more computationally expensive splitting procedure when we are not sure that the splitting errors are dominant over the errors caused by other sources. In other words, the choice of a more accurate splitting procedure should normally be connected with a choice of more accurate numerical methods if the numerical and the splitting errors are the only errors that appear in the computational process.

2.8.2 Only splitting errors appear in the computations

The splitting errors are normally not the only errors that appear during the computations. Other important sources of errors are the errors caused by the spatial discretization of the problem as well as the errors caused by the time discretization. Of course, there are many other sources for errors (errors due to uncertainties in the input data, errors due to uncertainties in the mechanisms by which the physical and chemical processes are described mathematically in the model, etc.). It is difficult to study the errors due to the splitting procedure when these are mixed with errors from other sources. Therefore, it is important to construct examples in which all other errors are equal to zero. The extra benefit of experiments with such examples is that one can study whether the behaviour of the errors when the splitting time-step size is reduced is the same as the behaviour that is expected. For example, if

- the splitting procedure is of order one, and

- the splitting stepsize is sufficiently small

then one should expect the error to be reduced by a factor approximately equal to two when the splitting time-step size is reduced by a factor of two. It should be stressed here that the same effect can be observed when other errors do appear during the computations, but the splitting errors are much larger than the other errors. However, in practice it is very difficult to establish that for the particular problem under consideration the splitting errors are much larger than the other errors. This is why a test, in which all errors excepting the splitting errors are vanishing, is very useful in the efforts to check the correctness of the implementation of the particular splitting procedure used in the model.

Consider the following problem:

$$\frac{\partial c}{\partial t} = y\frac{\partial c}{\partial x} - x\frac{\partial c}{\partial y} + 1 + x - y, \quad t \in (0,1], \qquad c(x,y,0) = x + y \tag{2.37}$$

with exact solution

$$c(x,y,t) = x + y + t. \tag{2.38}$$

Assume that a sequential splitting procedure is applied in the solution of (2.37) by using the following two simpler operators:

$$\frac{\partial c^{[1]}}{\partial t} = y\frac{\partial c^{[1]}}{\partial x} - x\frac{\partial c^{[1]}}{\partial y}, \tag{2.39}$$

$$\frac{\partial c^{[2]}}{\partial t} = 1 + x - y. \tag{2.40}$$

Linear finite elements are used in the semi-discretization of the first sub-problem (2.39). The resulting system of ODEs is treated by third-order predictor-corrector method. The second sub-problem (2.40) is solved exactly. Since the exact solution (2.38) of (2.37) is a linear function in x, y and t, this means that the numerical methods are not causing errors. Thus, the only error is caused by the splitting procedure.

Some results obtained by using different spatial discretizations and different time-step sizes (the time-step size Δt used in the time-integration method and the splitting time-step size τ are the same in this experiment) are given in Table 2.1 and Table 2.2. For this simple problem the results obtained without splitting were also calculated and presented in the tables.

Two important conclusions can be drawn using the results given in Table 2.1 and Table 2.2:

Table 2.1: The values of $ERRMAX$ obtained when (2.37) is solved without splitting and with splitting (the number of time-steps is 1500 for all grids, but the values of $ERRMAX$ are rather insensitive to variations of the number of the time-steps when no splitting is used). The values of $ERRMAX$ are close to the machine precision when no splitting is used. The values of $ERRMAX$ are of order $O(10^{-4})$ for all grids when the splitting time-step size is kept constant for all spatial grids.

Grid	Without splitting	With splitting
8×8	3.34E-14	1.70E-4
16×16	3.79E-14	1.80E-4
32×32	3.99E-14	1.80E-4
96×96	7.21E-14	1.97E-4
288×288	4.57E-14	1.99E-4
480×480	4.28E-14	1.99E-4

Table 2.2: The values of $ERRMAX$ obtained when (2.37) is solved using different numbers of time-steps without splitting and with splitting. The 16×16 grid is used in the discretization of the spatial derivatives. The experiment shows that the values of $ERRMAX$ does not depend on the number of time-steps when no splitting is used, while in the case where splitting is applied, the values of $ERRMAX$ are reduced when the number of steps is increased, because as stated in the text $\triangle t = \tau$.

Steps	Stepsize	Without splitting	With splitting
10	0.1	4.35E-14	2.42E-2
100	0.01	4.36E-15	2.39E-3
1000	0.001	5.76E-14	2.39E-4
10000	0.0001	7.24E-13	2.39E-5

- As mentioned above, there are no errors caused by the numerical algorithms. Therefore, all errors (excepting the rounding errors, which are negligible, because the computations were carried out in double precision) are caused by the use of the splitting procedure. This implies that refining the grids should have no effect on the accuracy of the approximate solution. This statement is fully confirmed by the results shown in Table 2.1.

- Since the sequential splitting procedure used is of first order, one should expect the errors to depend linearly on the splitting time-step size. The results shown in Table 2.2 indicate that this is true (reducing the splitting time-step size by a factor of ten is leading to a reduction of the errors of the numerical solution by a factor approximately equal to ten).

2.8.3 Splitting errors versus other numerical errors

The combined effect of splitting errors and errors caused

- by the numerical algorithms applied in the discretization of the spatial derivatives, and
- by the numerical methods applied in the treatment of the systems of ODEs arising after the semi-discretization

will be illustrated in this subsection.

Consider the following three-dimensional problem:

$$\frac{\partial c}{\partial t} = y\frac{\partial c}{\partial x} - x\frac{\partial c}{\partial y} + \frac{\partial c}{\partial z} + y(\cos z - \sin z) + x(\sin z + \cos z), \quad t \in [0,1], \tag{2.41}$$

with an initial condition

$$c(x,y,z,0) = e^z(x^2 + y^2). \tag{2.42}$$

The analytical solution of this problem is

$$c(x,y,z,t) = e^{z+t}(x^2 + y^2) + t[y(\cos z - \sin z) + x(\sin z + \cos z)]. \tag{2.43}$$

A sequential splitting procedure is used in the solution of this problem. This procedure is based on the sequential treatment of the following sub-problems:

$$\frac{\partial c^{[1]}}{\partial t} = y\frac{\partial c^{[1]}}{\partial x} - x\frac{\partial c^{[1]}}{\partial y}, \tag{2.44}$$

$$\frac{\partial c^{[2]}}{\partial t} = y(\cos z - \sin z) + x(\sin z + \cos z) \tag{2.45}$$

$$\frac{\partial c^{[3]}}{\partial t} = \frac{\partial c^{[3]}}{\partial z} \tag{2.46}$$

Results obtained with this splitting procedure for $\tau = \Delta t$ are given in Table 2.3 for different grid sizes and different values of Δt.

The results in Table 2.3 indicate that the error due to the spatial discretization is very quickly becoming dominant (excepting the results for the finest grid). Further reductions of the Δt do not lead to a reduction of the total error when the errors caused by the spatial discretization become dominant.

Table 2.3: Values of $ERRMAX$ obtained when the problem defined by (2.41) and (2.42) is solved with the splitting procedure (2.44) - (2.46) using different numbers of time-steps and different spatial grids. In this experiment, the splitting time-step size τ is equal to the time-step size $\triangle t$ used in the time-integration algorithm.

Time-step $\triangle t$	16x16x16	32x32x32	64x64x64	128x128x128
8.0E-3	1.58E-2	1.92E-2	unstable	unstable
4.0E-3	7.99E-3	9.03E-3	unstable	unstable
2.0E-3	4.03E-3	4.55E-3	5.07E-3	unstable
1.0E-3	2.22E-3	2.33E-3	2.53E-3	unstable
5.0E-4	1.32E-3	1.22E-3	1.28E-3	1.35E-3
2.5E-4	1.00E-3	6.73E-4	6.58E-4	6.74E-4
1.25E-4	1.02E-3	4.02E-4	3.47E-4	3.43E-4
6.25E-5	1.03E-3	2.96E-4	1.91E-4	1.77E-4
3.125E-5	1.03E-3	2.95E-4	1.14E-4	9.32E-5
1.5625E-5	1.03E-3	2.94E-4	8.11E-5	5.15E-5
7.8125E-6	1.03E-3	2.94E-4	8.09E-5	3.07E-5
3.90625E-6	1.03E-3	2.94E-4	8.08E-5	2.15E-5

The splitting time-step size and the time-step size used in the integration of the system of ODEs arising after the semi-discretization were, as stated above, equal when the results shown in Table 2.3 were calculated (i.e., $\tau = \triangle t$). This does not allow us to distinguish between errors due to the use of different splitting stepsizes τ and errors caused by use of different stepsizes $\triangle t$ in the solution of the system of ODEs. This is why some experiments in which τ is not equal to $\triangle t$ are also needed and were carried out. More precisely, the number of splitting steps was kept constant and the number of time-steps per a splitting step was varied. Some results obtained in such experiments are given in Table 2.4 for the $16 \times 16 \times 16$ grid. In the first column of Table 2.4 the values of $\triangle t$ (the time-step size used in the time-integration algorithm) are given. In the first row in the table the values of splitting time-step size τ are displayed. This means that both the splitting time-step size τ and and the the time-step size $\triangle t$, which is used in the time-integration algorithm, are varied in this experiment.

It is seen from the results presented in Table 2.4 that if splitting time-step size τ is fixed, then it does not matter how many time-steps with a stepsize $\triangle t$ will be carried out per splitting time-step. In other words, the results shown Table 2.4 indicate that the errors caused by the splitting time-step size τ are dominating over the errors caused by other sources in this example.

Table 2.4: Values of $ERRMAX$ obtained when the problem defined by (2.41) and (2.42) is solved using with different numbers of time-steps $\triangle t$ per splitting time-step τ (the values of τ are given in the first row of the table).

Δt	$8.0E-3$	$4.0E-3$	$2.0E-3$	$1.0E-3$	$5.0E-4$	$2.5E-4$
8.0E-3	1.59E-2	-	-	-	-	-
4.0E-3	1.59E-2	7.99E-3	-	-	-	-
2.0E-3	1.59E-2	7.99E-3	4.03E-3	-	-	-
1.0E-3	1.59E-2	7.99E-3	4.03E-3	2.22E-3	-	-
5.0E-4	1.59E-2	7.99E-3	4.03E-3	2.22E-3	1.32E-3	-
2.5E-4	1.59E-2	7.99E-3	4.03E-3	2.22E-3	1.33E-3	1.00E-3
1.25E-4	1.59E-2	7.99E-3	4.03E-3	2.22E-3	1.33E-3	1.01E-3
6.25E-5	1.59E-2	7.99E-3	4.03E-3	2.22E-3	1.32E-3	1.01E-3
3.125E-5	1.59E-2	7.99E-3	4.03E-3	2.22E-3	1.32E-3	1.00E-3
1.5625E-5	1.59E-2	7.99E-3	4.03E-3	2.22E-3	1.32E-3	1.01E-3
7.8125E-6	1.59E-2	7.99E-3	4.03E-3	2.22E-3	1.32E-3	1.01E-3
3.90625E-6	1.59E-2	7.99E-3	4.03E-3	2.22E-3	1.32E-3	1.01E-3

2.8.4 Vanishing splitting errors

There are no errors caused by the splitting procedure when all pairs of simpler operators in (2.2) are L-commuting (it should be reiterated here that if both operators in a given pair are linear, then L-commutativity is reduced to ordinary commutativity of the involved operators).

In the following example the two operators involved commute and, therefore there is no error caused by the splitting procedure:

$$\frac{\partial c}{\partial t} = y\frac{\partial c}{\partial x} - x\frac{\partial c}{\partial y} + 1, \quad t \in (0,1], \qquad c(x,y,0) = x^2 + y^2. \tag{2.47}$$

The exact solution of (2.47) is given by

$$c(x,y,t) = x^2 + y^2 + t. \tag{2.48}$$

Sequential splitting procedure based on the following two operators is used:

$$\frac{\partial c^{[1]}}{\partial t} = y\frac{\partial c^{[1]}}{\partial x} - x\frac{\partial c^{[1]}}{\partial y}, \tag{2.49}$$

$$\frac{\partial c^{[2]}}{\partial t} = 1. \tag{2.50}$$

Table 2.5: The values of *ERRMAX* obtained when the problem defined by (2.47) is solved without splitting and with splitting (the values for the same grid are equal when rounded to three significant digits). The same time-step size $\triangle t = \tau = 1.5625 \times 10^{-5}$ has been used in the runs on different grids.

Grid	Without splitting	With splitting
16×16	1.47E-3	1.47E-3
32×32	3.97E-4	3.97E-4
96×96	4.67E-5	4.67E-5
288×288	5.35E-6	5.35E-6
480×480	1.94E-6	1.94E-6

Since these operators are linear and commutative, there are no errors due to the splitting procedure. Thus, the results obtained by using the splitting procedure should be the same as those obtained without using any splitting (of course, this is strictly speaking only true if there are no rounding errors, but the rounding errors are in this case negligible, because double precision is used in the computations). The results shown in Table 2.5 and in Table 2.6 show that the results are practically the same when no splitting is used and when splitting is used. This is true both when different spatial grids are applied (see Table 2.5) and when the time-step size used in the solution of the semi-discretized problem is varied (see Table 2.6). Since the exact solution of this problem is linear in the time variable t, the numerical methods used in the time-integration part are not causing errors. This means that the errors shown in Table 2.5 and Table 2.6 are caused only by the particular finite discretization used to discretize the spatial derivatives in the right-hand-side of (2.49).

Table 2.6: The values of *ERRMAX* obtained when the problem defined by (2.47) is solved using different time-steps Δt without splitting and with splitting. The 16×16 grid is used in the discretization of the spatial derivatives. The experiment shows that the values of *ERRMAX* does not depend on the length of time-steps. As in the previous table, these values remain the same both when no splitting is used and when splitting is applied.

$\Delta t = \tau$	Without splitting	With splitting
$3.33E-3$	1.47E-3	1.47E-3
$6.67E-3$	1.47E-3	1.47E-3
$6.67E-4$	1.47E-3	1.47E-3

2.8.5 Splitting procedures versus numerical methods

In this subsection we shall show that there is an interaction between the numerical methods and the splitting results. This means that

- if the splitting procedure has been chosen, then some rules have to be applied when the numerical method is to be selected, and
- if the numerical method has been chosen, then, again, some rules have to be applied when the splitting procedure is to be selected.

A simple example will be used (only in order to facilitate the understanding of the main ideas). Consider the system of ordinary differential equation:

$$\frac{dc_1}{dt} = c_2, \quad \frac{dc_2}{dt} = \frac{c_2(c_2 - 1)}{c_1}, \tag{2.51}$$

where $t \in [0, 10]$ and the initial values are given by

$$c_1(0) = 0.5, \quad c_2(0) = 0.5. \tag{2.52}$$

The problem defined by (2.51) and (2.52) is very simple. Nevertheless, it has several properties that are typical for the problems arising in the chemical part of air pollution models:

- the problem is non-linear,
- the mass preservation law holds, and
- the components of the solution vector are non-negative on the whole time-integration interval.

Moreover, there is at least one clear advantage when this problems is used: the exact solution of this problem is known. The components of the exact solution of the initial value problems defined by (2.51) and (2.52) are given by

$$c_1(t) = 1.0 - 0.5\exp(-t), \quad c_2(t) = 0.5\exp(-t). \tag{2.53}$$

Rewrite (2.51) as

$$\frac{dc_1}{dt} = f_1(c_1, c_2), \quad \frac{dc_2}{dt} = f_2(c_1, c_2). \tag{2.54}$$

The right-hand-side vector in (2.54) can be represented as a sum of two simpler operators defined by

$$f_1^{[1]} = 0.0, \quad f_2^{[1]} = -\frac{c_2}{c_1} \tag{2.55}$$

and

$$f_1^{[2]} = c_2, \quad f_2^{[2]} = \frac{(c_2)^2}{c_1}. \tag{2.56}$$

It is clear that (a) $f_1^{[1]} + f_1^{[2]} = f_1$, while $f_2^{[1]} + f_2^{[2]} = f_2$ and (b) the four types of splitting procedures, which were discussed in the beginning of this chapter can be constructed for the particular problem (2.51) using the simpler operators (2.55) and (2.56). We shall solve (2.51) both without splitting and with the four splitting procedures:

- sequential splitting,
- symmetric splitting,
- weighted sequential splitting (with $\theta = 0.5$), and
- weighted symmetric splitting (with $\theta = 0.5$).

Three numerical methods will be used in the solution (both in the case where no splitting is used and in the case where the four splitting procedures are applied). If the problem (2.51) is rewritten in vector form as

$$\frac{dc}{dt} = f(c), \tag{2.57}$$

where c is a vector with components c_1 and c_2, then the numerical methods, which will be used in this subsection, can be represented by the following formulae:

(A) Backward Euler Method.

This is an implicit first-order numerical method for the solution of ordinary differential equations (see, for example, Butcher [37], Hairer and Wanner [127], Hundsdorfer and Verwer [142], Lambert [155]). The calculations are carried out step by step using the formula:

$$c_{n+1} = c_n + \triangle t f(c_{n+1}). \tag{2.58}$$

where c is the vector containing the two components of the particular problem (2.51) and the sub-scripts denote the number of the time-step (which means that having an approximation c_n of the solution at $t = t_n$ it is possible to calculate the next approximation c_{n+1} at the next mesh-point $t_{n+1} = t_n + \triangle t$ using the above formula).

(B) Implicit Mid-point Rule.

This is an implicit second-order numerical method for the solution of ordinary differential equations (see, again, Butcher [37], Hairer and Wanner [127], Hundsdorfer and Verwer [142], Lambert [155]). The calculations are carried out step by step using the formula:

$$c_{n+1} = c_n + \Delta t f(0.5(c_n + c_{n+1})). \tag{2.59}$$

where c_{n+1} and c_n are defined in the same way as the corresponding quantities are defined for the Backward Euler Method (see the previous paragraph).

(C) Fully Implicit Three-stage Runge-Kutta Method.

This is an implicit sixth-order numerical method for the solution of ordinary differential equations (more details can be found Butcher [37], Hairer and Wanner [127], Hundsdorfer and Verwer [142], Lambert [155]). The calculations are carried out step by step using the following formula:

$$c_{n+1} = c_n + \Delta t[\alpha_1 f(Y_1) + \alpha_2 f(Y_2) + \alpha_3 f(Y_3)], \tag{2.60}$$

the quantities Y_1, Y_2 and Y_3 being calculated by

$$Y_1 = c_n + \Delta t[\gamma_{11} f(Y_1) + \gamma_{12} f(Y_2) + \gamma_{13} f(Y_3)], \tag{2.61}$$

$$Y_2 = c_n + \Delta t[\gamma_{21} f(Y_1) + \gamma_{22} f(Y_2) + \gamma_{23} f(Y_3)], \tag{2.62}$$

$$Y_3 = c_n + \Delta t[\gamma_{31} f(Y_1) + \gamma_{32} f(Y_2) + \gamma_{33} f(Y_3)]. \tag{2.63}$$

The coefficients in formulae (2.60) - (2.63 (i.e., the coefficients α_i and γ_{ij}, where $i = 1, 2, 3$ and $j = 1, 2, 3$) are given below:

$$\alpha_1 = \frac{5}{18}, \quad \alpha_2 = \frac{8}{18}, \quad \alpha_3 = \frac{5}{18}, \tag{2.64}$$

$$\gamma_{11} = \frac{5}{36}, \quad \gamma_{12} = \frac{10 - 3\sqrt{15}}{45}, \quad \gamma_{13} = \frac{25 - 6\sqrt{15}}{180}, \tag{2.65}$$

$$\gamma_{21} = \frac{10 + 3\sqrt{15}}{72}, \quad \gamma_{22} = \frac{2}{9}, \quad \gamma_{23} = \frac{10 - 3\sqrt{15}}{72}, \tag{2.66}$$

$$\gamma_{31} = \frac{25 + 6\sqrt{15}}{180}, \quad \gamma_{32} = \frac{10 + 3\sqrt{15}}{45}, \quad \gamma_{33} = \frac{5}{36}. \tag{2.67}$$

The particular problem (2.51), which is considered in this subsection is an autonomous system of ordinary differential equations (i.e., the independent variable

t does not participate explicitly in the right-hand-side of the system). If the non-autonomous problem $dc/dt = f(t,c)$ is solved, then three additional coefficients are needed:

$$\beta_1 = \frac{5-\sqrt{15}}{10}, \quad \beta_2 = \frac{1}{2}, \quad \beta_3 = \frac{5+\sqrt{15}}{36}. \tag{2.68}$$

In this case, i.e., when the system of ordinary differential equations is non-autonomous, it is necessary to calculate values of the right-hand-side function f at times $t_n + \beta_i \Delta t$, where $i = 1, 2, 3$ (much more details about the application of Runge-Kutta type of methods for non-autonomous systems can be found in Butcher [37], Hairer and Wanner [127], Hundsdorfer and Verwer [142], Lambert [155]).

The transformation defined by

$$z_1 = Y_1 - y_n, \quad z_2 = Y_2 - y_n, \quad z_3 = Y_3 - y_n \tag{2.69}$$

is sometimes advocated (see, for example, Hairer and Wanner [127]). Our experiments indicate that this transformation is very useful when chemical schemes for large-scale air pollution models are treated by the Fully Implicit Three-stage Runge-Kutta Method. The transformation (2.69) has been used in the implementation of the Fully Implicit Three-stage Runge-Kutta Method for our tests.

(D) The notation used in the tables presented in this subsection.

Numerical results are calculated in the following way. For each numerical method and for each splitting procedure (considering here "no splitting" as a special case of splitting), we start with $\Delta t = 0.25$. In this case the numerical integration is completed in 40 time-steps. Then we are successively increasing the number of time-steps by a factor of two (increasing the number of time-steps by a factor of two implies a decrease of the time-step size Δt by a factor of two).

The notation used in connection with the tables given in this subsection can be explained as follows:

- *"Steps"* refers to the number of time-steps used in the run under consideration,
- *"Splitting 0"* refers to the case where no splitting is used,
- *"Splitting 1"* refers to the case where sequential splitting is used,
- *"Splitting 2"* refers to the case where symmetric splitting is used,
- *"Splitting 3"* refers to the case where weighted sequential splitting is used,
- *"Splitting 4"* refers to the case where weighted symmetric splitting is used,
- the global errors found during the whole integration process will simply be called *errors*

- the ratios between the errors obtained in two successive runs will simply be called *ratios.*

(E) What should be expected when the numerical methods are combined with different splitting procedures?

If no splitting is used, then the results should reflect the order of the numerical method used to solve the system of differential equations. In our study this means that the order of accuracy is

- one for the Backward Euler Method,
- two for the Implicit Mid-point Rule, and
- six for the Fully Implicit Three-stage Runge-Kutta Method,

when no splitting procedure is used.

Assume now that some splitting procedure has been applied. Assume also that the order of accuracy of the splitting procedure selected is q (q is equal to one for the sequential splitting, while q is equal to two for the remaining three splitting procedures). It is clear that, if the order of the numerical method used is p, then one should expect the order r of the combined method (i.e., the method obtained when the splitting procedure chosen is applied together with some numerical method) to satisfy the relationship

$$r = \min(q, p). \tag{2.70}$$

Numerical experiments indicate that the relationship (2.70) will normally hold, but some artificial examples where this relationship is not satisfied can be constructed. For example, one can construct particular problems for which

$$r > \min(q, p). \tag{2.71}$$

It is not very clear how this fact can be exploited in practice. If a particular problem has to be solved again and again, one can try to design, for this particular problem, a splitting procedure and a numerical method such that for the combined method the relationship $r > \min(q,p)$ holds for the particular numerical methods selected and for the particular splitting procedure applied. There is no guarantee that such an attempt will be successful, but if the attempt is successful then the remaining runs (which are presumably many) can be run in a more efficient manner. At the end of this section it will be demonstrated that for some particular problems the relationship $r > \min(q,p)$ is satisfied.

It will be assumed in this subsection that the relationship $r = \min(q,p)$ holds and it will be investigated what are the consequences of this fact for the results obtained with different splittings and the numerical methods described above. Assume that the time-stepsize Δt is successively decreased by a factor of two in several runs. Then there are three possibilities for the ratios of the errors obtained in two successive runs:

- If the combined method (formed by the splitting procedure used and by the numerical method applied) is of order one, then one should expect the ratios to be approximately equal to two.
- If the combined method (formed by the splitting procedure used and by the numerical method applied) is of order two, then one should expect the ratios to be approximately equal to four.
- If the combined method is of order six (this happens only for the Fully Implicit Three-stage Runge-Kutta Method when no splitting is used), then one should expect the ratios to be approximately equal to 64 (excepting the case where the error becomes very small and, therefore, the accuracy is limited not by the combined numerical method, but by other factors as, for example, rounding errors, tolerances used to stop iterations, etc.).

The numerical results indicate that the behaviour of the error obtained during the computations is in fact determined by the rules formulated above.

(F) Behaviour of the errors obtained when numerical methods are combined with different splitting procedures.

(F1) Backward Euler Method

It is appropriate to start the discussion of the results obtained by different numerical methods with the case where the simple Backward Euler Method is used with different splitting procedures (including here the case where no splitting procedure is applied).

Results from 15 runs (beginning with a run where 40 time-steps are performed, increasing the number of time-steps by a factor of two after each run and finishing with a run where 327 680 rime-steps were performed) are given

- in Table 2.7, showing the accuracy achieved for different time-step sizes, and
- in Table 2.8, showing the ratios of the errors between successive runs.

It is seen that the reduction of the stepsize by a factor of two is always resulting in a reduction of the error by a factor of two. In other words, the combined method behaves in this situation as a method of order one for all kinds of splitting. The conclusion is that in general **the use of splitting procedures of order higher than one is not advisable when the numerical method used is of order one.**

If the process of doubling the number of time-steps, and consequently reducing the time-step size by a factor of two, is continued (actually the time-steps were

Table 2.7: Accuracy results achieved when the Backward Euler Method is used with five splitting procedures and when problem (2.51) is solved. The abbreviations *Steps*, *"Splitting 0"*, *"Splitting 1"*, *"Splitting 2"*, *"Splitting 3"* and *"Splitting 4"* are used for (i) the number of time-steps and (ii) the cases where no splitting, sequential splitting, symmetric splitting, weighted sequential splitting and weighted symmetric splitting are used in the computations.

Steps	Splitting 0	Splitting 1	Splitting 2	Splitting 3	Splitting 4
40	2.08E-2	2.08E-2	1.40E-1	2.60E-1	1.81E-1
80	1.09E-2	1.09E-2	7.40E-2	1.26E-2	9.22E-2
160	5.60E-3	5.60E-3	3.80E-2	6.26E-2	4.64E-2
320	2.84E-3	2.84E-3	1.93E-2	3.12E-2	2.33E-2
640	1.43E-3	1.48E-3	9.70E-3	1.56E-2	1.17E-2
1280	7.16E-4	7.16E-4	4.86E-3	7.81E-3	5.85E-3
2560	3.59E-4	3.59E-4	2.44E-3	3.91E-3	2.93E-3
5120	1.79E-4	1.79E-4	1.22E-3	1.95E-3	1.46E-3
10240	8.98E-5	8.98E-5	6.10E-4	9.76E-4	7.31E-4
20480	4.49E-5	4.49E-5	3.05E-4	4.88E-4	3.66E-4
40960	2.24E-5	2.24E-5	1.52E-4	2.44E-4	1.83E-4
81920	1.12E-5	1.12E-5	7.63E-5	1.22E-4	9.15E-5
163840	5.61E-6	5.61E-6	3.81E-5	6.10E-5	4.58E-5
327680	2.81E-6	2.81E-6	1.91E-5	3.05E-5	2.28E-5

increased in this way to 20 971 520 in the experiments), then the same trend can be observed (i.e. the accuracy is improved twice every time when the number of time-steps is doubled). Accuracy of order $O(10^{-8})$ has been achieved in the last run; the run with 20 971 520 time-steps.

It should be mentioned that the accuracy achieved when no splitting is used is precisely the same as the accuracy achieved when a sequential splitting procedure is used. This is a surprising result. Istvan Faragó from the University of Budapest investigated the situation closely. He proved that for the particular problem (2.51) the sequential splitting procedure performed using first the operator from (2.55) and then the operator from (2.56) does not cause additional errors (in comparison with the case where no splitting is used). However, this result is no longer true if the order of the operators in the sequential splitting procedure is reversed. This example illustrates the fact that **the order in which the simpler operators are**

Table 2.8: The ratios of the errors between successive runs that are obtained when the Backward Euler Method is used with five splitting procedures and when problem (2.51) is solved. The abbreviations *Steps*, *"Splitting 0"*, *"Splitting 1"*, *"Splitting 2"*, *"Splitting 3"* and *"Splitting 4"* are used for (i) the number of time-steps and (ii) the cases where no splitting, sequential splitting, symmetric splitting, weighted sequential splitting and weighted symmetric splitting are used in the computations.

Steps	Splitting 0	Splitting 1	Splitting 2	Splitting 3	Splitting 4
40	-	-	-	-	-
80	1.908	1.908	1.894	2.061	1.966
160	1.951	1.951	1.947	2.015	1.983
320	1.975	1.975	1.974	2.004	1.992
640	1.987	1.987	1.987	2.001	1.996
1280	1.994	1.994	1.993	2.000	1.998
2560	1.997	1.997	1.997	2.000	1.999
5120	1.999	1.999	1.998	2.000	2.000
10240	1.999	1.999	1.999	2.000	2.000
20480	2.000	2.000	2.000	2.000	2.000
40960	2.000	2.000	2.000	2.000	2.000
81920	2.000	2.000	2.000	2.000	2.000
163840	2.000	2.000	2.000	2.000	2.000
327680	2.000	2.000	2.000	2.000	2.000

applied in the splitting procedure might be important in some situations.

(F2) Implicit Mid-point Rule

Let us consider now the application of the Implicit Mid-point Rule with different splitting procedures (including also here the case where no splitting procedure is applied).

Results from 15 runs (beginning again with a run where 40 time-steps are performed, increasing the number of time-steps by a factor of two after each run and finishing with a run where 327 680 rime-steps were performed) are given

- in Table 2.9, showing the accuracy achieved for different time-step sizes, and
- in Table 2.10, showing the ratios of the errors between successive runs.

It is seen that the reduction of the stepsize by a factor of two is resulting in a reduction of the error by a factor of four in all cases excepting the case where the sequential splitting procedure is used. In the latter case the reduction of the stepsize by a factor of two is resulting in a reduction of the error by a factor of two. In other words, the combined method behaves in this situation as a method of order two for all splitting procedures excepting the sequential splitting procedure. For the sequential splitting procedure the combined method is of order one. The conclusion is that **the use of splitting procedures of order one is not advisable when the numerical method used is of order two.**

It should be noted that, if the weighted sequential splitting procedure is used together with the Implicit Mid-point Rule, then the combined method is performing as a method of order approximately equal to three for large time-step sizes. However, the combined method behaves as a method of order two when the time-step size is sufficiently small. It is not very clear what is the reason for this behaviour of the combined method. Some cancellation of errors is probably taking place when this particular test problem is run.

Table 2.9: Accuracy results achieved when the Implicit Mid-point Rule is used with five splitting procedures and when problem (2.51) is solved. The abbreviations *Steps*, *"Splitting 0"*, *"Splitting 1"*, *"Splitting 2"*, *"Splitting 3"* and *"Splitting 4"* are used for (i) the number of time-steps and (ii) the cases where no splitting, sequential splitting, symmetric splitting, weighted sequential splitting and weighted symmetric splitting are used in the computations.

Steps	Splitting 0	Splitting 1	Splitting 2	Splitting 3	Splitting 4
40	9.65E-04	2.00E-01	2.12E-02	6.67E-03	3.70E-03
80	2.40E-04	1.11E-01	5.47E-03	8.16E-04	9.64E-04
160	5.99E-05	5.88E-02	1.38E-03	1.17E-04	2.43E-04
320	1.50E-05	3.03E-02	3.45E-04	2.13E-05	6.10E-05
640	3.74E-06	1.54E-02	8.64E-05	4.49E-06	1.53E-05
1280	9.36E-07	7.75E-03	2.16E-05	1.03E-06	3.81E-06
2560	2.34E-07	3.89E-03	5.40E-06	2.45E-07	9.52E-07
5120	5.85E-08	1.95E-03	1.35E-06	5.99E-08	2.38E-07
10240	1.46E-08	9.75E-04	3.38E-07	1.48E-08	5.96E-08
20480	3.65E-09	4.88E-04	8.44E-08	3.68E-09	1.49E-08
40960	9.14E-10	2.44E-04	2.11E-08	9.16E-10	3.72E-09
81920	2.28E-10	1.22E-04	5.27E-09	2.29E-10	9.30E-10
163840	5.11E-11	6.10E-05	1.32E-09	5.72E-11	2.33E-10
327680	1.43E-11	3.05E-05	3.30E-10	1.43E-11	5.83E-11

(F3) Fully Implicit Three-stage Runge-Kutta Method

Finally, we shall consider the application of the Fully Implicit Three-stage Runge-Kutta Method when this method is used with different splitting procedures

Table 2.10: The ratios of the errors between successive runs that are obtained when the Implicit Mid-point Rule is used with five splitting procedures when when problem (2.51) is solved. The abbreviations *Steps*, *"Splitting 0"*, *"Splitting 1"*, *"Splitting 2"*, *"Splitting 3"* and *"Splitting 4"* are used for (i) the number of time-steps and (ii) the cases where no splitting, sequential splitting, symmetric splitting, weighted sequential splitting and weighted symmetric splitting are used in the computations.

Steps	Splitting 0	Splitting 1	Splitting 2	Splitting 3	Splitting 4
40	-	-	-	-	-
80	4.021	1.800	3.879	8.181	3.835
160	4.005	1.889	3.968	6.944	3.962
320	4.001	1.941	3.992	5.502	3.991
640	4.000	1.970	3.998	4.751	3.998
1280	4.000	1.985	3.999	4.377	3.999
2560	4.000	1.992	4.000	4.189	4.009
5120	4.000	1.996	4.000	4.094	4.000
10240	4.000	1.998	4.000	4.047	4.000
20480	4.000	1.999	4.000	4.024	4.000
40960	4.000	2.000	4.000	2.012	4.000
81920	4.000	2.000	4.000	4.006	4.000
163840	4.000	2.000	4.000	4.003	3.999
327680	4.000	2.000	4.000	4.003	3.996

(including also here the case where no splitting procedure is applied).

As for the previous two numerical methods (the Backward Euler Method and the Implicit Mid-point Rule), results from 15 runs (beginning with a run where 40 time-steps are performed, increasing the number of time-steps by a factor of two after each run and finishing with a run where 327 680 rime-steps were performed) are given

- in Table 2.11, showing the accuracy achieved for different time-step sizes, and
- in Table 2.12, showing the ratios of the errors between successive runs.

Consider first the case where no splitting is used. It is seen from Table 2.12 that the reduction of the stepsize by a factor of two is resulting, when no other factors are limiting the accuracy of the results, in a reduction of the error by a factor of 64 when no splitting is used (which should be expected, because the Fully Implicit Three-stage Runge-Kutta Method is of order six). The results, which are presented in Table 2.11, show that in this case the accuracy is very quickly approaching the machine precision, and the results cannot be further improved when this happens.

Table 2.11: Accuracy results achieved when the Fully Implicit Three-stage Runge Kutta Method is used with five splitting procedures and when problem (2.51) is solved. The abbreviations *Steps*, *"Splitting 0"*, *"Splitting 1"*, *"Splitting 2"*, *"Splitting 3"* and *"Splitting 4"* are used for (i) the number of time-steps and (ii) the cases where no splitting, sequential splitting, symmetric splitting, weighted sequential splitting and weighted symmetric splitting are used in the computations.

Steps	Splitting 0	Splitting 1	Splitting 2	Splitting 3	Splitting 4
40	4.47E-10	1.90E-01	1.95E-02	1.66E-02	2.62E-03
80	6.96E-12	1.08E-01	5.12E-03	3.39E-03	6.49E-04
160	1.09E-13	5.82E-02	1.30E-03	7.49E-04	1.62E-04
320	1.99E-15	3.01E-02	3.25E-04	1.75E-04	4.07E-05
640	6.66E-16	1.53E-02	8.13E-05	4.22E-05	1.02E-05
1280	2.78E-15	7.74E-03	2.03E-05	1.03E-05	2.54E-06
2560	1.78E-15	3.89E-03	5.08E-06	5.08E-06	6.36E-07
5120	4.77E-15	1.95E-03	1.27E-06	6.39E-07	1.59E-07
10240	3.22E-15	9.75E-04	3.17E-07	1.59E-07	3.97E-08
20480	6.88E-15	4.88E-04	7.94E-08	3.98E-08	9.93E-09
40960	8.54E-15	2.44E-04	1.99E-08	9.94E-09	2.48E-09
81920	2.08E-14	1.22E-04	4.97E-09	2.48E-09	6.21E-10
163840	9.77E-15	6.10E-05	1.24E-09	6.21E-10	1.55E-10
327680	1.23E-14	3.05E-05	3.10E-10	1.55E-10	3.86E-11

The reduction of the stepsize by a factor of two is resulting in a reduction of the error by a factor of two when the sequential splitting is used. This should also be expected because the sequential splitting procedure is of order one, see also (2.70).

The reduction of the stepsize by a factor of two is resulting in a reduction of the error by a factor greater than four in the beginning when the weighted sequential splitting is used, however the reduction is by a factor of four when the stepsize become small. Similar results, even more pronounced, were, as mentioned above, observed when the Implicit Mid-point Rule is used. It is not very clear why the reduction of the error is by a factor greater than four when the time-step size is larger than four when the time-stepssizes are large (probably some kind of cancelation of errors caused by the numerical method and errors caused by the splitting procedure takes place for this particular problem when the time-step size is large).

The reduction of the stepsize by a factor of two is resulting in a reduction of the error by a factor of four when the symmetric and the weighted symmetric splitting procedures are used.

Table 2.12: The ratios of the errors between successive runs that are obtained when the Fully Implicit Three-stage Runge Kutta Method is used with five splitting procedures and when problem (2.51) is solved. The abbreviations *Steps*, *"Splitting 0"*, *"Splitting 1"*, *"Splitting 2"*, *"Splitting 3" "Splitting 4"*, *Error* and *Ratio* are used for (i) the number of time-steps, (ii) the cases no splitting, sequential splitting, symmetric splitting, weighted sequential splitting and symmetric splitting, (iii) the global error found during the integration and (iv) the ratio between two successive errors.

Steps	Splitting 0	Splitting 1	Splitting 2	Splitting 3	Splitting 4
40	-	-	-	-	-
80	64.116	1.753	3.815	4.892	4.037
160	64.002	1.866	3.950	4.531	3.995
320	54.458	1.930	3.987	4.283	3.997
640	3.000	1.964	3.997	4.145	3.999
1280	0.240	1.982	3.999	4.074	4.000
2560	1.562	1.991	4.000	4.037	4.000
5120	0.372	1.995	4.000	4.019	4.000
10240	1.483	1.998	4.000	4.009	4.000
20480	0.468	1.999	4.000	4.005	4.000
40960	0.805	1.999	4.000	2.002	4.000
81920	0.412	2.000	4.000	4.001	4.000
163840	2.125	2.000	4.000	4.001	3.999
327680	0.793	2.000	4.000	4.000	3.996

Two important conclusions can be drawn using the numerical results presented in Table 2.11 and in Table 2.12.

- The example, which is treated in this subsection indicates that **the use of a numerical method of very high order is justified only if no splitting is used.**

- It might be worthwhile, in some situations at least, to try **to avoid the use of splitting procedures by applying numerical methods of high order.** The computational work per time-step will be increased if such an action is taken. This increase will be very substantial for large-scale problems. On the other side, however, the number of time-steps can sometimes be reduced dramatically (as the results in Table 2.11 indicate). Indeed, better accuracy (than the accuracy achieved by the other two numerical methods) is achieved when only 320 time-steps are performed. Let us reiterate here that performing 20 971 520 time-steps with the Backward Euler Method results in numerical errors of order $O(10^{-8})$. It is also worthwhile to mention the computational times: 0.18 seconds for 320 time-steps with the Fully Implicit Three-stage Runge-Kutta Method and 31 745 seconds for 20 971 520 time-steps with the Backward Euler Method.

(G) Summarizing the results related to the order of the combined method

The interaction between the numerical methods and the splitting procedures is illustrated in Table 2.13. It should be stressed here that the results given in Table 2.13 are strictly speaking valid only for the particular problem (2.51). One should expect that these results are also valid for other problems. However, more experiments are necessary in order to verify such a conjecture.

Table 2.13: The order of combined method (splitting procedure and numerical method) observed when the problem (2.51) is solved with five splitting procedure (considering here the runs without splitting as a special case of splitting) and three numerical methods. The figure "2+" for the weighted sequential splitting procedure applied with the implicit mid-point rule reflects the fact that the combined method seems to be of order higher than two for large time-steps (see Table 2.9). The order of the combined method (the sequential splitting procedure + the fully implicit three-stage Runge-Kutta method) seems also to be slightly higher than two (see Table 2.12).

Type of splitting	Backward Euler	Mid-point	Three-stage Runge-Kutta
No splitting	1	2	6
Sequential	1	1	1
Symmetric	1	2	2
Weighted sequential	1	2+	2
Weighted symmetric	1	2	2

The results in Table 2.13 reflect the general case where the order of the combined method r satisfies the relationship (2.70), i.e., r is equal to the minimum of the order p of the numerical method and the order q of the splitting procedure. It was mentioned in the beginning of this section that for some particular problems the relationship $r > \min(q, p)$ can be observed. This statement will now be illustrated by a simple numerical example. Consider the problem:

$$\frac{dc_1}{dt} = c_2, \quad \frac{dc_2}{dt} = -c_1, \tag{2.72}$$

where $t \in [0, 10]$ and the initial values are given by

$$c_1(0) = 0, \quad c_2(0) = 0. \tag{2.73}$$

The exact solution of the problem defined by (2.72)-(2.73) is given by

$$c_1(t) = \sin(t), \quad c_2(t) = \cos(t). \tag{2.74}$$

The problem (2.72)-(2.73) was run in the same manner as the previous problem, i.e., in the same manner as the problem defined by (2.53)-(2.54) was run. Results obtained with the Backward Euler Method (of order $p = 1$) combined with different splitting procedures will be studied. The accuracy of the obtained solution (with different time-step sizes and different splitting procedures) is shown in Table 2.14, while the ratios are displayed in Table 2.15.

It is clearly seen that for the case where no splitting is used and for the sequential splitting the order of the combined method is one (which should be expected because $p = q = 1$). However, for the remaining three cases (i.e., for the symmetric splitting, the weighted sequential splitting and the weighted symmetric splitting) the order of the combined method is not one (which should be expected and which was observed in Table 2.7 and Table 2.8), but two. This means that the relationship $r > \min(q, p)$ holds for this particular problem when the three splitting procedures of order $q = 2$ are used.

Table 2.14: Accuracy results achieved when the Backward Euler Method is used with five splitting procedures and when problem (2.72) is solved. The abbreviations *Steps*, *"Splitting 0"*, *"Splitting 1"*, *"Splitting 2"*, *"Splitting 3"* and *"Splitting 4"* are used for (i) the number of time-steps and (ii) the cases where no splitting, sequential splitting, symmetric splitting, weighted sequential splitting and weighted symmetric splitting are used in the computations.

Steps	Splitting 0	Splitting 1	Splitting 2	Splitting 3	Splitting 4
40	6.83E-01	1.46E-01	2.51E-02	1.01E-01	2.51E-02
80	4.42E-01	6.77E-02	6.27E-03	2.49E-02	6.20E-03
160	2.55E-01	3.25E-02	1.57E-03	6.20E-03	1.54E-03
320	1.37E-01	1.60E-02	3.92E-04	1.55E-03	3.86E-04
640	7.13E-02	7.90E-03	9.80E-05	3.86E-04	9.64E-05
1280	3.62E-02	3.93E-03	2.45E-05	9.65E-05	2.41E-05
2560	1.83E-02	1.96E-03	6.13E-06	2.41E-05	6.03E-06
5120	9.21E-03	9.78E-04	1.53E-06	6.03E-06	1.51E-06
10240	4.62E-03	4.89E-04	3.83E-07	1.50E-06	3.77E-07
20480	2.31E-03	2.44E-04	9.57E-08	3.77E-07	9.42E-08
40960	1.16E-03	1.22E-04	2.39E-08	9.42E-08	2.35E-08
81920	5.78E-04	6.10E-05	5.98E-09	2.35E-08	5.88E-09
163840	2.89E-04	3.05E-05	1.50E-09	5.88E-09	1.47E-09
327680	1.45E-04	1.53E-05	3.74E-10	1.47E-09	3.68E-10

Table 2.15: The ratios of the errors between successive runs that are obtained when the Backward Euler Method is used with five splitting procedures and when problem (2.51) is solved. The abbreviations *Steps*, *"Splitting 0"*, *"Splitting 1"*, *"Splitting 2"*, *"Splitting 3"* and *"Splitting 4"* are used for (i) the number of time-steps and (ii) the cases where no splitting, sequential splitting, symmetric splitting, weighted sequential splitting and weighted symmetric splitting are used in the computations.

Steps	Splitting 0	Splitting 1	Splitting 2	Splitting 3	Splitting 4
40	-	-	-	-	-
80	1.544	2.215	4.004	4.038	4.053
160	1.735	2.082	3.999	4.019	4.013
320	1.858	2.041	4.000	4.009	4.002
640	1.926	2.020	4.000	4.005	4.001
1280	1.962	2.005	4.000	4.002	4.000
2560	1.981	2.003	4.000	4.001	4.000
5120	1.990	2.001	4.000	4.001	4.000
10240	1.995	2.001	4.000	4.000	4.000
20480	1.998	2.000	4.000	4.000	4.000
40960	1.999	2.000	4.000	4.000	4.000
81920	1.999	2.000	4.000	4.000	4.000
163840	2.000	2.000	4.000	4.000	4.000
327680	2.000	2.000	4.000	4.000	4.000

(H) Need for more experiments

The results given in Table 2.7 to Table 2.13 allowed us to draw some useful conclusions about the combined methods (splitting procedures with numerical methods). However, as mentioned above, these results are in fact valid only for the problem tested, the problem (2.51). It is expected that these conclusions are rather general and, therefore, hold also for many other problems. Nevertheless, much more experiments are needed. It is necessary to develop a powerful program which will facilitate the efforts related with performing in a systematic way many tests with different problems. Such program, EXAMINATOR, is in process of development. The basic principles used in the development of this program are sketched below.

- The same splitting procedures, as those discussed in this section, can be specified by setting an appropriate value to a special parameter $ISPLIT$. The following values of $ISPLIT$ are allowed at present:
  - $ISPLIT = 0$ if no splitting is to be used,
  - $ISPLIT = 1$ if a sequential splitting is to be used,

- $ISPLIT = 2$ if a symmetric splitting is to be used,
- $ISPLIT = 3$ if a weighted sequential splitting is to be used, and
- $ISPLIT = 4$ if a weighted symmetric splitting is to be used.

- The numerical methods used in this subsection can be specified. However, three additional methods can also be used. The desired numerical method can be selected by choosing an appropriate value of parameter $METHOD$. The following six values of parameter $METHOD$ can be used at present:
  - $METHOD = 1$ for the Backward Euler Method (see Butcher [37], Hairer and Wanner [127], Hundsdorfer and Verwer [142], Lambert [155]),
  - $METHOD = 2$ for the Implicit Mid-point Rule (see Butcher [37], Hairer and Wanner [127], Hundsdorfer and Verwer [142], Lambert [155]),
  - $METHOD = 3$ for the Modified Diagonally Implicit Runge-Kutta Method (see Zlatev [273]),
  - $METHOD = 4$ for the Fully Implicit Three-stage Runge-Kutta Method (see Butcher [37], Hairer and Wanner [127], Hundsdorfer and Verwer [142], Lambert [155]),
  - $METHOD = 5$ for the Two-stage Rosenbrock Method (see Rosenbrock [206], Hundsdorfer and Verwer [142], Lambert [155]), and
  - $METHOD = 6$ for the Trapezoidal Rule (see Hairer and Wanner [127] or Lambert [155]).

All possible combinations of the five splitting procedures and the six numerical methods can be used. Thus, it is possible to carry out many different experiments when this program is used. In Chapter 10, it will be demonstrated that the program EXAMINATOR can be used to examine the performance of different splitting procedures and numerical methods not only when the problems are solved directly, but also when some variational data assimilation technique is used.

The input data which have to be provided can be described as follows (it is clear that the parameters $ISPLIT$ and $METHOD$, which were described above, are also input parameters):

- The integration interval $[a, b]$ should be defined.
- Subroutines, which calculate the functions f from (2.54), $f^{[1]}$ from (2.55) and $f^{[2]}$ from (2.56), should be prepared.
- Subroutines, which calculate the Jacobian matrices of the functions f from (2.54), $f^{[1]}$ from (2.55), and $f^{[2]}$ from (2.56), should be prepared.
- A subroutine, which calculates the exact solution of the problem $dc/dt = f(t, c)$ has to be prepared.

It is easy to define the integration interval. Two parameters, $TSTART$ and $TEND$, are be given in order to show where the computations are to be started and where the computational process will be finished.

The subroutines for the calculation of f, $f^{[1]}$ and $f^{[2]}$ are normally very simple and there will normally be no problems in the preparation of these subroutines.

In the present version of the program EXAMINATOR the exact Jacobian matrices are required. Differences might be used to prepare approximate Jacobian matrices. Approximate Jacobian matrices might cause difficulties in some situations.

If the exact solution of the problem studied, $dc/dt = f(t,c)$, is not known, then a reference solution (see, for example, Alexandrov et al. [6]) can be calculated, saved in a file and used to evaluate the behaviour of the errors obtained when different combinations of splitting procedures and numerical methods are tested.

The present version of the program EXAMINATOR is dealing with problems of the type $dc/dt = f(t,c)$. There are plans to extend the class of the problems which can be handled by this program by adding options for the treatment of some other operators which are participating in advanced air pollution models. The problem is that if this is done, then it becomes difficult to check the behaviour of the errors in the same manner as the procedure used in the preparation of the results shown in Table 2.7 to Table 2.13.

At the end of this section some numerical results will be presented (see Table 2.16, Table 2.17, and Table 2.18) in order to demonstrate the efficiency of the program EXAMINATOR described above in the efforts

- to explain the results obtained by different splitting procedures and different numerical methods

as well as

- to facilitate the selection of the most suitable combination of a splitting procedure and a numerical method.

The problem defined by (2.51) and ((2.52) is again considered. Sixteen runs were performed for each splitting procedure and for each numerical method and the results are summarized into the three tables given below (i.e., the total number of runs was 480, but it should be emphasized here that the conclusions were drawn using much more experiments). For each combination of a numerical method and a splitting procedure, the first run is performed with a time-step size $\Delta t = 0.25$. After that the stepsize is successively reduced (15 times) by a factor of two. Reducing the time-step size by a factor of two leads to an increase of the number of time-steps by a factor of two (the number of time steps for the first run is 40, while it becomes 5242880 for the last run).

The smallest and the largest errors obtained in the runs with 16 different time-step sizes (or, in other words, the left-hand-end-points and the right-hand-end-points of the intervals in which the errors vary) are given, for each combination of a splitting procedure and a numerical method, in Table 2.17 and Table 2.16 respectively. Taking the corresponding figures in these two table, one can obtain the interval in which the errors are varied for a given combination of a numerical method and a splitting procedure.

Table 2.16: The largest errors obtained in the runs with 16 different time-step sizes (i.e., the right-hand-end-points of the intervals in which the errors vary) are given in this table for different splittings and different numerical methods. The abbreviations *"Splitting 0"*, *"Splitting 1"*, *"Splitting 2"*, *"Splitting 3"* and *"Splitting 4"* are used for the different splitting, while the abbreviations *"BE"*, *"IMR"*, *"RK2"*, *"RK6"*, *"ROS2"* and *"TR"* are denoting the Backward Euler Method, the second-order RK method, the six-order RK method, the second-order Rosenbrock method and the Trapezoidal Rule, respectively.

Method	Splitting 0	Splitting 1	Splitting 2	Splitting 3	Splitting 4
BE	2.1E-02	2.1E-02	1.4E-01	2.6E-1	1.8E-01
IMR	9.6E-04	2.0E-01	2.1E-02	6.7E-3	3.7E-03
RK2	8.7E-04	1.4E-01	3.7E-02	2.1E-2	2.7E-02
RK6	4.5E-10	1.9E-01	2.0E-02	1.7E-2	2.6E-02
ROS2	8.6E-03	1.4E-01	3.7E-02	2.1E-2	2.7E-02
TR	9.6E-04	2.0E-01	2.1E-02	6.7E-3	3.7E-03

The largest errors (i.e., the right-hand-end-points of the intervals in which the errors vary) are normally obtained when the largest time-step size, $\Delta t = 0.25$, is used. It was expected that the smallest errors will be obtained when the smallest stepsizes are used. However, this is not always true for the smallest errors (i.e., the left-hand-end-points of the intervals in which the errors vary), because if the errors become less than $O(10^{-11})$ and millions of time-steps are to be carried out, then the rounding error start to interfere with the other errors. Accuracy of order $O(10^{-11})$ is achieved very quickly, when the RK6 method is used without splitting).

The rates of decreasing the errors when the time-step sizes are decreased are given in Table 2.18. These values are approximate values. For some combinations "numerical method with splitting procedure", the rates are lower in the beginning. If the errors become very small, then rates start to oscillate around 1. However, the figures in Table 2.18 give quite adequate information about the rates which can be achieved by different combinations of a numerical method and a splitting procedure.

Several conclusions can be drawn using the results from the 480 runs (as well as results from some other runs, including here runs performed by other test-problems):

- If the sequential splitting is used, then the errors caused by this splitting (which is of order one) are dominating. Therefore, in this case it does not

Table 2.17: The smallest errors obtained in the runs with 16 different time-step sizes (i.e., the left-hand-end-points of the intervals in which the errors vary) are given in this table for different splittings and different numerical methods. The abbreviations *"Splitting 0"*, *"Splitting 1"*, *"Splitting 2"*, *"Splitting 3"* and *"Splitting 4"* are used for the different splitting, while the abbreviations *"BE"*, *"IMR"*, *"RK2"*, *"RK6"*, *"ROS2"* and *"TR"* are denoting the Backward Euler Method, the second-order RK method, the six-order RK method, the second-order Rosenbrock method and the Trapezoidal Rule, respectively.

Method	Splitting 0	Splitting 1	Splitting 2	Splitting 3	Splitting 4
BE	1.8E-07	1.8E-07	1.2E-06	1.9E-06	1.4E-06
IMR	8.7E-14	1.9E-06	1.2E-12	2.2E-12	1.7E-12
RK2	9.2E-13	1.9E-06	6.2E-13	1.3E-13	5.9E-12
RK6	6.7E-16	1.9E-06	1.1E-12	2.8E-12	1.3E-12
ROS2	9.5E-13	1.9E-06	6.6E-13	1.2E-11	7.4E-12
TR	8.7E-14	1.9E-06	1.2E-12	2.2E-13	1.7E-13

Table 2.18: The approximate rates of convergence (measured by the ratio of the error found at run i and the error found at run $i-1$, where $i = 2, 3, \dots, 16$). The abbreviations *"Splitting 0"*, *"Splitting 1"*, *"Splitting 2"*, *"Splitting 3"* and *"Splitting 4"* are used for the different splitting, while the abbreviations *"BE"*, *"IMR"*, *"RK2"*, *"RK6"*, *"ROS2"* and *"TR"* are denoting the Backward Euler Method, the second-order RK method, the six-order RK method, the second-order Rosenbrock method and the Trapezoidal Rule, respectively.

Method	Splitting 0	Splitting 1	Splitting 2	Splitting 3	Splitting 4
BE	2	2	2	2	2
IMR	4	2	4	4	4
RK2	4	2	4	4	4
RK6	64	2	4	4	4
ROS2	4	2	4	4	4
TR	4	2	4	4	4

matter too much what numerical method is selected. In fact, practically the numerical method does not have any influence on the results (excluding here the Backward Euler Method, which has, as mentioned before, some special properties for this particular test-problem)

- As mentioned before, the combination of a second-order numerical method and a second-order splitting procedure results in a second-order of the combined method. However, one should be careful when the second-order Rosenbrock methods is used for non-autonomous problems. In this case one should first transform the problem, by adding an extra equation, to an autonomous problem. This transformation was not needed here, because the problem (2.51) is autonomous.

- There is, in general, no sense to use a combination of a high-order splitting procedure with a low-order numerical method and vice-versa. This is prescribed by the relationship (2.70), but it is immediately seen that the numerical results are in a very good agreement with this relationship. However, let us re-iterate that for a particular problem one can also design combinations in which the relationship (2.71) holds instead of (2.70). An example, where this happens, was given in Table 2.14 and Table 2.15.

2.9 Using splitting procedures in connection with other applications

The presentation of the splitting procedures made in this chapter is very general. Therefore, it is quite clear that all splitting procedures that were discussed in the previous sections of this chapter can be applied with some obvious modifications when the applications discussed in Section 7 of Chapter 1 or other large-scale applications are to be treated on high-speed computers.

2.10 Conclusions and plans for future research

The examples given in Section 8 of this chapter show that even in very simple cases the splitting procedure will in general cause splitting errors. Therefore, it is worthwhile to carry out work on the solution of the following four important problems:

- to attempt to avoid the use of splitting procedures when this is possible (it should be emphasized, however, that at present this is not always possible),
- to construct splitting procedures of order higher than two,
- to search for optimal combinations of splitting procedures and numerical methods such that the order of the combined methods is as higher as possible, and
- to find reliable and robust methods for evaluating the splitting errors.

These four problems are still open (or, at least, not satisfactorily well resolved) both when general large-scale air pollution models are treated and when other large-scale mathematical problems are to be handled. Success in the solution of any of these four problems is a very desirable, but also a very challenging task. It should be noted that even a partial solution of any of these four problems will lead to a very considerable progress in the numerical treatment of large-scale air pollution models, as well as in the numerical treatment of many other large-scale computational problems arising in science and engineering.

Better understanding of the interference of errors due to the splitting procedures and errors that are caused by other reasons (numerical methods, uncertainties of the input data, etc.) is highly desirable. Research efforts in this directions are needed.

Chapter 3

Treatment of the advection-diffusion phenomena

The advection-diffusion phenomena in a large-scale air pollution model can be divided (for example, by using an appropriate splitting procedure; see the previous chapter) in three parts

- the mathematical terms describing the horizontal advection,
- the mathematical terms describing the horizontal diffusion, and
- the mathematical terms describing the vertical exchange (the vertical exchange is a combination of vertical advection and vertical diffusion).

Several numerical methods, which can be applied to treat the advection-diffusion phenomena, will be discussed in this chapter. It should be emphasized here that we are mainly interested in the computational aspect of the treated problems. Therefore, we shall try to emphasize the fact that the application of different numerical methods leads to the same type of computational problems and the resulting computational problems will be the main topic of the discussions in this chapter. More details about

- the choice of particular numerical methods, and
- the properties of different classes of numerical methods

can be found, for example, in the monograph *"Numerical Solution of Time-Dependent Advection-Diffusion-Reaction Equations"* written by Hundsdorfer and Verwer [142].

It should also be emphasized, before the beginning of the discussion of different computational problems, that the use of some splitting procedure is assumed in this chapter. This is a natural assumption, because all operationally used large-scale air pollution models are treated, as mentioned in the previous chapters, by applying some splitting procedure. Some of the computational problems, which arise when no splitting procedure is used, will be shortly described in Chapter 6.

3.1 Treatment of the horizontal advection

We shall start by assuming that the splitting procedure described by (2.5) - (2.9) is used. The horizontal advection, which is described by the system of PDEs (2.5), is separated from the other physical processes when this assumption is made. It is immediately seen that the system of PDEs (2.5) consists of **independent** partial differential equations (one equation per each chemical compound when a two-dimensional version of the model is studied, while the number of equations is equal to the product of the number of chemical compounds and the number of layers when the horizontal transport in a three-dimensional version of the model is to be handled). This means that in many cases (as, for example, the important case where a fast and reliable numerical method is to be found) it is quite sufficient to consider only one of these equations. Under this assumption, the horizontal transport of any pollutant in the atmosphere can be described mathematically by the following first-order partial differential equation:

$$\frac{\partial c}{\partial t} = -\frac{\partial (uc)}{\partial x} - \frac{\partial (vc)}{\partial y}. \tag{3.1}$$

Let us introduce the parameter z_k (where $k = 1$ in the two dimensional case, while $k \in \{1, 2, \ldots, 10\}$ when a three-dimensional version of the particular model discussed in the Chapter 1, DEM, is used). Assume that the value of k is fixed. Then the notation used in (3.1) can (as in the previous chapters) be explained as follows:

- the unknown function $c = c(x, y, z_k, t)$ is the concentration of some pollutant in the kth layer of the model, and
- the known functions $u = u(x, y, z_k, t)$ and $v = v(x, y, z_k, t)$ are wind velocities (again in the kth layer of the model) along the Ox and Oy axes respectively.

This means that the quantity c in (3.1) is representing the concentration of any of the chemical species both in the two-dimensional case and in the three-dimensional case. It is clear that in the two-dimensional case there are $q = N_s$ (q being the number of the chemical species studied by the model) independent equations of type (3.1). In the three-dimensional case, the number of independent equations of type (3.1) is equal to the product of the number of chemical species q and the number of layers k. As mentioned above, it is sufficient to consider only one equation of type (3.1) when a good numerical method for the treatment of the advection sub-model is to be found. However, sometimes it is better to combine several equations and to consider the enlarged problem obtained by the combination chosen (for example, when parallel computers are used this is done in order to obtain bigger parallel tasks; see Chapter 7).

The numerical solution of (3.1) causes some great difficulties in the case when this equation is a part of a large-scale air pollution model, because the computational tasks can become **very large**. This can be explained by the following example (see also Chapter 1). Assume that a three-dimensional air pollution model is to be handled and that the number of pollutants is N_s. Assume also that the model is discretized on a $(N_x \times N_y \times N_z)$ grid, where N_x, N_y and N_z are the numbers of grid-points along the coordinate axes. Then the application of some splitting procedure followed by some kind of discretization of the spatial derivatives leads to the treatment of $N_s \times N_z$ systems of ordinary differential equations (ODEs) during many (typically several hundred thousand) time-steps. Each of these ODE systems contains $N_x \times N_y$ equations. If $N_x = 480$, $N_y = 480$, $N_z = 10$, $N_s = 35$, then 350 ODE systems each of them containing 230400 equations have to be treated at every time-step. It is difficult to treat such a model on the available supercomputers when the time interval is long (say, several years) or when many different scenarios have to be run. This is why the two-dimensional version of DEM is often used in this case (i.e. $N_z = 1$ is chosen). The number of ODE systems that are to be handled at every time-step is reduced considerably, from 350 to 35, when the two-dimensional version is used. However, even under this assumption it is not possible to handle

- long-term runs (by using meteorological data for many, at present up to ten, years), or
- simulations consisting of many (several hundred or even several thousand) scenarios.

Coarser grids have to be used in these two cases ($N_x = N_y = 96$). Of course, this is a compromise. It is much more desirable to use finer grids (avoiding in this way the use of non-physical simplifications; i.e. avoiding introducing such simplifications only in order to be able to treat large-scale air pollution models on the computers that are available at present).

The actual situation is in fact more complex, when (3.1) is arising from air pollution models, because similar systems of ODEs have also to be treated in connection with the other physical processes (chemistry, diffusion, deposition and vertical exchange). The five major physical and chemical processes are treated successively at every time-step in the older version of DEM; see (2.5)-(2.9) in Chapter 2. In the latest version of DEM the horizontal advection is combined with the horizontal diffusion as described in (2.12); the remaining processes are treated by (2.11) and (2.13).

The short description of the computational difficulties, which has been sketched above, explains clearly that (a) the treatment of the horizontal advection is a very challenging problem and (b) it is very important to handle efficiently PDEs of type (3.1). Three major tasks have to be resolved in the efforts to make the treatment of (3.1) more efficient:

- Fast, but sufficiently accurate, numerical methods have to be selected and carefully implemented. The use of the scalar equation (3.1) in the solution of this task is quite sufficient.

- The selected numerical methods have to be adjusted for efficient runs on different high-speed computers. The organization of the computations is very important when this task is handled. It is normally necessary to treat simultaneously several equations (3.1) in order to efficiently resolve this task.

- Finally, if (3.1) is a part of an large-scale air pollution model, then the fact that (3.1) is, as mentioned above, a system of PDEs that has to be treated together with several other PDE systems (arising from the other physical and chemical processes) must be taken into account. The solution of this task will be discussed in Chapter 7.

The solution of these three tasks is important. The main part of this chapter is devoted to the choice of efficient numerical algorithms. However, the requirements implied from the other two tasks will also be taken into account in order to synchronize the efficient solution of the first task with the requirement for efficient treatment of the whole air pollution model of which (3.1) is an essential part. Therefore, we shall assume that either a single equation (3.1) is considered or that several equations (3.1) are combined and treated simultaneously in an attempt to organize better the computational process. It will either be stated directly which of these two cases is discussed or it will be clear from the context whether (3.1) is considered as a single equation or several equations (3.1) are combined.

As in many other places in this book, it should be emphasized that the problem described by (3.1) is rather general. Therefore many of the results presented in this chapter are also valid if (3.1) arises from some other fields of science or engineering.

3.2 Semi-discretization of the advection equation

The choice of numerical methods for the solution of (3.1) has been discussed in many publications (many references on this topic can be found in Zlatev [282]). The following numerical methods are often used in the treatment of the horizontal advection in large-scale air pollution models:

(a) Numerical methods, which are based on the direct use of finite differences (many details about the use of finite differences in the solution of different scientific problems can be found in Chock [43], [44], Chock and Dunker [45], Pepper and Baker [192], Pepper et al. [193], Kreiss [152], Richtmyer and Morton [201]). Some kind of smoothing is often used when there are sharp gradients. More details about the use of finite differences in the numerical solution of partial differential equations can also be found in the recently published, by Strikwerda, text-book on this subject; see [237].

(b) Numerical methods, which are based on the direct use of finite elements (many details about the use of finite elements in the solution of different scientific problems can be found in Axelsson and Barker, [13], Chock [43], [44], Chock and Dunker [45], Pepper and Baker [192], Pepper et al. [193] Segerlind [216], Strang and Fix [236], Zienkiewicz [271]). Also in this case, smoothing procedures are often used. More details about the use of finite elements in the numerical solution of partial differential equations can also be found in the recently published, by Solin, text-book on this subject; see [229].

(c) Numerical methods, which are based on the use of either finite elements or finite differences combined with some kind of limiting (see Hundsdorfer and Verwer [142], LeVeque, [162], [163]). Limiting can be viewed as a special smoothing procedure.

(d) Semi-Lagrangian methods (see Loon [167]), Rancic [198], Rancic and Williamson [199], Robert et al. [204]). Some kind of interpolation must be used together with the semi-Lagrangian methods. Finding a good interpolation scheme, which can efficiently be used together with the implemented semi-Lagrangian method, can be difficult.

(e) The method proposed by A. Bott is very popular in the field of large-scale air pollution modelling (see Bott [24]). In this method, an attempt is made to avoid negative concentrations (this is why the method is sometimes called *positive definite*). There have been made several attempts to improve this method; see, for example, Holm and Erbes [137], Syrakov [242], Syrakov and Galperin [243], [244]).

(f) Pseudo-spectral methods (different applications of this numerical technique can be found in Fornberg [103], [104], Kreiss and Oliger [153], Lee [160], Merilees [182], Roache [202], Zlatev [276], [282]). The pseudo-spectral methods are based on an expansion of the unknown function in Fourier series and truncating this series after some number of terms. This means that sines and cosines are used instead of the polynomials used when some of the previous methods is selected. The methods requires periodic boundary conditions. If this requirement is satisfied, then the results are as a rule very good (see again the above references). If the requirement for periodic boundary conditions is satisfied, then some special procedures can be applied in order to achieve periodicity on the boundaries of the space domain. Such special procedures for achieving periodicity can be based on ideas that are discussed in Lyness [170] and Roache [202], [203].

Comparisons of different numerical methods, which are used in the case where (3.1) arises from some air pollution models, are given, for example, in Chock, [43], [44] and Chock and Dunker [45].

As mentioned in the previous section, it is quite sufficient to consider a single equation (3.1) when a numerical algorithm is to be chosen, but in many practical cases it is useful to combine several independent equations in order to organize better the computational process. To be specific, we shall assume here that the chemical compound is fixed, and the equations (3.1) are combined (in the three-dimensional case) along the layers (of course, other combinations can also be considered).

It will also be assumed that finite elements are used in the discretization of the spatial derivatives in (3.1), but all results are also valid, with some small modifications, when finite differences are used instead of finite elements. Assume that the three-dimensional version of the model is handled by using ten layers in the vertical direction. Then ten independent equations of the type (3.1) have to be considered instead of the single equation (3.1). Thus, the single equation (3.1) is considered with the two-dimensional version of the model, while ten independent equations of type (3.1) are considered when the three-dimensional version of the model is used. Under these assumptions, the application of any finite element method leads (both for the two-dimensional version and for the three-dimensional version of the model) to a system of ODEs of the following type:

$$P\frac{dg}{dt} = Hg, \quad g \in R^{N_{xyz}}, \quad P \in R^{N_{xyz} \times N_{xyz}}, \quad H \in R^{N_{xyz} \times N_{xyz}}, \tag{3.2}$$

where $N_{xyz} = N_x \times N_y \times N_z$ is the total number of grid-points in the space domain (with $N_z = 1$ in the two-dimensional version of the model). The components of the function g are some approximations to the concentrations at the grid-points of the space domain.

Matrix P is a block-diagonal matrix in which each block corresponds to one layer in the three-dimensional model. If the two-dimensional model is used, then there is only one block. Each block is in general a banded matrix. If finite differences are used, then matrix P is the identity matrix (i.e. $P = I$). The elements of matrix P are constants.

Matrix H is also a block-diagonal matrix in which each block corresponds to one layer in the three-dimensional model. If the two-dimensional model is used, then again there is only one block. Each block is a banded matrix. The elements of matrix H depend on the wind velocity vectors (consisting of values of u and v at the grid-points). This means that in general matrix H depends both on the spatial variables and on the time variable.

Both the structure of matrices P and H and their non-zero elements depend on the particular finite element method or finite difference method that has been selected.

If splitting to one-dimensional sub-models is used (see, for example, McRae et al. [181]), then both P and H are tri-diagonal matrices. In our case, one dimensional sub-models appear after the discretization of the vertical exchange module (2.11), see Section 5 of this chapter.

Thus, the application of any finite element method or finite difference method in the discretization of the spatial derivatives in the right-hand-side of (3.1) leads to replacing the PDE (3.1) with a system of ODEs of type (3.2). In fact, the application of some other numerical methods (as, for example, the pseudo-spectral method; see, for example, Zlatev [282] or Zlatev et al. [289]) will also result in a system of type (3.2).

The next problem is to decide how to handle numerically the ODE system which is represented by (3.2).

3.3 Time integration of the semi-discretized advection equation

Assume that matrix P from the equation (3.2) is a non-singular matrix and denote $f(t,g) = P^{-1}Hg$. Then (3.2) can be re-written as

$$\frac{dg}{dt} = f(t,g), \quad g \in R^{N_{xyz}}, \quad f \in R^{N_{xyz}}. \tag{3.3}$$

The following two questions are important and the answers to these questions will be discussed in this section.

- How to select efficient methods for the solution of (3.3)?
- How to implement them taking into account the fact that (3.3) is a part of a bigger computational process?

Very often simple numerical methods for solving systems of ODEs are used in the treatment of (3.3). Moreover, these methods are as a rule **explicit**. Some examples for such numerical methods that can be used in connection with the solution of (3.3) are

- the leap-frog method (which is well-known, especially in the theory for the numerical solution of systems of ODEs as the explicit mid-point rule; see, for example, Lambert [155]), and
- the forward Euler method (this method is studied in nearly all text books for numerical methods for ordinary differential equation; see, for example, the classical book written by Henrici [132]).

The computational requirements are low when simple explicit numerical methods are used. However, the use of such methods is also connected with some restrictions on the time-step size, which have to be imposed in order preserve the stability of the computations. Normally, the simple explicit methods are used with some sufficiently small time-step size, which is experimentally found and for which the computations stay stable for the particular problem under consideration. It is intuitively clear that it is much better to incorporate some rules for **controlling automatically the stability of the computational process**. Methods, for which such rules can be used, will be discussed in this section.

3.3.1 Predictor-corrector schemes with several correctors

Predictor-corrector schemes with several different correctors can be used in the solution of equation (3.3). This can be done in the following way. Consider a set of predictor-corrector schemes

$$\mathbf{F} = \{F_1, F_2, \dots, F_m\}. \tag{3.4}$$

Assume that $m \geq 1$ and F_j, ($j \in M$, $M = \{1, 2, \dots, m\}$) is a predictor-corrector (PC) scheme $PEC_1EC_2 \dots C_{q_j}E$ in which

- P is some predictor formula,
- E denotes an evaluation of the right-hand-side of (3.3), and
- C_r, $r = 1, 2, \dots, q_j$, are correctors, which **can be different**.

Assume also that

- the time-step size Δt is a positive constant,
- approximations g_k of the exact solution $g(t_k)$ of (3.3) are to be found at the points t_k (which satisfy $t_k - t_{k-1} = \Delta t$ for $k = 1, 2, \dots, K$) starting with a given initial approximation $g_0 = g(t_0)$, and
- $f_k = f(t_k, g_k)$.

Then the predictor P and the q_j correctors C_r, which are involved in the predictor-corrector scheme F_j, can be defined by the two formulae (3.5) and (3.6), which are given below (see also Zlatev [279]). More precisely, the predictor is defined below by (3.5), while the correctors are defined by (3.6):

$$g_k^{[0]} = \sum_{i=1}^{\mu_j^{[0]}} \alpha_{ji}^{[0]} g_{k-i} + \Delta t \sum_{i=1}^{\nu_j^{[0]}} \beta_{ji}^{[0]} f_{k-i}, \tag{3.5}$$

$$g_k^{[r]} = \sum_{i=1}^{\mu_j^{[r]}} \alpha_{ji}^{[r]} g_{k-i} + \Delta t \beta_{j0}^{[r]} f_k^{[r-1]} + \Delta t \sum_{i=1}^{\nu_j^{[r]}} \beta_{ji}^{[r]} f_{k-i}. \tag{3.6}$$

where $r = 1, 2, \dots, q_j$, $f_k^{[0]} = f(t_k, g_k^{[0]})$ and $f_k^{[r]} = f(t_k, g_k^{[r]})$.

If $q_j = 0$, then only the predictor formula (3.5) is used in the computations. The usually used schemes with one predictor and one corrector can be obtained for $q_j = 1$. This means that in fact

- single integration formulae,

- the usual predictor-corrector scheme with one predictor and one corrector, and
- more advanced predictor-corrector schemes containing several, not necessarily the same) correctors

can be considered when the above formulae, (3.5) and (3.6), are used.

Index j is used in the above two formulae only to show that (3.5) and (3.6) represent element F_j from set $\mathbf{F}$, which is defined by (3.4). Assume that F_j is used **at every time-step** when (3.3) is solved. Let

$$s_j = max(\mu_j^{[0]}, \nu_j^{[0]}, \mu_j^{[1]}, \nu_j^{[1]}, \ldots, \mu_j^{[q_j]}, \nu_j^{[q_j]}). \tag{3.7}$$

Assume furthermore that $g_1, g_2, \ldots, g_{s_j-1}$ have already been obtained in some way. Then the calculations can successively be carried out for $k = s_j, s_{j+1}, \ldots, K$ by using (3.5) and (3.6) for $r = 1, 2, \ldots, q_j$. At the end of step k it is necessary to set $g_k = g_k^{[q_j]}$ and $f_k = f_k^{[q_j]}$.

This is the classical way of solving numerically (3.3): the computational process is carried out by using at every time-step the **same** PC scheme F_j and the **same** time-step size Δt. If the computations are very time-consuming, then it is desirable to generalize the classical approach by allowing variations of both the PC schemes and the time-step size. This leads to variable stepsize variable formula methods (VSVFMs) based on PC schemes. Such methods can formally be defined in the following way (see more details in Zlatev [275] and [279]). Introduce the non-equidistant grid G_K^* and the vector Δt_{kj}^* by the two formulae that are given below:

$$G_K^* = \{t_k | t_0 = a, t_k = t_{k-1} + \Delta t_k, \Delta t_k > 0, k = 1(1)K, t_K = b\}, \tag{3.8}$$

$$\Delta t_{kj}^* = \left\{ \frac{\Delta t_{k-1}}{\Delta t_k}, \frac{\Delta t_{k-2}}{\Delta t_k}, \ldots, \frac{\Delta t_{k-s_j+1}}{\Delta t_k} \right\}. \tag{3.9}$$

The following PC scheme can be obtained by using the introduced above quantities G_K^* and Δt_{kj}^*:

$$g_k^{[0]} = \sum_{i=1}^{\mu_j^{[0]}} \alpha_{ji}^{[0]}(\Delta t_{kj}^*) g_{k-i} + \sum_{i=1}^{\nu_j^{[0]}} \Delta t_{k-i} \beta_{ji}^{[0]}(\Delta t_{kj}^*) f_{k-i}, \tag{3.10}$$

$$\begin{aligned} g_k^{[r]} &= \sum_{i=1}^{\mu_j^{[r]}} \alpha_{ji}^{[r]}(\Delta t_{kj}^*) g_{k-i} + \Delta t_k \beta_{j0}^{[r]}(\Delta t_{kj}^*) f_k^{[r-1]} \\ &+ \sum_{i=1}^{\nu_j^{[r]}} \Delta t_{k-i} \beta_{ji}^{[r]}(\Delta t_{kj}^*) f_{k-i}. \end{aligned} \tag{3.11}$$

A VSVFM for calculating approximations to the solution of (3.3) at the points of G_K^* can be based on PC schemes defined by (3.10)-(3.11), the coefficients of which depend on vector $\triangle t_{kj}^*$ (see Zlatev [275], [276], [279]). It is said, Zlatev [279], that *the PC scheme (3.10)-(3.11) is corresponding to* F_j *with regard to* $\triangle t_{kj}^*$. The most important features of such a scheme are:

- its coefficients are determined so that (3.10) and (3.11) have the same order as the corresponding formulae (3.5) and (3.6) of the PC scheme F_j, and
- if all components of $\triangle t_{kj}^*$ are equal to 1 (i.e. if the same time-step size has been used during the last $s_j - 1$ steps), then the PC scheme (3.10) - (3.11) reduces to the PC scheme (3.5) - (3.6).

The elements of set **F**, defined by (3.4), are called *basic* PC schemes for the VSVFM (Zlatev [279]).

It is desirable to find a sub-class of the class defined by (3.5) - (3.6) such that all VSVFMs based on PC schemes (3.10) - (3.11)

- are consistent, zero-stable and convergent (the properties consistency, zero-stability and convergence of predictor-corrector schemes used as constant stepsize constant formula methods are studied, for example, in Henrici [132] and Lambert [155]), and
- have good absolute stability properties on the imaginary axis (absolute stability properties have been first studied by Dahlquist [56] in connection with the test-equation $y' = -\lambda y$, where λ is a positive constant; see also Hairer et al. [126], Hairer and Wanner [127], Lambert [155] or Shampine and Gordon [218]).

It can be proved that the well-known and commonly used VSVFMs that are based on Adams PC schemes (see Shampine [217]) satisfy the first of these two requirements but not the second one. If the following conditions are imposed on the coefficients of formulae (3.10)-(3.11), then both requirements can be satisfied for the resulting PC schemes:

$$\triangle t = \max_{1 \le k \le K} (\triangle t_k) \quad and \quad \triangle t K \le c < \infty \quad for \ \forall K \tag{3.12}$$

$$(i.e. \quad K \to \infty \quad implies \quad \forall \triangle t_k \to 0);$$

$$0 < \overline{\alpha} \le \frac{\triangle t_k}{\triangle t_{k-1}} \le \overline{\beta} < \infty \quad for \quad \forall k \in \{1, 2, \dots, K\} \stackrel{\text{def}}{=} K^*; \tag{3.13}$$

$$\alpha_{j1}^{[r]}(\triangle_{kj}^*) = \alpha_{j1}^{[r]} = \alpha_j^{[r]}, \qquad \alpha_{j2}^{[r]}(\triangle_{kj}^*) = \alpha_{j2}^{[r]} = 1 - \alpha_j^{[r]}; \tag{3.14}$$

$$\alpha_{js}^{[r]}(\triangle_{kj}^*) = \alpha_{js}^{[r]} = 0 \ \ (s = 3(1)\mu_j, r = 0(1)q_j, j \in M, k \in K^*); \tag{3.15}$$

the PC scheme used at step $k \in K^*$ *is such that* $s_j \leq k$. (3.16)

The time-step size selection strategy is said to be *stable* when (3.12) and (3.13) are satisfied. If (3.14) and (3.15) hold and if the order of the predictor is $\nu_j^{[0]}$ while the order of the rth corrector is $\nu_j^{[r]} + 1$, $r = 1, 2, \ldots, q_j$, then it is said that F_j is *two-ordinate*. The order requirements lead to the solution of $q_j + 1$ systems of algebraic equations the unknowns of which are the coefficients $\beta_{ji}^{[r]}(\Delta t_{kj}^*)$. It can be proved that each of these systems has a unique solution. The VSVFM is *self-starting* if (3.16) is satisfied. The desired properties of the VSVFM (consistency, zero-stability and convergence) follow from the following theorem (this theorem has been proved in Zlatev [276]).

Theorem 3.1 *If a self-starting VSVFM, which is based on two-ordinate PC schemes corresponding to two-ordinate basic PC schemes of some set* $\boldsymbol{F}$ *defined as in (3.4), is applied on a grid* G_K^* *which determines a stable time-step size selection strategy, then the VSVFM is consistent, zero-stable and convergent when* $0 \leq \alpha_j^{[q_j]} < 2$ *for* $\forall j \in M$.

3.3.2 Absolute stability properties

While the requirement for constructing numerical schemes which are consistent, zero-stable and convergent is absolutely necessary, it is not sufficient in the efforts to ensure an efficient computational process. It is also necessary to select methods with good absolute stability properties along the imaginary axis (Zlatev [276]). If PC schemes are to be used, then there are some barriers of the length h_{imag} of the absolute stability interval on the positive part of the imaginary axis that can be achieved. More precisely, the following theorem holds (see Jeltsch and Nevanlinna [146] or Zlatev [276]).

Theorem 3.2 *The length* h_{imag} *of the absolute stability interval on the positive part of the imaginary axis cannot exceed* $q_j + 1$ *when a PC scheme of type (3.5)-(3.6) with* q_j *correctors is used.*

Theorem 3.2 indicates that the absolute stability properties along the imaginary axis can be improved by selecting PC schemes that contain more correctors. However, increasing the number of correctors does not automatically lead to a PC scheme with better absolute stability properties. This is why the two-ordinate PC schemes were introduced. This class of PC schemes ensures the important properties consistency, zero-stability and convergence. Moreover, there are free parameters ($\alpha_j^{[r]}$, $r = 0, 1, \ldots, q_j$, $j \in M$) that can be used to search for particular PC schemes with good absolute stability properties. The PC schemes listed in Table 3.1 have been obtained (see Zlatev [276]) by organizing an optimization process based on the use of subroutines from Fletcher [102].

Table 3.1: Absolute stability intervals on the imaginary axis of three PC schemes which are actually used in DEM. The figures in the brackets are showing how close (in percent) the absolute stability intervals are to the barriers from Theorem 3.2.

PC scheme	Values of the parameters ($\alpha_j^{[r]}$)	h_{imag}
$F_1 = PEC_1EC_2E2EC_3E$	-0.3412, 0.3705, 0.5766, 0.4584	3.26 (81.5%)
$F_2 = PEC_1EC_2E$	-0.65, 1.5, 1.0	2.51 (83.7%)
$F_3 = PEC_1E$	-0.90, 1.6	1.62 (81.0%)

3.3.3 Restrictions on the time-step size

Several additional assumptions are needed in the efforts to apply the absolute stability intervals derived in the previous section in the solution of our problems, because these intervals are valid when matrix $P^{-1}H$ is a constant matrix with distinct eigenvalues, which are purely imaginary numbers. If matrix $P^{-1}H$ is a constant matrix with distinct eigenvalues, then there exists a decomposition $P^{-1}H = Q\Lambda Q^T$, where Λ is a diagonal matrix the diagonal elements of which are the eigenvalues of $P^{-1}H$, while Q is an orthogonal matrix. This means that (3.2) can be transformed into

$$\frac{dh}{dt} = \Lambda h \tag{3.17}$$

by applying the substitution $Q^T g = h$. It is clear that (3.17) consists of independent equations. Assume that

- all eigenvalues of $P^{-1}H$ (or, in other words, all diagonal elements of Λ) lie on the imaginary axis, and
- the largest in absolute value eigenvalue of $P^{-1}H$ is λ.

It should be noted that the second of these two requirements implies that λ satisfies the following inequality

$$\lambda = \max_{1 \le i \le N_{xyz}} \{| \lambda_i |\} \tag{3.18}$$

The following conclusion can be drawn when these two requirements are satisfied. If a numerical method with an absolute stability interval on the imaginary axis h_{imag} is applied in the solution of (3.17), then the computations will be stable when the condition

$$\lambda \Delta t \le h_{imag} \tag{3.19}$$

is satisfied. Some further assumptions are needed in order to demonstrate the role of parameter λ from (3.18) in the efforts to preserve the stability of the computational process. Consider the one-dimensional equation

$$\frac{\partial c}{\partial t} = -u\frac{\partial c}{\partial x} \tag{3.20}$$

where u is a positive constant. Assume that the finite element discretization is performed on an equidistant spatial grid with an increment $\triangle x$. The result is a system of type (3.2) with $N_{xyz} = N_x$.

The first assumption for matrix $P^{-1}H$ that was made after formula (3.17) (i.e. the assumption that all eigenvalues of $P^{-1}H$ lie on the imaginary axis) is in general not satisfied when different finite element or finite difference algorithms are used. It will be assumed here that the finite element method from Pepper and Baker [192] and Pepper et al. [193] is applied. This method is currently used in DEM (see Zlatev [282], Georgiev and Zlatev [117]). Both P and H are tri-diagonal matrices when finite element methods are used to disscretize (3.20). The particular finite element method proposed by Pepper and Baker [192] and Pepper et al. [193] leads to a matrix P that has elements $2/3$ on the main diagonal and $1/6$ on the two adjacent diagonals for all rows excepting the first and the last rows. The first row of P has elements $p_{1,1} = 1/3$ and $p_{1,2} = 1/6$. The last row of P has elements $p_{N_x,N_x} = 1/3$ and $p_{N_x-1,N_x} = 1/6$. Matrix H has elements 0 on the main diagonal, $u/(2\triangle x)$ on the diagonal over the main diagonal and $-u/(2\triangle x)$ on the diagonal under the main diagonal for all rows excepting the first and the last rows. The first row of H has elements $h_{1,1} = u/(2\triangle x)$ and $h_{1,2} = -u/(2\triangle x)$. The last row of H has elements $h_{N_x,N_x} = -u/(2\triangle x)$ and $h_{N_x-1,N_x} = u/(2\triangle x)$.

It is relatively easy to compute the eigenvalues for the particular matrix $P^{-1}H$ when P and H are defined as above, because the size of matrix $P^{-1}H$ is not very large when the selected numerical method is applied to discretize the right-hand-side of (3.20). This has actually been done by using standard eigenvalue subroutines from LAPACK (see Anderson et al. [11]). All eigenvalues were purely imaginary numbers in this particular case. It is not clear if this will be true for all finite elements methods. However, if the discretization of (3.20) is based on the use of finite differences, then the eigenvalues will be purely imaginary numbers (because in the latter case $P = I$ and, thus, $P^{-1}H = H$, where H is as above, an anti-symmetric matrix).

Having the eigenvalues of matrix $P^{-1}H$, one can proceed by calculating λ; see (3.18). The experiments show that, for the particular matrix $P^{-1}H$ arising in our case, the value of λ can be expressed by the following formula:

$$\lambda \approx \frac{1.73u}{\triangle x}. \tag{3.21}$$

The substitution of λ from (3.21) in (3.19) leads to the following relationship:

$$\frac{\lambda^* u \triangle t}{\triangle x} < h_{imag}, \quad \lambda^* \approx 1.73. \tag{3.22}$$

Formula (3.22) depends on three parameters: $\triangle x$, u and h_{imag}. Assume that only one of these three quantities (i.e. either only $\triangle x$ or only u or only h_{imag}) is varied (while the remaining two quantities are kept fixed). The impact of these variations on the time-step size $\triangle t$ can be summarized in the following three conclusions:

- If the spatial resolution is improved (i.e. if $\triangle x$ is reduced), then $\triangle t$ has also to be reduced in order to preserve stability of the computations.
- If the wind velocity u is increased, then $\triangle t$ has to be reduced in order to keep the computations stable.
- If a numerical method with better stability properties (i.e. with a larger h_{imag}) is selected, then the time-step size $\triangle t$ can be increased if the accuracy requirements are not stringent.

The factor λ^* depends on the numerical method used in the discretization of the spatial derivatives. For the particular method used in DEM (based on the finite elements from Pepper and Baker [192] and Pepper et al. [193]) this factor is, as seen in (3.22), approximately equal to 1.73. If a pseudo-spectral method is used in the discretization of $\partial c/\partial x$ from (3.20), see Zlatev [276], [282] and [289], then $\lambda^* = \pi(N_x - 1)/N_x$, which is about 1.8 times larger when N_x is sufficiently large.

If the numerical method is fixed and the number of grid-points N_x is varied under the condition that $\triangle x$ is kept unchanged (by decreasing or increasing the spatial interval), then the computations show that λ^* from (3.22) remains practically the same; some results are given in Table 3.2. This factor can probably be determined analytically for the particular finite element discretization applied in DEM (note that the numbers in Table 3.2 are very good approximations of $\sqrt{3}$). However, this is not very important, because the results are obtained under an assumption that the wind velocity u is a positive constant. It is much more important to emphasize the fact that experiments indicate that the stability of the computations seems to be preserved very well when the requirement for a constant wind velocity u is removed provided that some norm U of the wind velocity vector (formed by the values of the wind velocity at all grid-points and at the time-point under consideration) is used in the determination of the time-step size $\triangle t$ instead of the constant wind velocity u. More precisely, the experiments indicate that the inequality

$$\triangle t < \alpha \frac{h_{imag}}{\lambda^* U} \triangle x \tag{3.23}$$

Table 3.2: Values of the factor λ^* when the number N_x of grid-points is varied (the variation of N_x implies variations of the size of matrix $P^{-1}H$).

Number of grid-points	Size of matrix $P^{-1}H$	λ^*
96	(96x96)	1.73078850874
288	(288x288)	1.73191245476
480	(480x480)	1.73200113778
1024	(1024x1024)	1.73203991785
2048	(2048x2048)	1.73204808779
3072	(3072x3072)	1.73204959917
4096	(4096x4096)	1.73205012795

can successfully be used to decide whether the computations should be expected to be stable or not ($0 < \alpha < 1$ being used, as in many ODE codes, to increase the reliability of the experimentally derived criterion).

Consider again the two-dimensional case described by (3.1). Assume that U and V are some norms of the wind velocity vectors (formed by the values of the wind velocities at the grid-points). Then the inequality

$$\Delta t < \alpha \frac{h_{imag}}{\lambda^*(U+V)} \Delta x \tag{3.24}$$

can be used, instead of (3.23), to decide whether the computations that are carried out with a time-step size Δt will be stable or not. It is necessary to mention here that, although it is not possible to justify rigorously the above criterion, it is based on some plausible assumptions and it works very well in practice.

3.3.4 Implementation of the PC schemes in the advection module of DEM

As mentioned in the previous subsection, experiments indicate that (3.24) is a rather efficient tool in the efforts to check in a reliable way whether the stability of the computations is preserved or not. This formula can be used in two ways:

1. to build a fully VSVFM code in which both the time-step size and the formula can be varied in an attempt to optimize the computational process (as those discussed in Zlatev [276], [277]),
2. to try to choose a more stable formula when the test (3.24) indicates that the stability requirements will be violated if the formula currently used is not

replaced with a more stable formula (or, in other words, we try to prevent reductions of the time-step size by selecting a more stable formula).

If only the advection problem (3.1) is to be solved, then it will be relatively easy to solve the first task. The same is true in the case where only a few simple linear chemical reactions are used in a two-dimensional model. In this case, the time-step size, which is used in the advection part, can also be applied in the other parts of the model (diffusion, deposition and chemistry). Thus, the determination of a time-step size and a PC scheme for the advection part by using (3.24) will not affect the efficiency of the computations in the other parts of the model. The use of VSVFMs for such simple air pollution models is discussed in Zlatev [277], [278] and [290] (it should also been mentioned that a pseudo-spectral method is used in the advection part of the codes described in Zlatev [277], [278] and [290]).

The solution of the first task becomes rather difficult when (3.1) is a part of much more complicated air pollution models, in which a large number of non-linear chemical reactions are involved. In this case it is not clear anymore whether the changes of the time-step size in the advection module will cause problems in the other parts of the model (the critical part being the stiff chemical module). On the other hand, the problem of keeping the time-step size unchanged (the second task) is very important for the successful computer treatment of such models. It turns out that this second task can efficiently be handled by using the powerful stability criterion (3.24). There arise two major cases in the efforts to keep the time-step size unchanged.

1. **Need for a more stable formula.** Assume that the stability condition (3.24) is not satisfied at some time-step and that the PC scheme used is F_3 from Table 3.1. Then PC scheme with a longer stability interval on the imaginary axis (either F_2 or F_1: if both F_2 and F_1 satisfy the stability condition, then F_2 should be chosen) has to be selected. If the same situation occurs but the PC scheme currently used is F_2, then F_1 has to be selected. The situation will be critical if F_1 is used and if the test (3.24) fails. However, this could occur only if the wind velocity is extremely high, and in practice this never happens (in order to achieve this, Δt_k must be kept smaller than Δt_{max} which is such that the PC-scheme F_1 is stable also when the wind velocities are very high).

2. **Using more economical formulae.** If a PC scheme has better stability properties, then it is also more expensive (because more formulae have to be used in such a scheme, see Table 3.1). Therefore, every time when it becomes possible to change to a more economical PC scheme (a PC schemes using smaller number of formulae), such a change is performed. One should take care to perform the change safely, with regard to the stability control, i.e. requiring (3.24) to be satisfied for the PC scheme which is to be chosen.

The major part of the additional computational work that has to be carried out in order to use the algorithm, which was sketched above, is the calculation of the norms of the two wind velocity vectors U and V. This additional work is normally fully compensated because

- the computational process is much safer with a stability control, and
- reductions of the time-step size, caused by the fact that the stability condition (3.24) is not satisfied, are fully avoided.

The stability check sketched above is currently used to keep the time-step size unchanged in the operational versions of DEM. This increased considerably the efficiency of the computations and, thus, it was possible to run the model

- over very long-time periods (until now up to 10 years), and
- with many scenarios (up to several hundred scenarios in some of the comprehensive studies that were performed with DEM).

The ability of DEM to solve such big computational tasks is demonstrated in the papers where the comprehensive studies performed until now are described Zlatev et al. [296], [298]. The studies discussed in Chapter 8 and Chapter 9 are another illustration of the ability of DEM to resolve very challenging computational tasks.

3.4 Numerical treatment of the horizontal diffusion

As mentioned in Subsection 2.2.2, the horizontal advection sub-model and the horizontal diffusion sub-model are united in the current splitting (2.11) - (2.13) used in DEM. It is easy to add horizontal diffusion in the finite element discretization (see Georgiev and Zlatev [117]). When the spatial derivatives in the united horizontal advection-diffusion part are discretized, the same time-integration methods as the time-integration methods discussed in the previous sections of this chapter can be applied. Moreover, the addition of the horizontal diffusion part is not affecting the stability of the time-integration scheme, because the horizontal advection process is dominating (this fact is exploited by some modellers to neglect totally the horizontal diffusion when they build up their large-scale air pollution models).

3.5 Numerical treatment of the vertical exchange

The vertical exchange sub-model is given by the system of PDEs (2.11). The equations in this system are independent. Exploiting this fact, it is sufficient to consider in this section the following scalar equation (where also the indices induced by the splitting procedure are omitted) instead of (2.11):

$$\frac{\partial c}{\partial t} = -\frac{\partial (wc)}{\partial z} + \frac{\partial}{\partial z}\left(K_z \frac{\partial c}{\partial z}\right) \tag{3.25}$$

The semi-discretization of the spatial derivatives leads to a system of ODEs which is of the same type as (3.2). However, there is a substantial difference. The system of ODEs (3.2), which appear in the horizontal transport sub-model, is non-stiff. Therefore, it was possible to use explicit methods for numerical integration in the numerical solution of this system of ODEs (see the methods discussed in the previous sections of this chapter). The system of ODEs which appear when the spatial derivatives in (3.25) are discretized is **stiff**. The stiffness is caused by the vertical diffusion term (in this case the vertical diffusion is dominating over the vertical advection). This implies the use of implicit methods in the solution of the system of ODEs arising after the discretization of the spatial derivatives in (3.25).

The same finite element discretization as that used in the previous sections is used to discretize the spatial derivatives in (3.25). After that the $\theta-$method (see Lambert [155]) is used in the treatment of the resulting system of ODEs. The implementation of these methods is straight-forward.

3.6 Applicability to other large-scale models

It was established in Section 7 of Chapter 1 that many other large scale models can be described with a system of PDEs that is similar to the system of PDEs (1.1), which has been obtained, in Chapter 1, for the particular air pollution model, DEM, which is discussed in this book. This means that the methods discussed in this chapter will be applicable also to the systems of PDEs that were discussed in Section 7 of Chapter 1 (assuming that some splitting procedure has been applied to separate the transport processes from the other physical and chemical processes).

It is necessary to point out that it might be worthwhile to take into account some specific properties of the particular model under consideration in order to improve the efficiency. This is especially true when the model under consideration leads to very big computational tasks.

Furthermore, one should be careful when combined advection-diffusion sub-models are treated. The approach sketched in Section 3 of this chapter is only applicable if the advection process is dominating over the diffusion process. If this is not the case, then one might consider a splitting procedure in which the advection and the diffusion are treated as separate sub-models.

3.7 Concluding remarks and plans for future research

The implementation of a truly VSVFM in the advection part is a challenging task when large-scale air pollution models with advanced chemical schemes are to be treated. Such an implementation requires a closer investigation of the impact of a time-step size changes in the horizontal advection module on the computations in the other parts of the air pollution model.

Finding formulae with better stability properties, for example formulae of Runge-Kutta type (different properties of Runge-Kutta methods are fully described in Butcher [37], Hairer et al. [126], Hairer and Wanner [127] and Lambert [155]), may lead to improvements of the performance of the horizontal advection part. However, note that the computational work per time-step for these formulae is much more expensive than computational work per time-step for the formulae proposed in this chapter. Therefore, improvements will be achieved only if it becomes possible to run the horizontal advection sub-model with a sufficiently large stepsize when Runge-Kutta formulae are used.

It should be mentioned here that if the requirement for a constant wind velocity is imposed (as in Subsection 3.3.3) and if $P = I$, then all the eigenvalues of matrix $P^{-1}H = H$ lie on the imaginary axis. If these requirements are removed, then there is no guaranty that this will still be the case. Therefore, the numerical methods should have good stability properties not only on the imaginary axis, but also in a sufficiently large stability region in the part of the complex plane to the left of the imaginary axis. It should be mentioned here that many Runge-Kutta methods have good stability properties in this part of the complex plane. Thus, this is another reason to try to apply Runge-Kutta formulae (as, for example, some of the Runge-Kutta methods studied in Butcher [37] Hairer et al. [126], Hairer and Wanner [127] and Lambert [155]) in the horizontal advection part.

Only some basic ideas, which are related to the particular finite elements that are used in the discretization of the spatial derivatives in the horizontal transport sub-model, have been sketched in this chapter. These methods are rather simple. More details about the implementation of such methods in large-scale air pollution models can be found in Georgiev and Zlatev [117]i; see also Pepper and Baker [192], Pepper et al. [193]. However, it should be emphasized here that several other methods have also been tested and used in the runs with DEM. The use of the pseudo-spectral method is discussed in detail in Zlatev [282] and in Zlatev et al. [289], [290].

It is very important (in our opinion) to explain the principle that can be applied in order to carry out in an reliable way the computations without changing the time-step size when explicit time-integration algorithms are selected for the horizontal advection sub-model. The algorithms that are used in the efforts to avoid changing the time-step size have been fully described in Section 3.3.

Note that the algorithms from Section 3.3 do not depend very much on the particular numerical methods that are used for discretization of the spatial derivatives. The algorithms from Section 3.3 were successfully tested in connections with several traditionally used methods for discretization of the spatial derivatives (both methods based on the use of finite differences and the pseudo-spectral method).

Furthermore, the algorithms presented in Section 3.3 were used with success in a previous version of our air pollution model, DEM, in which a pseudo-spectral discretization of the spatial derivatives in the advection sub-model was applied. The pseudo-spectral discretization is based on an expansion of the unknown function in a Fourier series.

The possibility of using the algorithm from Section 3.3 in connection with different algorithms for the discretization of the spatial derivatives in the horizontal advection sub-model shows that this algorithm can be viewed as a **template** by which an attempt to avoid changing the time-step size during the run is made.

Other numerical methods can also be applied in the vertical exchange sub-model. However, this is not urgently needed, because this part of the computations is much cheaper than the horizontal advection-diffusion part, which will be demonstrated in Chapter 7.

Chapter 4

Treatment of the chemical part: general ideas and major numerical methods

The numerical treatment of the chemical part of a large-scale air pollution model is very often **the most time consuming part** of the computational work. Therefore, it is important to select fast and sufficiently accurate numerical methods for the treatment of the chemical part of a large-scale air pollution model.

The application of appropriate discretization and splitting procedures, as those described in Chapter 2, reduces the chemical sub-model to a large number of independent and relatively small systems of ODEs (ordinary differential equations). One system of ODEs per each grid-point is to be handled at each time-step. Therefore, in the process of searching for efficient numerical algorithms for the chemical sub-models one can carry out experiments by using only one such ODE system in order to facilitate the work.

The above approach has been used in connection with a particular chemical scheme, the condensed CBM IV scheme developed by Gery et al. [118]. CBM stands for carbon-bond-mechanism. All reactions of the version of this chemical scheme, which is currently used in model discussed in this book, DEM, are listed in Zlatev [282]. The CBM IV chemical scheme is used in several other large-scale air pollution models. The same approach (i.e. the use of only one system of ODEs, corresponding to an arbitrary grid-point, in order to facilitate the search for good numerical methods) can also be used when any other chemical scheme is used instead of the CBM IV scheme.

Six algorithms for solving systems of ODEs have been tested on a set of typical scenarios (consisting of different starting concentrations and/or of different values of the emissions). The advantages and the disadvantages of the numerical algorithms, which are tested in this chapter, are also discussed.

The use of a small system of ODEs, which represent the chemical reactions at a given grid-point (sometimes considering additionally several typical situations at the grid-point chosen by applying several scenarios) is only possible when an appropriate splitting procedure is properly implemented in the air pollution model under consideration. Therefore, it should again be mentioned that in this chapter it will always be assumed (as in the previous chapter) that some splitting procedure has been applied. Let us also reiterate here that some details about the case where no splitting procedure is used will be discussed in Chapter 6 (this has also been mentioned in the beginning of Chapter 3).

4.1 The chemical sub-model

The chemical sub-model can formally be obtained from (2.7). Note that the chemical reactions are considered together with the emissions in (2.7). This means that the treatment of the chemical sub-model leads, as the treatment of the other four sub-models in Subsection 2.1 in Chapter 2, to the solution of a very large system of $N_x \times N_y \times N_z \times N_s$ ODEs. However, it can easily be established that the chemical sub-model consists (after the implementation of appropriate discretization and splitting procedures) of $N_x \times N_y \times N_z$ **independent** ODE systems. Each of these systems will be called a *chemical ODE system*. The size of each chemical ODE system is equal to the number N_s of the chemical species studied by the model. The physical interpretation of the fact that the chemical ODE systems are independent is the following: *the chemical species at each grid-point react with each other, but not with chemical species from other grid-points.*

The above observation can be exploited in different ways. For example, the treatment of a chemical ODE system at each grid-point can be considered as a parallel task. Thus, the total number of parallel tasks in the treatment of the chemical sub-model at a given time-step is equal to the number of grid-points. Parallel computations will be discussed in detail in Chapter 7.

The treatment of the chemical sub-model is often, as mentioned above, the most time consuming part of the computational work. Therefore, it is crucial to apply efficient numerical methods when the chemical part is handled. The fact that the chemical ODE systems at every grid-point are independent can easily be exploited when the efficiency of different numerical methods is tested, i.e. one can test different numerical methods by using a small system of ODEs representing the chemical reactions at an arbitrary grid-point. This approach will consistently be used in this chapter. This means that we shall select the chemical ODE system arising at an arbitrary grid-point and use this system in the experiments (the term *box model* is commonly used when a chemical ODE system is considered at an arbitrary grid-point). Sometimes several scenarios will be used in order to carry out the experiments in different typical situations that occur often in practice when large-scale air pollution models are used in different studies.

The system of ODEs at any point of the grid can be written in the form traditionally used when ODE systems are treated numerically:

$$y' = f(t, y), \qquad y \in R^{N_s}, \qquad f \in R^{N_s}, \tag{4.1}$$

where

- the components of vector y from (4.1) are the concentrations of the chemical species at the grid-point under consideration, and
- the components of vector f in the right-hand-side of (4.1) are formed by the chemical reactions in which the corresponding components of y are involved and by appropriate emission terms.

A remark about the dependence of function $f(t, y)$ on the independent variable (the time variable) t is needed here. This function can, also in the general case, be easily rewritten in an autonomous form, i.e. the system of ODEs can be represented in the form

$$y' = f(y). \tag{4.2}$$

This can be done by increasing the number of equations by one (adding the equation $t' = 1$). However, the function $f(t, y)$ from (4.1) depends indirectly on the time variable t. The term *indirectly* can be explained as follows. The chemical rate coefficients depend on several parameters (such as temperature, pressure, humidity, etc.). These parameters vary in time, but it is not possible to express this dependence by analytical terms; all these parameters are read from huge input files (the input data files needed when large-scale air pollution models are to be run were discussed in Chapter 1). Therefore, one must be careful when numerical methods, which require $\partial f / \partial t$, are used in the chemical part of the model, because it is not very easy to evaluate this quantity. The class of the Rosenbrock methods (see Rosenbrock [206], Hundsdorfer and Verwer [142], Lambert [155]) can be given as an example where the partial derivative $\partial f / \partial t$ is needed.

4.2 Why is it difficult to handle the chemical sub-model?

Several difficulties occur when the chemical sub-model of a large air pollution model is to be treated numerically. The main difficulties are discussed below.

4.2.1 The number of relatively small ODE systems that are to be solved is very large

For example, if the space domain of the air pollution model is discretized by using a $(96 \times 96 \times 10)$ grid and if the number of chemical species involved in the model is 35, then 92160 ODE systems, each of them containing 35 equations, are to be treated during many time-steps (in conjunction with the other physical processes). If $(480 \times 480 \times 10)$ grid is used, then the size of the ODE systems remains the same (35 equations), but the number of the ODE systems that have to be treated at every time-step is increased from 92160 to 2304000.

4.2.2 The chemical ODE systems are normally very stiff

Some of the chemical reactions are very fast, others are rather slow. As a rule, this fact implies stiffness.

There are periods (sun-rises and sun-sets) when certain chemical processes (the so-called photochemical reactions) are activated or deactivated. This also causes quick changes of some of the chemical species and, thus, introduces extra difficulties. Therefore, one should be very careful in the selection of numerical methods for the chemical sub-models.

4.2.3 The chemical ODE systems are very badly scaled

The concentrations of the chemical species involved in an air pollution model, which are often measured in $molecules/cm^3$, vary in different ranges. An example, which demonstrates this fact, is given in Table 4.1. It is seen that

- the unknown quantities (i.e. the concentrations of the different chemical species) have different magnitudes, and
- some chemical species vary very slowly (as CO), while other chemical species vary in very large intervals (which is easily seen by comparing the minimal concentrations with the maximal concentrations for NO and $O^{(1)}D$ in Table 4.1).

Table 4.1: Variations of some chemical species in the case where Scenario 2 (see Subsection 4.4.2) is used.

Species	Starting concentration	Min. concentration	Max. concentration
NO	10^{+10}	$3.6 * 10^{+03}$	$1.8 * 10^{+10}$
NO_2	10^{+11}	$8.8 * 10^{+08}$	10^{+11}
CO	$3.8 * 10^{+14}$	$3.7 * 10^{+14}$	$3.8 * 10^{+14}$
$O^{(1)}D$	10^{+03}	$2.1 * 10^{-43}$	$1.3 * 10^{+03}$

4.2.4 Coupling with the other atmospheric processes

The use of a splitting procedure allows us to treat the chemical sub-model with numerical methods that are specially designed for this sub-model. However, the chemical sub-model has to be coupled with the other physical processes:

- horizontal transport (advection),
- horizontal diffusion,
- deposition, and
- vertical exchange between the different layers.

This implies that the chemical ODE systems should be integrated over a series of very short time-intervals (see also Chapter 2 where the coupling of the sub-models, which are obtained after the implementation of an appropriate splitting procedure, is discussed). After the integration of the chemical ODE system over a short time-interval (consisting of one or several chemical time-steps), the other physical processes are handled (which may lead to considerable changes of the concentrations) and the modified concentrations are used as starting values in the chemical sub-model for the next short time-interval. This means that the difficult problem of starting the integration process with the selected numerical method occurs very often during the treatment of the chemical part of a large air pollution model (in most of the cases, this happens at every time-step). Moreover, this means that some difficulties may arise when linear multistep methods are applied (different linear multistep methods are discussed in Hairer et al. [126], Hairer and Wanner [127], Lambert [155] and Shampine and Gordon [218]). The well-known and commonly used (when stiff systems of ODEs are solved) backward differentiation formulae form a sub-class of the class of linear multistep methods.

4.2.5 The use of efficient integration techniques is crucial

The difficulties discussed above show that the choice of numerical methods for the solution of the ODE systems in the chemical sub-model is very important. In many atmospheric models, the treatment of the chemical sub-models is the most time consuming part. The problem of finding the optimal numerical method the treatment of the chemical ODEs, with regard to both

- the requirement to reduce the computational time as much as possible, and
- the requirement to achieve the required accuracy

is still open.

Several algorithms that can be applied in the treatment of the chemical sub-model of a large-scale air pollution model will be discussed in the following sections of this chapter.

4.3 Algorithms for the numerical integration of the chemical ODE systems

Consider the chemical system (4.1). Numerical experiments with several algorithms have been carried out. A short description of these algorithms for the treatment of (4.1) is given in this section. Some of these algorithms together with some other algorithms have been studied in Brandt et al. [29], Chock et al. [46], Hertel et al. [134], Jay et al. [145], Odman et al. [187], Sandu et al. [211], [213], [212], Shieh et al. [221], Verwer and van Loon [259], Verwer and Simpson [260], Verwer et al. [258].

The case where the chemical part is combined with the vertical exchange is treated together with appropriate splitting procedures and with applications on parallel computers in Botchev et al. [23].

The interest in developing efficient numerical algorithms for ODE systems arising in atmospheric chemistry seems to be increasing during the last decade.

4.3.1 The original QSSA algorithm

QSSA stands for quasi-steady-state-approximation. Methods of this type have been used (and are still used) in many areas of science and engineering. This algorithm has originally been suggested for usage in the area of air pollution modelling in Hesstvedt et al. [135]. The QSSA algorithm, which has been introduced in Hesstvedt et al. [135], can be described as follows. Let us rewrite (4.1) in the following form:

$$\frac{dy_s}{dt} = P_s(t, y_1, y_2, ..., y_q) - L_s(t, y_1, y_2, ..., y_q) y_s, \quad s = 1, 2, ..., q, \ q = N_s, \quad (4.3)$$

where

$$P_s = P_s(t, y_1, y_2, ..., y_q) \quad and \quad L_s = L_s(t, y_1, y_2, ..., y_q) \quad (4.4)$$

are non-negative functions which are called production terms and loss terms respectively.

Assume that some approximations y_s^n to the solutions $y_s(t_n)$ of (4.3) at $t = t_n$ have been found for all values of s, where $s = 1, 2, \ldots, q$. Let $\triangle t$ be a given increment such that $t_{n+1} = t_n + \triangle t$. Then the QSSA algorithm can be applied to calculate approximations y_s^{n+1} to the solutions $y_s(t_{n+1})$ of (4.3) at $t = t_{n+1}$. In Hesstvedt et al. [135] the QSSA is defined by using (for $s = 1, 2, ..., q$) the following three formulae:

$$y_s^{n+1} = \frac{P_s}{L_s} \qquad for \qquad \triangle t L_s > 10, \quad (4.5)$$

$$y_s^{n+1} = \frac{P_s}{L_s} + (y_s^n - \frac{P_s}{L_s}) e^{-\triangle t L_s} \qquad for \qquad 0.01 < \triangle t L_s \leq 10, \quad (4.6)$$

$$y_s^{n+1} = y_s^n + \triangle t (P_s - L_s y_s^n) \qquad for \qquad \triangle t L_s \leq 0.01, \quad (4.7)$$

where it is assumed that the functions P_s and L_s are calculated for $t = t_n$. We shall refer to the version of the algorithm which is defined by the formulae (4.5) - (4.7) as **the original QSSA algorithm**.

It is seen that any of the three formulae (4.5) - (4.7) is, in fact, an explicit formula. However, the original QSSA algorithm is normally combined with some iterative process. The following actions are to be carried out at each iteration (until certain stopping criteria are satisfied) when an iterative procedure is used:

- Step 1. Calculate approximations y_s^{n+1} for $s = 1, 2, ..., q$ $(q = N_s)$.
- Step 2. Update the values of P_s and L_s for $s = 1, 2, ..., q$ $(q = N_s)$ by using the approximation found in the previous step.
- Step 3. Check the selected stopping criteria (if these are satisfied, then stop, else go to Step 1).

Such an iterative process is often used in practice (it is called either the functional iteration or the simple iteration), but there is no guarantee that it will converge. Moreover, very often no stopping criteria are specified (i.e. the third step in the algorithm given above is simply omitted, and the iteration is stopped after some prescribed number of iterations; five iterations are often used). The original QSSA algorithm is in fact an explicit algorithm when it is applied with a fixed number of iterations. The chemical species are often divided into two groups: slowly varying species and quickly varying species. Iterations are carried out only for the quickly varying species. Finally, the lumped mass principle is often used in an attempt to improve the accuracy (much more details about the practical implementation of the original QSSA algorithm, which is defined by the three formulae (4.5) - (4.7), can be found in Hesstvedt et al. [135]).

Many experiments, which were carried in Hesstvedt et al. [135], indicated that the original QSSA algorithm (implemented as sketched above) will often give good results when the time-steps are small ($\triangle t \leq 30$ seconds is recommended in Hesstvedt et al. [135]).

The original QSSA algorithm has at least two important advantages:

- the formulae used in this algorithm are very simple, and
- the algorithm can very efficiently be implemented in the treatment of large-scale air pollution models on high-speed computers which will be demonstrated in Chapter 7 (see also Alexandrov et al. [5], Dimov et al. [70], [71], Georgiev and Zlatev [116], [117] and Owczarz and Zlatev [188], [189], [190]).

The original QSSA algorithm has also some drawbacks. It is desirable to increase further the speed of computations (although the algorithm as mentioned above is faster than most of the other algorithms). Moreover, the algorithm produces inaccurate results in cases where the concentrations are both high and rapidly varying (see Table 4.6 and Table 4.7).

Because of the use of equation (4.5), the original QSSA method can be viewed as an attempt to transform dynamically, during the process of integration, the system of ordinary differential equation (4.3) into two systems: a system of ordinary differential equations and a system of non-linear algebraic equations. The approach used in the original QSSA method is rather crude. There are more advanced and, therefore, more accurate methods. The main idea can be sketched as follows. The two systems, the system of ordinary differential equations and the system of non-linear algebraic equations, are written in the following generic form:

$$\frac{dg_1}{dt} = f_1(t, g_1, g_2), \tag{4.8}$$

$$0 = f_2(t, g_1, g_2), \tag{4.9}$$

and these two systems have to be treated simultaneously at every time-step.

In this way we arrive at a system of differential-algebraic equations (DAEs). There are special methods for treating such systems as, for example, the code DASSL (see Brenan et al. [30]). Problem-solving environments (such as MATLAB or Simulink) can be used in the preparation stage (where a small chemical systems at one grid-point only is used in the tests). More details about the use of such problem solving environments can be found in Shampine et al. [220].

Solvers for systems of DAEs are not very popular in the area of the air pollution models. However, some results in this direction were recently reported in Djouad and Sportisse [76].

4.3.2 An improved QSSA algorithm

Several attempts to improve the original QSSA algorithm has been carried out (see, for example, Jay et al. [145], Verwer and van Loon [259], Verwer and Simpson [260]). An improvement based on ideas proposed in [259] and [260]) will be discussed in this subsection.

It is not possible to achieve a very high performance on some computers by using the original QSSA algorithm because:

1. Three questions, are to be asked at every grid-point and for every species. These questions are expressed by logical operations (normally by IF constructs when the code is written in FORTRAN) and used in very long loops. This means that the questions are to be asked up to 10^6 times at each time-step when the chemical part of a large air pollution model is handled (it should be mentioned here that several hundred thousand time-steps are often needed when large-scale air pollution models are run).

2. The calculation of the exponential function which appears in (4.6) is expensive on some computers.

Therefore, it is worthwhile to try to improve the performance of the original version of QSSA algorithm by performing the following two actions:

- using only one formula, formula (4.6), in the numerical integration of the system of ODEs defined by (4.1), and
- replacing the exponential function in formula (4.6) with the following approximation:

$$e^{-\Delta t L_s} \approx \frac{1}{1 + \Delta t L_s + 0.5(\Delta t L_s)^2}. \tag{4.10}$$

If the approximation given by (4.10) is used, then it can easily be proved that the following formula can be obtained from formula (4.6) and this single formula can be used to calculate approximations to y_s^{n+1}, where $s = 1, 2, ..., q$ and $n = 1, 2, ...$ (assuming here that some starting approximations y_s^0 for $s = 1, 2, ..., q$ are given):

$$y_s^{n+1} = \frac{y_s^n + [1 + 0.5\Delta t L_s]\Delta t P_s}{1 + \Delta t L_s + 0.5(\Delta t L_s)^2}. \tag{4.11}$$

The use of the last formula leads to a new QSSA algorithm, which will be called the **improved QSSA algorithm**. The improved QSSA algorithm can be combined with all devices that are usually used together with the original QSSA algorithm. For example, the lumped mass principle can be applied and/or the chemical species can be separated into two groups (slowly varying species and quickly varying species). Furthermore, this formula can be applied both as

- an explicit formula (inserting in P_s and L_s the values of the concentrations which are obtained at time-step n)

or as

- an implicit formula (inserting in P_s and L_s the values of the concentrations which are obtained at time-step $n + 1$).

An iterative method **must** be used in the latter case. One can use again the functional iteration, but a more advanced iterative method (some modification of the classical Newton method, methods based on the use of a Krylov subspace or the Gauss-Seidel iteration) can also be applied.

The improved QSSA algorithm based on (4.11) is implemented and used in the same way as the original QSSA algorithm proposed in Hesstvedt et al. [135]. This will allow us to see directly the effects of using

- only one formula, (4.11), in the improved QSSA algorithm instead of the three formulae (4.5) - (4.7), which are used in the original QSSA algorithm, and
- the rational approximation of the exponential function, which is defined by formula (4.10).

4.3.3 Using the Backward Euler Method

Consider again the ODE system defined by (4.1). It is worthwhile to change the notation when any of the classical numerical methods for solving systems of ODEs is to be used. In this and in the following sections the lower indices will show the number of the time-step under consideration, while the upper indices in square brackets (when used) will show the number of the current iteration. Let us also introduce the following abbreviations under the assumption that y_n is an approximation to the exact solution $y(t_n)$ of the system of ODEs under consideration at $t = t_n$:

$$f = f(t, y) \qquad and \qquad f_n = f(t_n, y_n) \qquad for \qquad n = 0, 1, 2, \ldots . \tag{4.12}$$

Then the Backward Euler Method (see, for example, Hairer and Wanner [127] or Lambert [155]) can be defined as follows:

$$y_{n+1} = y_n + \Delta t f_{n+1}, \qquad for \quad n = 0, 1, 2, \ldots, \quad with \;\; y_0 \;\; given. \tag{4.13}$$

The last equation is in general non-linear in y_{n+1}, because $f_{n+1} = f(t_{n+1}, y_{n+1})$ depends also on y_{n+1}. Therefore, some iterative method must be applied when the Backward Euler Method is used (the case where f is a linear, in y, function being an exception). The classical Newton method is often used. This method can be introduced as follows. Consider the quantities:

$$F(y_{n+1}) = y_{n+1} - y_n - \Delta t f_{n+1} \tag{4.14}$$

and

$$J_{n+1} = \left. \frac{\partial f(t, y)}{\partial y} \right|_{t=t_{n+1}, y=y_{n+1}} . \tag{4.15}$$

The relationship (I being the identity matrix):

$$\frac{\partial F(y_{n+1})}{\partial y_{n+1}} = I - \Delta t J_{n+1} \tag{4.16}$$

follows from (4.14) and (4.15).

Assume that the classical Newton method is used to solve (approximately, according to some prescribed accuracy) the system:

$$F(y_{n+1}) = 0, \tag{4.17}$$

which is in fact the same as (4.13).

The major formulae that are needed in the classical Newton method can be written in the following form (assuming that the iteration numbers are given as superscripts in square brackets):

$$(I - \Delta t J_{n+1}^{[i-1]}) \Delta y_{n+1}^{[i]} = -y_{n+1}^{[i-1]} + y_n + \Delta t f(t_{n+1}, y_{n+1}^{[i-1]}) \tag{4.18}$$

and

$$y_{n+1}^{[i]} = y_{n+1}^{[i-1]} + \triangle y_{n+1}^{[i]}. \tag{4.19}$$

Some initial approximation $y_{n+1}^{[0]}$ is needed to start the iterative process defined by (4.18)-(4.19). The following two choices are often used in practice:

$$y_{n+1}^{[0]} = y_n, \tag{4.20}$$

$$y_{n+1}^{[0]} = y_n + \frac{\triangle t_{n+1}}{\triangle t_n}(y_n - y_{n-1}), \tag{4.21}$$

where $\triangle t_{n+1}$ and $\triangle t_n$ are the last two time-step sizes applied in the integration process. The second choice is always used in the experiments described in Section 4.4.

Consider an arbitrary iteration step i (where $i = 1, 2, \ldots, \mu$, where μ is the last iteration step, i.e. the iteration step at which the iterative process will be stopped) of the classical Newton method used to calculate an approximation y_{n+1} to the exact solution $y(t_{n+1})$ of the problem defined by (4.1). Such an iteration step consists of six major parts:

1. **Part 1 - Function evaluation**. Calculate the components of the right-hand side vector $f(t_{n+1}, y_{n+1}^{[i-1]})$.

2. **Part 2 - Jacobian evaluation**. Compute the elements of the Jacobian matrix $J_{n+1}^{[i-1]}$.

3. **Part 3 - Form a system of linear algebraic equations**. Calculate the elements of the shifted Jacobian matrix $I - \triangle t J_{n+1}^{[i-1]}$ and the components of vector $-y_{n+1}^{[i-1]} + y_n + \triangle t f(t_{n+1}, y_{n+1}^{[i-1]})$.

4. **Part 4 - Solve the system of linear algebraic equations formed in the previous part**. Use the well-known Gaussian elimination method to solve the system of linear algebraic equations defined by (4.18).

5. **Part 5 - Update the solution**. Use formula (4.19) to calculate the components of vector $y_{n+1}^{[i]}$.

6. **Part 6 - Stopping checks**. Apply some set of stopping criteria and

   - If all stopping criteria are satisfied, then set y_{n+1} equal to $y_{n+1}^{[i]}$ and stop the iteration.

   - If there are stopping criteria that are not satisfied and the iterative process is converging sufficiently fast, then set $i := i + 1$ and **go to Part 1** to start the next Newton iteration.

- If there are stopping criteria that are not satisfied and the iterative process is either divergent or the convergence rate is very slow, then set $i := 0$, reduce $\triangle t$ and restart the Newton iteration.

Some modifications of the classical Newton iteration are often used in practice. One of the following two procedures, which are described in Remark 4.1 and Remark 4.2, can be applied in the efforts to reduce the amount of computational work.

Remark 4.1. The iterative algorithm described above can be carried out under the assumption that the calculations needed in Part 2 are only performed during the first Newton iteration at time step $n+1$. If the algorithm is implemented in this way, then the elements of the coefficient matrix are also calculated only during the first Newton iteration at time step $n+1$. Finally, an LU decomposition of $I - \triangle t J_n^{[0]}$ is calculated only during the first iteration and used in all consecutive iterations at time step $n+1$.

Remark 4.2. One can try to apply a Jacobian matrix calculated in a previous integration step (say, step m with $m \leq n$). It is not necessary to calculate the Jacobian matrix at the first Newton iteration of step m; we shall assume that the Jacobian matrix has been calculated at some iteration j, where $j \geq 0$. If this is done, then the elements of the coefficient matrix are also calculated only at iteration j of step m. Finally, an LU decomposition of $I - \triangle t J_m^{[j]}$ is calculated, kept and used at all iterations from iteration j of step m to the current iteration of step n.

The use of any of the modified Newton procedures sketched above may lead to some considerable reduction in the computational work per iteration (Part 2 is not carried out at many iterations and the computations at Part 3 and Part 4 are considerably reduced). The reduction tends to be greater when the problem is larger. On the other hand, while the convergence rate of the classical Newton method is of order two, the convergence rate of the modified versions is linear (see, for example, Zlatev [273]). Thus, the number of iterations may be increased when the latter versions are used.

If any of the modified Newton procedures is applied, then some rules, by which a decision whether a new Jacobian matrix should be calculated or not, are to be given. The simplest way to do this is to insert an additional condition in **Part 6**:

If the convergence rate is judged (by the code) to be slow, then recalculate the Jacobian matrix.

It should be noted here that a reduction of the step-size $\triangle t$ should be carried out only if the recalculation of the Jacobian matrix (performed when the condition in the additional rule introduced above is satisfied) does not accelerate the convergence rate.

4.3.4 Using the Trapezoidal Rule

Assume that y_0 is given. Then the well-known and commonly used Trapezoidal Rule (see again Hairer and Wanner [127] or Lambert [155]) can be defined in the following way:

$$y_{n+1} = y_n + 0.5\triangle t(f_n + f_{n+1}), \qquad for \qquad n = 0, 1, 2, \ldots . \tag{4.22}$$

The non-linear system that has to be solved when the Trapezoidal Rule is used and its Jacobian matrix are given by the following formulae (which are rather similar to the corresponding formulae which were used with the Backward Euler Method in the previous subsection):

$$F(y_{n+1}) = y_{n+1} - y_n - 0.5\triangle t(f_n + f_{n+1}) \tag{4.23}$$

and

$$\frac{\partial F(y_{n+1})}{\partial y_{n+1}} = I - 0.5\triangle t J_{n+1}. \tag{4.24}$$

The major formulae that are needed in the classical Newton method applied for the Trapezoidal Rule can now be written in the following form (assuming again that the iteration numbers are given as superscripts in square brackets):

$$A_{n+1}^{[i-1]}\triangle y_{n+1}^{[i]} = -y_{n+1}^{[i-1]} + y_n + 0.5\triangle t\left[f(t_n, y_n) + f(t_{n+1}, y_{n+1}^{[i-1]})\right] \tag{4.25}$$

where $A_{n+1}^{[i-1]} = I - 0.5\triangle t J_{n+1}^{[i-1]}$ and

$$y_{n+1}^{[i]} = y_{n+1}^{[i-1]} + \triangle y_{n+1}^{[i]}. \tag{4.26}$$

The rules for stopping the iterative process in the classical Newton method can be defined as in the previous subsection, but these rules are now related to the above four formulae, (4.23)-(4.26). Moreover, modifications of the classical Newton method that are similar to those discussed in the previous subsection (based on Remark 4.1 and Remark 4.2) can also be introduced for the Trapezoidal Rule, in an attempt to reduce the computational work.

4.3.5 Partitioning the chemical ODE system: using a dense matrix technique

As mentioned before, some of the chemical reactions are very fast, while other chemical reactions are relatively slow. The fast reactions introduce stiffness to the system (4.1). Therefore, it is worthwhile to try to separate these reactions, and to treat the equations in which they appear independently. This can be done by partitioning both the vector y_{n+1} and the matrix $I - \triangle t J_{n+1}$ ($n = 1, 2, \ldots, N-1$).

Let us assume that the components of vector y_n have been partitioned in some way into $NBLOCKS$ blocks, where $NBLOCKS \le q = N_s$. The partitioning of vector y_{n+1} implies a unique partitioning of matrix $I - \triangle t J_{n+1}$ into two groups of blocks. Consider one of the blocks of vector y_{n+1}; say the block that contains the components $r_1, r_2, \ldots, r_p$, $p \le q$. Consider the sub-matrix of $I - \triangle t J_{n+1}$ formed by the rows $r_1, r_2, \ldots, r_p$. This sub-matrix consists of two blocks. The first one is formed by the columns $r_1, r_2, \ldots, r_p$, while the second block contains the remaining columns. The first block will be called a *strong* block, while the name *weak* block will be used for the second one. It is clear that each block of vector y_{n+1} induces a pair of blocks (one strong and the other weak) in an appropriate sub-matrix of matrix $I - \triangle t J_{n+1}$. We shall assume here that $r_1, r_2, \ldots, r_p$ are consecutive integers (i.e. $r_2 = r_1 + 1,\ r_3 = r_2 + 1,\ \ldots,\ r_p = r_{p-1} + 1$). It is clear that this is not a restriction, because it can be achieved by permuting the components of vector y_{n+1}. It should be mentioned here that the permutation of the components of vector y_{n+1} induces row and column permutations in matrix $I - \triangle t J_{n+1}$.

Denote

$$A_{n+1} = I - \triangle t J_{n+1}. \tag{4.27}$$

Matrix A_{n+1} can be represented as

$$A_{n+1} = S_{n+1} + W_{n+1}, \tag{4.28}$$

where matrix S_{n+1} contains only elements of the strong blocks, while W_{n+1} contains only elements of the weak blocks. It can easily be verified that matrix S_{n+1} is a block-diagonal matrix under the assumption that the permutations mentioned above have been performed.

Assume again that the classical Newton method is used. By using superscripts and (4.27), we can rewrite (4.18) as follows:

$$A_{n+1}^{[i-1]} \triangle y_{n+1}^{[i]} = -y_{n+1}^{[i-1]} + y_n + \triangle t f(t_{n+1}, y_{n+1}^{[i-1]}) \tag{4.29}$$

By applying (4.28), the last equation can be expressed by the strong and the weak blocks of matrix $A_{n+1}^{[i-1]}$:

$$S_{n+1}^{[i-1]} \triangle y_{n+1}^{[i]} = -y_{n+1}^{[i-1]} + y_n + \triangle t f(t_{n+1}, y_{n+1}^{[i-1]}) - W_{n+1}^{[i-1]} \triangle y_{n+1}^{[i]}. \tag{4.30}$$

The expression on the left-hand-side of (4.30) is very nice; it is a block-diagonal matrix (or, at least, can be permuted to a block-diagonal matrix under the assumption made above). The trouble is that the unknown vector $\triangle y_{n+1}^{[i]}$ appears also on the right-hand-side. This problem can be solved by using an inner iteration loop in order to obtain an approximate solution of (4.30). This means that we have to carry out successively the following computations (the second superscript in the square brackets denotes the iteration number in the inner iteration loop):

$$S_{n+1}^{[i-1]} \triangle y_{n+1}^{[i,k]} = -y_{n+1}^{[i-1]} + y_n + \triangle t f(t_{n+1}, y_{n+1}^{[i-1]}) - W_{n+1}^{[i-1]} \triangle y_{n+1}^{[i,k-1]}. \tag{4.31}$$

Consider again the ith iteration of the classical Newton method. By using the splitting of the matrix to strong and weak blocks and an inner iteration, we can rewrite the algorithm for performing the ith iteration, which has be presented in Subsection 4.3.3, as follows:

1. **Part 1 - Function evaluation.** Calculate the components of the right-hand side vector $f(t_{n+1}, y_{n+1}^{[i-1]})$.

2. **Part 2 - Jacobian evaluation.** Compute the elements of the Jacobian matrix $J_{n+1}^{[i-1]}$.

3. **Part 3 - Form a set of systems of linear algebraic equations.** Calculate the elements of matrix $S_{n+1}^{[i-1]}$ and the components of vector $-y_{n+1}^{[i-1]} + y_n + \triangle t f(t_{n+1}, y_{n+1}^{[i-1]}) - W_{n+1}^{[i-1]} \triangle y_{n+1}^{[i,0]}$.

4. **Part 4 - Solve the systems of linear algebraic equations formed in previous part.** Gaussian elimination is used to factorize every block of matrix $S_{n+1}^{[i-1]}$. Then an inner iteration loop is used in order to get a sufficiently accurate approximation to vector $\triangle y_{n+1}^{[i]}$ (for $k + 1, 2, \ldots, K$). If the inner iteration loop converges in K iterations, then vector $\triangle y_{n+1}^{[i]}$ is set equal to $\triangle y_{n+1}^{[i,K]}$. If the inner iteration loop is not convergent, then reduce the stepsize and restart the Newton iteration process.

5. **Part 5 - Update the solution.** Use formula (4.19) to calculate the components of vector $y_{n+1}^{[i]}$.

6. **Part 6 - Stopping checks.** Apply some set of stopping criteria and

   - If all stopping criteria are satisfied, then set y_{n+1} equal to $y_{n+1}^{[i]}$ and stop the iteration.
   - If there are stopping criteria that are not satisfied and the iterative process is converging sufficiently fast, then set $i := i + 1$ and **go to Part 1** to start the next Newton iteration.
   - If there are stopping criteria that are not satisfied and the iterative process is either divergent or the convergence rate is very slow, then set $i := 0$, reduce $\triangle t$ and restart the Newton iteration.

Comparing the algorithm given above with the algorithm in the previous section, it is seen that changes only appear in Part 3 and in Part 4.

The computational process can be further simplified by removing $W_{n+1}^{[i-1]} \triangle y_{n+1}^{[i,k-1]}$ from the right-hand side of (4.31). This is equivalent to decoupling the ODE system (4.1) into sub-systems (the number of equations in any of these sub systems is equal to the order of the corresponding diagonal block of $S_{n+1}^{[i-1]}$). It is clear that such decoupling of the ODE system can only be carried out if the problem studied has some appropriate physical properties. In the next chapter it will be justified that the term $W_{n+1}^{[i-1]} \triangle y_{n+1}^{[i,k-1]}$ can be omitted for ODE systems arising in the chemical part of air pollution models. Therefore, in the numerical examples, which will be given in the second part of this chapter, the term $W_{n+1}^{[i-1]} \triangle y_{n+1}^{[i,k-1]}$

will be removed. This leads to the solution of systems of the following type (the sub-index r, $r = 1, 2, \ldots, NBLOCKS$, shows the number of the block):

$$S^{[i-1]}_{n+1,r} \triangle y^{[i]}_{n+1,r} = -y^{[i-1]}_{n+1,r} + y_{n,r} + \triangle t \, g \tag{4.32}$$

with

$$g = f(t_{n+1}, y^*_{n+1,1}, y^*_{n+1,2}, \ldots, y^*_{n+1,r-1}, y^{[i-1]}_{n+1,r}, y^*_{n+1,r+1}, \ldots, y^*_{n+1,NBLOCKS}), \tag{4.33}$$

where $S^{[i-1]}_{n+1,r}$ is the r'th diagonal block of matrix $S^{[i-1]}_{n+1}$ evaluated at iteration i and the quantities $y^*_{n+1,1}, y^*_{n+1,2}, \ldots, y^*_{n+1,r-1}, y^*_{n+1,r+1}, \ldots, y^*_{n+1,NBLOCKS}$ are assumed to be known. The following two choices are often used for these quantities:

$$y^*_{n+1,r} = y_{n,r}, \qquad r = 1, 2, \ldots, NBLOCKS, \tag{4.34}$$

or

$$y^*_{n+1,r} = y_{n,r} + \frac{\triangle t_{n+1}}{\triangle t_n}(y_{n,r} - y_{n-1,r}), \qquad r = 1, 2, \ldots, NBLOCKS. \tag{4.35}$$

where $\triangle t_{n+1}$ and $\triangle t_n$ are the last two time-step sizes used in the integration process. The second choice will always be used in the experiments.

It is clear that (4.32)-(4.33) can be handled by the Newton iterative procedure was used used to handle (4.18) in Subsection 4.3.3.

The algorithm sketched above, and based on the use of (4.32)-(4.33), is very convenient for computations on parallel computers, because the computations for each r, $r = 1, 2, \ldots, NBLOCKS$, can be carried out concurrently. However, in our case this is not needed, because we have much more and much larger parallel tasks. The number of parallel tasks in an air pollution model is equal to the number of grid-points (see Subsection 4.1) and a parallel task contains all $NBLOCKS$ systems (4.32)-(4.33). Therefore, it is natural to use in the right hand side of (4.33) the approximations already calculated. Let us denote the calculated approximations by $\tilde{y}_{n+1,k}$, $k = 1, 2, \ldots, r-1$. By inserting these values in (4.33), the following formulae can be obtained:

$$S^{[i-1]}_{n+1,r} \triangle y^{[i]}_{n+1,r} = -y^{[i-1]}_{n+1,r} + y_{n,r} + \triangle t g \tag{4.36}$$

with

$$g = f(t_{n+1}, \tilde{y}_{n+1,1}, \tilde{y}_{n+1,2}, \ldots, \tilde{y}_{n+1,r-1}, y^{[i-1]}_{n+1,r}, y^*_{n+1,r+1}, \ldots, y^*_{n+1,NBLOCKS}), \tag{4.37}$$

The last two formulae will be used in the experiments described in Section 4.4. Normally, it will be necessary to introduce an outer "Gauss-Seidel-type" iteration by using the following two formulae (where m is the iteration index for the outer iteration loop):

$$\tilde{y}^{[m]}_{n+1,r} = y_{n,r} + \Delta t g \tag{4.38}$$

with

$$\begin{aligned} g \;=\; & f(t_{n+1}, \tilde{y}^{[m]}_{n+1,1}, \tilde{y}^{[m]}_{n+1,2}, \ldots, \tilde{y}^{[m]}_{n+1,r-1}, \tilde{y}^{[m]}_{n+1,r}, \\ & \tilde{y}^{[m-1]}_{n+1,r+1}, \ldots, \tilde{y}^{[m-1]}_{n+1,NBLOCKS}), \end{aligned} \tag{4.39}$$

where $\tilde{y}^{[m]}_{n+1,1}, \tilde{y}^{[m]}_{n+1,2}, \ldots, \tilde{y}^{[m]}_{n+1,r-1}$ have already been calculated at the current outer iteration m, while $\tilde{y}^{[m-1]}_{n+1,r+1}, \ldots, \tilde{y}^{[m-1]}_{n+1,NBLOCKS}$ are known from the previous outer iteration $m-1$. Thus, one has to calculate, in an inner iteration loop, $\tilde{y}^{[m]}_{n+1,r}$. When this is done, one proceeds with the calculation of $\tilde{y}^{[m]}_{n+1,r+1}$. The m'th outer iteration is finished when $\tilde{y}^{[m]}_{n+1,NBLOCKS}$ is calculated. After that outer iteration $m+1$ is to be started if the accuracy required is not achieved.

If the iteration process (4.38)-(4.39) converges for all r, then the final approximation y_{n+1} is formed (as a combination of the sub-vectors $\tilde{y}_{n+1,r}$) and one proceeds with the next step; step $n+2$. If the process does not converge or converges slowly, then the stepsize Δt is reduced and the iteration is restarted with the reduced stepsize.

The algorithm described in this section is based on ideas discussed in Skelboe and Zlatev [227].

The Backward Euler Method has been applied in this section. However, other numerical methods can also be selected and used.

More details about partitioning and splitting of systems of ODEs can be found in the monograph of Burrage, see [36]. It is shown in [36] that partitioning is a rather general procedure, which can be applied in many fields of science and engineering.

Some theoretical aspects of the partitioning procedures will be studied in detail in the next chapter. We shall proceed now with a sparse matrix implementation of the partitioning procedure and with numerical examples.

4.3.6 Partitioning the chemical ODE system: using a sparse matrix technique

The partitioning techniques applied in Subsection 4.3.5 leads to the solution of several systems of linear algebraic equations during the Newton iteration (or the modified Newton iteration). Moreover, it has been assumed that some dense matrix technique is to be used in Subsection 4.3.5. The diagonal blocks of matrix S are sparse matrices. However, these matrices are normally rather small when air pollution models are to be handled. This is why the straight forward application of a general sparse matrix code (as, for example, the codes discussed in Duff et al. [81] or Zlatev [281]) will be inefficient. Therefore, a special sparse matrix technique has been developed. This technique is based on the following steps:

1. A preliminary reordering procedure based on the application of a Markowitz pivotal strategy (see Zlatev [281]) is performed.

2. The positions, in which new non-zero elements will be created, are determined and locations for these elements are reserved in the arrays where the LU factorization of the diagonal block under consideration is stored.

3. A loop-free code for the numerical calculation of the LU factorization of the diagonal block under consideration is used.

4. A loop-free code for the back-substitution with the LU factorization computed in the previous step is used.

Moreover, it must be emphasized that no implicit addressing is used during the calculations.

The advantage of this approach is that the use of sparse matrix techniques becomes more efficient than the use of dense matrix techniques. On vector machines one can achieve high speed computations by vectorizing across the grid-points (see Zlatev [282], [278]).

The special sparse matrix technique based on the ideas sketched above has several disadvantages (compared with the general sparse matrix codes studied in Duff et al. [81] or Zlatev [281]; it should be mentioned here that some algorithms for treating general sparse matrices will be discussed in Section 6):

- There is no pivoting for numerical stability and this may sometimes cause problems. However, the problems will normally be detected (because the Newton iterative procedure will not converge). The code will then automatically reduce the stepsize, which will lead to an improvement of the numerical stability of the computations during the LU factorization.

- Loop-free codes have been developed in the 60's (see Willoughby [268]). It became very quickly clear that if the matrix is very large, then these codes become too long and the compilers cannot deal with them. However, the number of chemical species that are to be handled in a single grid-point remains reasonably small when large air pollution models are handled. This means that the sparse matrices involved are also reasonably small and, thus, the use of a loop-free code in the sparse subroutines does not cause problems in this particular situation.

It should be mentioned here that an approach based on the application of similar ideas has been used in Sandu et al. [213] and Swart and Bloom [240]; see also some of the references in these papers.

Sometimes the use of general-purpose methods for solving linear algebraic equations with large and sparse coefficient matrices is nevertheless more profitable. This is especially true when no splitting procedure is used in the discretization of the large-scale air pollution model under consideration. As mentioned above, some methods for the treatment of large and very large general sparse matrices will be introduced and discussed in Chapter 6.

4.4 Numerical results

The condensed CBM IV scheme has been used in the experiments, results of which will be presented in this section. The scheme has been run over a time period of 42 hours (starting at 6 o'clock in the morning and finishing at 24 o'clock the next day). This period contains several changes from day to night or from night to day. It is important to include such changes because the photochemical reactions cause quick changes of some concentrations in these periods (around sunrises and sunsets). Therefore, it is absolutely necessary to check the performance of the numerical algorithms used in the treatment of the chemical part of a large-scale air pollution model when changes

- from day-time to nigh-time, and
- from night-time to day-time

take place during the computations.

The rate coefficients of most of the chemical reactions depend on the temperature. Therefore, it is necessary to introduce some temperature variation during the experiments. A diurnal variation of the temperature has been simulated by using a simple cosine function.

4.4.1 Implementation of the different algorithms

The implementation of the QSSA algorithms is quite straight-forward. For the original QSSA, we just followed the recommendation in Hesstvedt et al. [135]. The species have been divided into two groups (slowly varying and quickly varying). An iterative process with a fixed number of iterations (five) has been carried out for the quickly varying species; this means that the method is in fact explicit. The same division and the same type of iteration has been applied with the improved QSSA. This is done in order to be able to compare directly the effect of using (4.10) and (4.11) instead of (4.5) - (4.7). It should be emphasized here that the implementation of the improved QSSA based on (4.10)-(4.11) could be made in a more efficient way (which will lead to the achievement of more accurate results; while it is not very clear whether one can also achieve better computational times).

The two classical numerical ODE methods, the Backward Euler Method and the Trapezoidal Rule, are implemented in a rather traditional way. The basic idea is to reduce the number of evaluations of the Jacobian matrix and the number of factorizations of the shifted Jacobian matrix $I - \triangle tJ$. This is done by using the ideas sketched in Remark 4.2 in Subsection 4.3.3.

The partitioning algorithms, which were introduced and discussed in the previous sections of this chapter, have been used with

- one big block containing 13 species, and
- 22 small blocks, each of them containing only one species.

The treatment of the small blocks is carried out by the Newton method applied for a scalar equation. The Jacobian matrix of the large block is treated as the Jacobian matrix in the Backward Euler Method and the Trapezoidal Rule. The ideas discussed in Subsection 4.3.6 are used in the sparse matrix version of the partitioned algorithm.

Two starting time-step sizes, a large one ($\triangle t = 30$ sec.) and a small one $\triangle t = 1$ sec., have been used.

The rules, which are used in order to determine the time-step size at a given time-step n can be described as follows:

- The time-step size is kept fixed (equal to the starting time-step size) when the two QSSA algorithms are used.
- The time-step size is varied in the Backward Euler Method, the Trapezoidal Rule and the two partitioning algorithms. If the iterative method is slowly convergent or diverges, then first the Jacobian matrix is re-evaluated (when a Jacobian matrix calculated at a previous step or at a previous iteration is used; see Remark 4.2 in Subsection 4.3.3) and a new LU-factorization is calculated. If the iterative process is still slowly convergent or divergent, then the time-step size is reduced (and again a new Jacobian matrix as well as a new LU-factorization is calculated). This process can be repeated several times (until sufficiently fast convergence is achieved). The time-step size can also be increased when the convergence is fast (but the chemical time-step size cannot become larger than the stepsize which is used in the other sub-models obtained when an appropriate splitting procedure, see Chapter 2, has been applied).

4.4.2 The scenarios used in the experiments

Six scenarios, which are related to the starting concentrations and/or to the emissions, have been used in the runs. In the first three scenarios there are no emissions. Emissions are used in the last three scenarios. A qualitative characterization of the six scenarios, which have been selected by us, is given in Table 4.2.

Table 4.2: Qualitative descriptions of the six scenarios used in the experiments.

Scenario	Short description
1	Small starting concentrations (remote rural areas)
2	Medium starting concentrations (areas close to big cities)
3	Large starting concentrations (big cities)
4	As Scenario 1 but also emissions are added
5	As Scenario 2 but also emissions are added
6	As Scenario 3 but also emissions are added

Table 4.3: The nitrogen oxide starting concentrations and the NO_x emissions for the six scenarios used in the experiments. The concentrations are measured in $molecules/cm^3$, while $molecules/(cm^2s)$ are used for the emissions.

Scenario	Nitrogen oxide	NO_x emissions
1	1.0E+08	0.0
2	1.0E+10	0.0
3	2.6E+12	0.0
4	1.0E+08	3.3E+06
5	1.0E+10	3.3E+06
6	2.6E+12	3.3E+06

An example is given in Table 4.3 to illustrate how different some of the starting concentrations are and how big some of the emissions used are in the different scenarios. Concentrations of nitrogen oxide (NO) and emissions of nitrogen oxides (NO_x) are used in Table 4.3 to show the differences in the six scenarios.

4.4.3 Computational times for the different algorithms

The computational times (measured in seconds), which were obtained for the six scenarios by the different algorithms that are used in this study, are given

- in Table 4.4 (with a starting time-step size $\Delta t = 30$ sec.), and
- in Table 4.5 (with a starting time-step size $\Delta t = 1$ sec.).

All experiments have been run on a SUN workstation with a 170 MHz processor. All subroutines have been compiled by using option *"-fast"* of the SUN FORTRAN 77 compiler. No other jobs have been run on the workstation during the experiments.

4.4.4 Accuracy of the results obtained by the six algorithms

Three chemical species have been chosen when the accuracy of the six numerical algorithms was tested:

- nitrogen dioxide (NO_2),
- ozone (O_3), and
- sulphur dioxide (SO_2).

The first two species participate in many chemical reactions and vary in large intervals (for Scenario 3, NO_2 varies in the interval from 4.463E+09 to 1.665E+12, while O_3 varies in the interval from 4.443E+10 to 2.972E+12). The third species is participating only in two reactions and varies in a small interval (in the interval from 1.796E+11 to 2.647E+11 for Scenario 3).

The accuracy achieved by the different algorithms for these species is shown in

- Table 4.6 for nitrogen dioxide,
- Table 4.7 for ozone, and
- Table 4.8 for sulphur dioxide.

The accuracy of the concentration of the ith species at a given time t_N (where $t_0 = 21600, t_N = t_0 + 900 * N, N = 1, 2, \ldots, 168$) is measured by the quantity:

$$\varepsilon_{i,N} = \frac{\mid y_{i,N}^{calc} - y_{i,N}^{ref} \mid}{max(y_{i,N}^{ref}, 10^{-6} y_i^{background})} \tag{4.40}$$

where

- $y_{i,N}^{calc}$ is the value calculated by the numerical method under consideration,
- $y_{i,N}^{ref}$ is the corresponding value of the reference solution, and
- $y_i^{background}$ is some small background concentration of the ith species.

The factor $y_i^{background}$ is only used to prevent division by zero when the concentrations become very small and could be set by the computer to zero when underflows take place (the results given in Table 4.1 indicate that this is a real danger for some species on some computers). The reference solution, $y_{i,N}^{ref}$, has been obtained by running the Backward Euler Method with a very small time-step size ($\triangle t = 0.001$). The maximal values (taken over the whole time-integration interval of 42 hours) of the relative errors $\varepsilon_{i,N}$ for the three selected species are given in Table 4.6 to Table 4.9.

Table 4.4: Computational times obtained by running six numerical algorithms with $\triangle$ t = 30 sec. on a SUN workstation. "QSSA1" is the original QSSA, "QSSA2" is the improved version of QSSA, "Euler" is the Backward Euler Method, "Trapez" is the Trapezoidal Rule, "Dense" and "Sparse" refer to the dense and sparse versions of the partitioned algorithm.

Scenario	QSSA1	QSSA2	Euler	Trapez	Dense	Sparse
1	0.42	0.38	0.52	0.58	0.39	0.32
2	0.43	0.38	0.53	1.03	0.39	0.32
3	0.45	0.39	0.55	0.69	0.37	0.31
4	0.45	0.38	0.61	0.68	0.39	0.32
5	0.45	0.39	0.62	0.85	0.41	0.33
6	0.45	0.39	0.55	0.73	0.37	0.32

Table 4.5: Computational times obtained by running six numerical algorithms with $\triangle$ t = 1 sec. on a SUN workstation. "QSSA1" is the original QSSA, "QSSA2" is the improved version of QSSA, "Euler" is the Backward Euler Method, "Trapez" is the Trapezoidal Rule, "Dense" and "Sparse" refer to the dense and sparse versions of the partitioned algorithm.

Scenario	QSSA1	QSSA2	Euler	Trapez	Dense	Sparse
1	12.85	11.72	15.11	15.94	10.09	9.21
2	12.38	11.72	15.10	16.69	10.08	9.22
3	12.78	11.78	15.27	16.10	10.28	9.34
4	12.59	11.79	15.67	16.90	10.29	9.33
5	12.64	11.77	15.12	15.98	10.08	9.22
6	12.86	11.79	15.28	16.12	10.40	9.41

4.4.5 Conclusions from the experiments

Six scenarios have been used in the experiments. These scenarios are typical examples for situations that arise when large air pollution models are to be treated over a large space domain and a long time-interval. Several main conclusions can be drawn from these experiments:

- The QSSA algorithms (both the original and the improved) are faster than the classical ODE algorithms (the Backward Euler Method and the Trapezoidal Rule) when these are implemented in the traditional manner. However, note that the difference is not very large and if an attempt to exploit the sparsity of the Jacobian matrix is made, then these methods may become competitive

Table 4.6: Accuracy of the nitrogen dioxide concentrations obtained by running six numerical algorithms with $\triangle$ t = 30 sec. on a SUN workstation. "QSSA1" is the original QSSA, "QSSA2" is the improved version of QSSA, "Euler" is the Backward Euler Method, "Trapez" is the Trapezoidal Rule, "Dense" and "Sparse" refer to the dense and sparse versions of the partitioned algorithm.

Scenario	QSSA1	QSSA2	Euler	Trapez	Dense	Sparse
1	3.08E-03	1.16E-03	1.78E-03	5.17E-04	1.69E-3	1.69E-3
2	2.01E-02	5.33E-02	1.08E-02	2.91E-03	9.76E-3	9.79E-3
3	4.05E-00	1.18E-00	1.03E-02	2.78E-02	1.03E-2	1.09E-2
4	1.87E-02	2.11E-02	6.01E-03	6.01E-03	6.03E-3	6.02E-3
5	1.38E-02	4.97E-02	1.85E-03	1.36E-03	2.03E-3	2.01E-3
6	2.86E-00	8.38E-01	7.50E-03	2.97E-03	8.03E-3	7.96E-3

with the QSSA algorithms.

- The use of partitioning in conjunction with the Backward Euler Method leads to an improvement of the performance. The two partitioned versions of the Backward Euler Method (the dense version and the sparse version) are at least competitive with the QSSA algorithms with regard to the computational time (see the results given in Table 4.4 and Table 4.5). However, this is true when a sequential computer is used. It is not clear at present whether these two versions of the partitioned algorithm will be as efficient as the QSSA algorithms on modern high-speed computers. In this chapter we have shown that the partitioned algorithms are good candidates for methods that could be applied instead of the QSSA algorithms. Such a replacement seems to be necessary, because in some situations the QSSA algorithms are rather inaccurate (see Table 4.6 and Table 4.7).

- The accuracy is an important issue. While a low accuracy is normally quite sufficient when large air pollution models are treated numerically, it is also important **to achieve** the required accuracy. The experiments indicate that the QSSA algorithms may have severe accuracy problems for the chemically active species in the important case, from a practical point of view, where the concentrations are very high. As a matter of fact, the QSSA algorithms may produce completely wrong results in this situation (see the results for Scenario 3 in Table 4.6 and Table 4.7).

- It is desirable to be able to improve the accuracy by reducing the time-step size. The reduction of the time-step size from 30 seconds to 1 second resulted

Table 4.7: Accuracy of the ozone concentrations obtained by running six numerical algorithms with $\triangle$ t = 30 sec. on a SUN workstation. "QSSA1" is the original QSSA, "QSSA2" is the improved version of QSSA, "Euler" is the Backward Euler Method, "Trapez" is the Trapezoidal Rule, "Dense" and "Sparse" refer to the dense and sparse versions of the partitioned algorithm.

Scenario	QSSA1	QSSA2	Euler	Trapez	Dense	Sparse
1	9.73E-05	1.44E-04	3.58E-05	3.43E-05	3.59E-5	3.59E-5
2	3.20E-03	3.39E-03	5.78E-04	5.78E-04	5.72E-4	5.73E-4
3	4.01E-01	3.88E-01	3.56E-03	3.56E-03	3.25E-3	3.25E-3
4	9.22E-03	1.22E-02	7.47E-04	7.13E-04	7.89E-4	67.89-4
5	5.55E-03	5.79E-03	7.82E-04	7.82E-04	7.53E-4	7.53E-4
6	4.39E-01	3.86E-01	3.57E-03	3.57E-03	3.26E-3	3.26E-3

Table 4.8: Accuracy of the sulphur dioxide concentrations obtained by running six numerical algorithms with $\triangle$ t = 30 sec. on a SUN workstation. "QSSA1" is the original QSSA, "QSSA2" is the improved version of QSSA, "Euler" is the Backward Euler Method, "Trapez" is the Trapezoidal Rule, "Dense" and "Sparse" refer to the dense and sparse versions of the partitioned algorithm.

Scenario	QSSA1	QSSA2	Euler	Trapez	Dense	Sparse
1	3.09E-05	3.48E-05	2.99E-05	3.49E-05	2.30E-3	2.30E-3
2	5.35E-05	6.59E-05	1.76E-05	1.76E-05	1.04E-5	1.04E-5
3	1.96E-02	1.49E-02	2.58E-04	1.18E-04	2.98E-4	2.98E-4
4	8.41E-04	3.81E-04	5.70E-03	5.70E-03	2.30E-4	2.30E-4
5	9.17E-05	9.13E-05	2.28E-04	2.28E-04	3.41E-5	3.41E-5
6	1.32E-02	1.11E-02	2.24E-04	8.32E-05	2.84E-4	2.83E-4

in a considerable improvement of the accuracy of all classical numerical methods. For the QSSA algorithms this is not true. Some results, which illustrate the above statements are given in Table 4.9 (the results in Table 4.9 should be compared with the results in Table 4.6). It should be noted here that we measure the accuracy by the largest error found during the integration. Therefore the fact that the largest errors obtained when the QSSA algorithms are run with $\Delta t = 30$ sec. and $\Delta t = 1$ sec. are nearly the same does not mean that the accuracy achieved is nearly the same in **all** sub-intervals of the period of 42 hours. This means only that there are some critical sub-intervals (where the concentrations change very quickly) for which the reduction of the

Table 4.9: Accuracy of the nitrogen dioxide concentrations obtained by running six numerical algorithms with △ t = 1 sec. on a SUN workstation. "QSSA1" is the original QSSA, "QSSA2" is the improved version of QSSA, "Euler" is the Backward Euler Method, "Trapez" is the Trapezoidal Rule, "Dense" and "Sparse" refer to the dense and sparse versions of the partitioned algorithm.

Scenario	QSSA1	QSSA2	Euler	Trapez	Dense	Sparse
1	2.49E-03	7.65E-04	3.30E-05	1.09E-05	3.28E-5	3.28E-5
2	2.23E-02	2.89E-02	2.88E-04	2.90E-05	2.88E-4	2.88E-3
3	3.90E-00	3.74E-00	2.56E-04	8.50E-05	5.49E-4	5.49E-4
4	2.02E-02	8.94E-03	6.02E-05	3.35E-05	3.70E-5	3.70E-5
5	1.47E-02	3.45E-03	6.22E-05	3.77E-05	6.25E-5	6.25E-3
6	2.75E-00	2.66E-00	1.63E-04	6.61E-05	5.00E-4	5.00E-4

stepsize does not lead to an improvement of the accuracy.

4.5 Treatment of the deposition

The deposition process can be treated as an independent sub-model, see (2.8) in Chapter 2. In this case the deposition process is described mathematically by a **linear** system of ODEs. Moreover, each equation of this system of ODEs is independent. This means that the system of ODEs is split to $N_x \times N_y \times N_z \times N_s$ independent equations each of which is linear. These equations can be solved exactly by using the analytical formula for the solution of scalar linear ODEs.

In the last version of DEM, the deposition process is treated, as mention in Subsection 2.2.2, together with the chemical problems. This means that some linear terms are added to the non-linear terms by which the chemical reactions are described. It is obvious that the same methods, as the methods discussed in the previous sections of this chapter, can be used for the combined chemistry-deposition sub-model.

The short description of the treatment of the deposition process shows that there are no numerical difficulties related to this process. This is true both in the case where the deposition is considered as a separate sub-model and in the case where the deposition terms are added to the chemical sub-model.

4.6 Applicability to other large-scale models

Other models (and, first and foremost, models for studying other large-scale environmental and ecological problems; see Section 7 of Chapter 1) lead, after applying both splitting and discretization, to similar mathematical tasks. Therefore, the numerical methods, which were discussed in the previous sections of this chapter, can successfully be applied also when other large-scale models are treated numerically on high-speed computers. It is necessary to point out, as in the previous chapter, that it might be worthwhile to take into account some specific properties of the particular model under consideration in order to improve the overall efficiency. This is especially true when the model under consideration leads to very big computational tasks.

4.7 Plans for future work

The work on improving the performance of the algorithms used in the chemical schemes must be continued. Faster, but also sufficiently accurate numerical procedures are still needed. Methods of Runge-Kutta type (Butcher [37], Hairer et al. [126], Hairer and Wanner [127]) as well as methods of Rosenbrock type (Rosenbrock [206], Hundsdorfer and Verwer [142], Lambert [155]) must be systematically investigated. These methods are more expensive (require more arithmetic operations per time-step), but they are more accurate. If the later property can successfully be applied in order to carry out the computations with large stepsizes, then the number of time-steps will be reduced. This will, of course, lead to a reduction of the computational time, which might make these methods competitive (in terms of computational time) with the numerical methods that were discussed in this chapter.

It should be mentioned here that the program discussed in Section 2.8 of Chapter 2 can be used in the experiments. Solvers based both on Runge-Kutta methods (including the Fully Implicit Three-stage Runge-Kutta Method of order six) and on Rosenbrock methods are already included into this program. Some results were presented in Chapter 2, but more experiments and some systematical investigations using this program might give useful information about the ability of the different methods to run efficiently when these are included into large-scale mathematical models.

Moreover, it is necessary to investigate the performance of the algorithms when these are combined with the horizontal transport sub-model. In the latter case, it is again necessary both

- to achieve a good accuracy, and
- a fast computational speed.

Some results, which were obtained when the whole model is treated on high-speed computers will be presented in Chapter 7.

Chapter 5

Error analysis of the partitioning procedures

Numerical results, which are obtained when partitioning procedures are applied to treat the particular ODE systems arising in air pollution modelling, were discussed in the previous chapter. It will be shown in this chapter that the partitioning procedures are rather general tools which are applicable for a much broader class of problems. Some algebraic conditions, under which such procedures can successfully be applied, will be derived. Then it will be shown that these conditions are satisfied when the system of ODEs is arising after the discretization of an air pollution model.

All ideas discussed in the first part of this chapter are applicable, under an assumption that certain requirements (stated in the theorems that are proven in this chapter) are satisfied, for many mathematical models, arising in different fields of science and engineering.

In the second part of this chapter, it will be shown that the requirements for a successful application of some partitioning procedures are satisfied for the chemical sub-model of a large-scale air pollution model.

It is assumed in the second part of this chapter (as in the previous two chapters) that some splitting procedure has been applied. Under this assumption, a chemical system of ODEs arising at an arbitrary grid-point (a box model) will be treated using the same partitioning procedure as that considered in the previous chapter. Moreover, the same scenarios (as those used in the previous chapter) will be applied in the experiments.

5.1 Statement of the problem

Systems of ordinary differential equations (ODEs) appear often in practical computations. In fact, the discretization of the spatial derivatives in systems of partial differential equations (PDEs) leads to the solution of ODE systems. Many models arising in science and engineering are described mathematically by systems of

PDEs; see, for example, Burrage [36] or Zlatev [282]). Partitioning can sometimes successfully be used in the numerical treatment of these ODE systems. The major principles used in the implementation of partitioning procedures were discussed in the previous chapter.

Some theoretical properties, which are related to the partitioning procedures, will be studied in this chapter. Only one numerical method, the Backward Euler Method, which was described in Subsection 4.3.5, will be used in this chapter (but this is not a restriction: the same ideas can also be applied when other numerical methods are used). We shall assume that the modified Newton iterative method (see Remark 4.1 in the previous chapter), which is defined by

$$(I - \triangle t J_{n+1}^{[0]}) \triangle y_{n+1}^{[i]} = -y_{n+1}^{[i-1]} + y_n + \triangle t f(t_{n+1}, y_{n+1}^{[i-1]}), \tag{5.1}$$

is used. Let us denote

$$A_{n+1}^{[0]} = I - \triangle t J_{n+1}^{[0]}. \tag{5.2}$$

Matrix $A_{n+1}^{[0]}$ can be represented as

$$A_{n+1}^{[0]} = S_{n+1}^{[0]} + W_{n+1}^{[0]}, \quad where \quad S_{n+1}^{[0]} = I - \triangle t \tilde{S}_{n+1}^{[0]}. \tag{5.3}$$

Matrix $S_{n+1}^{[0]}$ can, as in the previous chapter, be reordered to a block-diagonal matrix by using appropriate permutations to a block-diagonal matrix. It will always be assumed in this chapter that the permutations, by which matrix $S_{n+1}^{[0]}$ is transformed into a block-diagonal matrix, are performed. Thus, matrix $S_{n+1}^{[0]}$ has non-zero elements only in its diagonal blocks; these blocks were called strong blocks in Chapter 4.

Assume that the modified version (5.1) of the Newton method produces, in μ iterations, some acceptable approximation $y_{n+1}^{[\mu]}$ when the classical Backward Euler Method (i.e. when the Backward Euler Method is applied without partitioning) is used. By this assumption and by applying (5.3), vector y_{n+1}^{μ} is obtained by solving the following system of linear algebraic equations in the last iteration in the modified Newton process (5.1):

$$A_{n+1}^{[0]} \triangle y_{n+1}^{[\mu]} = -y_{n+1}^{[\mu-1]} + y_n + \triangle t f(t_{n+1}, y_{n+1}^{[\mu-1]}). \tag{5.4}$$

We should like to use matrix $S_{n+1}^{[0]}$ instead of $A_{n+1}^{[0]}$ in an attempt to save some computations. When this replacement is done, we have to apply

$$S_{n+1}^{[0]} \triangle z_{n+1}^{[i]} = -z_{n+1}^{[i-1]} + z_n + \triangle t f(t_{n+1}, z_{n+1}^{[i-1]}) \tag{5.5}$$

instead of (5.1) or (5.4). Note that (5.5) consists of several independent ODE systems. The number of these systems is equal to the number of blocks into which vector y, in (4.1), is partitioned. Note too that the weak blocks, which are contained in matrix $W^{[0]}_{n+1}$, are not used when (5.5) is used instead of (5.1); the weak blocks were also discussed in the previous chapter. In other words, the matrix $W^{[0]}_{n+1}$, which contains the weak blocks, is neglected when (5.5) is used instead of (5.4). The conditions under which (5.5) can successfully be used in the treatment of systems of ODEs (not necessarily systems of ODEs arising from the chemical sub-model of a large-scale air pollution model) will be studied in this chapter.

The use of a partitioned version (5.5) of the Backward Euler Method has several advantages:

- it is not necessary to calculate the non-zero elements of the weak blocks,
- several small matrices are to be factorized when (5.5) is used instead of the factorization of only one big matrix in (5.1), because matrix $S^{[0]}_{n+1}$ is block-diagonal, and
- several small systems of linear algebraic equations are to be solved at each Newton iteration (instead of the solution of only one large system of linear algebraic equations).

Assume that (5.5) produces in ν iterations some approximation $z^{[\nu]}_{n+1}$ and that $\|\triangle z^{[\nu]}_{n+1}\|$ is sufficiently small. It is clear that vector $z^{[\nu]}_{n+1}$ may differ very much from $y^{[\mu]}_{n+1}$ even when $\|\triangle z^{[\nu]}_{n+1}\|$ is very small. Therefore, not only should $\|\triangle z^{[\nu]}_{n+1}\|$ be sufficiently small, but one must also require that some norm of the difference $y^{[\mu]}_{n+1} - z^{[\nu]}_{n+1}$ is sufficiently small. The conditions under which the latter requirement is satisfied will be studied in the next section.

5.2 When is $\|y^{[\mu]}_{n+1} - z^{[\nu]}_{n+1}\|$ small?

Assume again that the modified version (5.1) of the Newton method produces in μ iterations some acceptable approximation $y^{[\mu]}_{n+1}$ when the classical Backward Euler Method is used. In this section we shall first derive two expressions for the difference $y^{[\mu]}_{n+1} - z^{[i]}_{n+1}$, where $z^{[i]}_{n+1}$ is obtained from (5.5) after performing i iterations, $i \in \{1, 2, \ldots, \nu\}$, with the modified Newton iterative method. After that we shall study $\|y^{[\mu]}_{n+1} - z^{[i]}_{n+1}\|$ under different assumptions.

5.2.1 Derivation of two expressions for $y^{[\mu]}_{n+1} - z^{[i]}_{n+1}$

Theorem 5.1. *Assume that $(S^{[0]}_{n+1})^{-1}$ exists. Then the difference $y^{[\mu]}_{n+1} - z^{[i]}_{n+1}$ can be expressed in the following form:*

$$y^{[\mu]}_{n+1} - z^{[i]}_{n+1} = B^{[i-1]}_{n+1}(y^{[\mu]}_{n+1} - z^{[i-1]}_{n+1}) + C_{n+1}(y_n - z_n) - D_{n+1}\triangle y^{[\mu]}_{n+1}, \quad (5.6)$$

where

$$C_{n+1} = (S^{[0]}_{n+1})^{-1}, \tag{5.7}$$

$$B^{[i-1]}_{n+1} = \triangle t (S^{[0]}_{n+1})^{-1} \, (T^{[i-1]}_{n+1} - \tilde{S}^{[0]}_{n+1}), \tag{5.8}$$

$$D^{[i-1]}_{n+1} = (S^{[0]}_{n+1})^{-1} \, W^{[0]}_{n+1} + B^{[i-1]}_{n+1}, \tag{5.9}$$

with

$$T^{[i-1]}_{n+1} = \int_0^1 \frac{\partial f(t_{n+1}, \lambda y^{[\mu-1]}_{n+1} + (1-\lambda) z^{[i-1]}_{n+1})}{\partial y} \, d\lambda. \tag{5.10}$$

Proof. Subtract (5.5) from (5.4). Apply (5.3). Move the term containing $W^{[0]}_{n+1}$ to the right-hand-side. The result is:

$$\begin{aligned} S^{[0]}_{n+1}(\triangle y^{[\mu]}_{n+1} - \triangle z^{[i]}_{n+1}) \;&=\; -\,(y^{[\mu-1]}_{n+1} - z^{[i-1]}_{n+1}) + (y_n - z_n) \\ &+\; \triangle t \left[f(t_{n+1}, y^{[\mu-1]}_{n+1}) - f(t_{n+1}, z^{[i-1]}_{n+1}) \right] \\ &-\; W^{[0]}_{n+1} \, \triangle y^{[\mu]}_{n+1}. \end{aligned} \tag{5.11}$$

Apply (4.19) to eliminate $\triangle y^{[\mu]}_{n+1}$ and $\triangle z^{[i]}_{n+1}$ from the left-hand-side of (5.11). Move the terms calculated at iterations $\mu - 1$ and $i - 1$ to the right-hand-side. Use the integral mean theorem (this theorem has been used in connection with error estimations of the numerical solutions of ODE systems in Losinskii [168], see Lemma 3 on pp. 65-66 in [168]; it should also be noted here that this theorem is called *"Lagrange theorem for vector functions"* in [168]) to eliminate the difference $f(t_{n+1}, y^{[\mu-1]}_{n+1}) - f(t_{n+1}, z^{[i-1]}_{n+1})$ from the right-hand-side of (5.11). The following formula will be obtained when all these modifications are done:

$$\begin{aligned} S^{[0]}_{n+1}(y^{[\mu]}_{n+1} - z^{[i]}_{n+1}) \;&=\; (S^{[0]}_{n+1} - I)\,(y^{[\mu-1]}_{n+1} - z^{[i-1]}_{n+1}) + (y_n - z_n) \\ &+\; \triangle t \left[\int_0^1 g(\lambda) d\lambda \right] (y^{[\mu-1]}_{n+1} - z^{[i-1]}_{n+1}) \\ &-\; W^{[0]}_{n+1} \, \triangle y^{[\mu]}_{n+1} \end{aligned} \tag{5.12}$$

with

$$g(\lambda) = \frac{\partial f(t_{n+1}, \lambda y^{[\mu-1]}_{n+1} + (1-\lambda) z^{[i-1]}_{n+1})}{\partial y}. \tag{5.13}$$

Multiply from the left with $(S_{n+1}^{[0]})^{-1}$. Group the first and the third terms in the right hand side. Use, in the term so obtained, (5.3) and (5.10). These transformations lead to

$$\begin{aligned} y_{n+1}^{[\mu]} - z_{n+1}^{[i]} &= \Delta t (S_{n+1}^{[0]})^{-1} (T_{n+1}^{[i-1]} - \tilde{S}_{n+1}^{[0]}) (y_{n+1}^{[\mu-1]} - z_{n+1}^{[i-1]}) \\ &+ (S_{n+1}^{[0]})^{-1} (y_n - z_n) \\ &- (S_{n+1}^{[0]})^{-1} W_{n+1}^{[0]} \, \triangle y_{n+1}^{[\mu]}. \end{aligned} \tag{5.14}$$

It is clear now that (5.6) can be obtained by substituting (5.7)-(5.9) in (5.14) and by adding and subtracting $B_{n+1}^{[i-1]} y_{n+1}^{[\mu]}$. This proves the theorem. □

Remark 5.1. The equality (5.6) tells us that the "local" error due to the application of $S_{n+1}^{[0]}$ instead of $A_{n+1}^{[0]}$ contains three terms. The first of them depends on the error in the previous iteration. The second term depends on the error in the previous time-step. The third term depends on difference of the last two iterates, $\triangle y_{n+1}^{[\mu]}$, in the Newton iterative method carried out with the classical Backward Euler Method. Assume that

- there is no error from the previous time-step (i.e. $y_n - z_n = 0$), and
- there is no error made in the solution of (4.13) when the classical Backward Euler Method is used (i.e. $\triangle y_{n+1}^{[\mu]} = 0$).

If these two requirements are satisfied, then the iterative process (5.5) will converge and, moreover, $z_{n+1}^{[i]} \rightarrow y_{n+1}^{[\mu]}$ as $i \rightarrow \infty$ when $\|B_{n+1}^{[i]}\| < 1$ for all values of i. Of course, these two requirements are not realistic assumptions. Nevertheless, we shall show that the condition imposed on matrix $B_{n+1}^{[i]}$ ensures convergence also under some much weaker assumptions. Further transformations are needed to confirm this statement.

Theorem 5.2. *Use the notation introduced in Theorem 5.1 and assume that $(S_{n+1}^{[0]})^{-1}$ exists. Then (5.6) can be rewritten as follows:*

$$\begin{aligned} y_{n+1}^{[\mu]} - z_{n+1}^{[i]} &= \left(\prod_{j=1}^{i} B_{n+1}^{[i-j]} \right) (y_{n+1}^{[\mu]} - z_{n+1}^{[0]}) \\ &+ \left[\sum_{k=0}^{i-1} \left(\prod_{j=1}^{k} B_{n+1}^{[i-j]} \right) \right] C_{n+1}(y_n - z_n) \end{aligned} \tag{5.15}$$

$$- \left[\sum_{k=0}^{i-1} \left(\prod_{j=1}^{k} B_{n+1}^{[i-j]} \right) D_{n+1}^{[i-1-k]} \right] \triangle y_{n+1}^{[\mu]},$$

where it is assumed that a product is equal to the identity matrix if the lower index is greater than the upper one.

Proof. Write (5.6) for $j = i, i-1, i-2, \ldots, 1$. Multiply the first of these equalities with I, the second with $B_{n+1}^{[i-1]}$, the third with $B_{n+1}^{[i-1]} B_{n+1}^{[i-2]}$, ... , the last one with $\prod_{j=1}^{i-1} B_{n+1}^{[i-j]}$. Sum all the equalities so found. The result is (5.15), which proves the theorem. □

Remark 5.2. The second and the third terms in the right-hand-side of (5.15) depend on the same quantities as the second and the third terms in right-hand-side of (5.6): the error at the previous time-step and the difference of the last two iterates in the Newton procedure (5.1) carried out with the classical Backward Euler Method. The first term in the right-hand-side of (5.15) depends on the starting approximation used at time-step $n+1$ in the partitioned method. If $\|B_{n+1}^{[i]}\| < 1$ for all indices i, then this term can be made arbitrarily small if sufficiently many iterations have been carried out. This fact will be exploited in some of the next theorems.

5.2.2 Evaluation of the "local" error due to the use of $S_{n+1}^{[0]}$

Assume that the errors made in the previous time-steps ($k = 1, 2, \ldots, n$) are sufficiently small. Assume also that the Newton iterative method converges for the classical (non-partitioned) Backward Euler Method. Then it could be shown that the error made at time-step $n+1$ is also small if the number of iterations i carried out by (5.5) is sufficiently large and if

$$B_{n+1} < 1, \tag{5.16}$$

where

$$B_{n+1} \geq \max_{0<j<i-1} \left(\|B_{n+1}^{[j]}\| \right) \quad \text{for all values of} \quad i. \tag{5.17}$$

Some bounds of the errors caused by the use of the iterative process defined by (5.5), instead of that defined by (5.1), will be derived in this and in the following subsections. Two cases will be considered:

- **Idealized case.** The exact solution y_{n+1} of (4.13) is used with the classical (non-partitioned) Backward Euler Method.

- **More realistic situation.** The Newton iterative method (5.1), which is used in the treatment of the classical (non-partitioned) Backward Euler Method, converges. Moreover, the calculations with (5.1) are carried out until certain accuracy stopping criteria are satisfied.

Both cases are important. Theorems based on the assumption made in the idealized case are telling us that under certain conditions the Newton iterative method used in the partitioned algorithm converges to the exact solution of (4.13) (although we are normally not able to find this solution in practice). Theorems based on the second case are telling us that under certain conditions the Newton iterative method used in the partitioned algorithm will produce as accurate results as those obtained when the calculations in the Newton iterative method (5.1) applied in the classical (non-partitioned) Backward Euler Method are carried out until certain stopping criteria are satisfied.

Theorem 5.3. *Assume that the exact solution y_{n+1} of (4.13) is used in the computations with the classical (non-partitioned) Backward Euler Method. Assume also that B_{n+1} from (5.17) satisfies (5.16). Then for any choice of a real number $\varepsilon > 0$ the inequality*

$$\|y_{n+1} - z_{n+1}^{[i]}\| < \varepsilon \tag{5.18}$$

will be satisfied if

- *the number of iterations that are carried out in (5.5) is sufficiently large, and*
- *the error from the previous time-step, $\|y_n - z_n\|$, is sufficiently small.*

Proof. It is clear that if y_{n+1} is the exact solution of (4.13), then (5.4) is satisfied with $y_{n+1}^{[\mu]} = y_{n+1}^{[\mu-1]} = y_{n+1}$ and $\triangle y_{n+1}^{[\mu]} = 0$. This means that all transformations carried out to obtain (5.15) can also be carried out when y_{n+1} is used instead of $y_{n+1}^{[\mu]}$. Moreover, the last term of (5.15) vanishes when y_{n+1} is used (because $\triangle y_{n+1}^{[\mu]} = 0$). Thus, (5.15) can be replaced by

$$\begin{aligned} y_{n+1} - z_{n+1}^{[i]} &= \left(\prod_{j=1}^{i} B_{n+1}^{[i-j]}\right)(y_{n+1} - z_{n+1}^{[0]}) \\ &+ \left[\sum_{k=0}^{i-1}\left(\prod_{j=1}^{k} B_{n+1}^{[i-j]}\right)\right] C_{n+1}(y_n - z_n). \end{aligned} \tag{5.19}$$

The following bound can be obtained from the last equality by using (a) inequality (5.16) and (b) some basic properties of the norms:

$$\|y_{n+1} - z_{n+1}^{[i]}\| < (B_{n+1})^i \, \|y_{n+1} - z_{n+1}^{[0]}\| + \frac{\|C_{n+1}\|}{1 - B_{n+1}} \|y_n - z_n\|. \tag{5.20}$$

Consider the first term in the right-hand-side of (5.20). Since $\|y_{n+1} - z_{n+1}^{[0]}\|$ is a constant and $B_{n+1} < 1$, it is clear that there exists an integer i_0 such that for all $i \geq i_0$ we have:

$$(B_{n+1})^i \, \|y_{n+1} - z_{n+1}^{[0]}\| < \frac{\varepsilon}{2}. \tag{5.21}$$

Consider now the second term in the right-hand-side of (5.20). Assume that the error from the previous time-step satisfies the following inequality:

$$\|y_n - z_n\| < \frac{1 - B_{n+1}}{\|C_{n+1}\|} \frac{\varepsilon}{2}. \tag{5.22}$$

Use the last two inequalities, (5.21) and (5.22), in (5.20). The result is inequality (5.18), which proves Theorem 5.3 □

Remark 5.3. The requirement for finding the exact solution of (4.13) can be replaced with a requirement that (4.13) is solved in a sufficiently accurate way by using the modified Newton iterative method (5.1). More precisely, the following theorem can be proved.

Theorem 5.4. *Assume that the Newton iterative method (5.1) converges and let i be the iteration number. Assume also that B_{n+1} from (5.17) satisfies (5.16). Finally, assume that*

$$D_{n+1} > \max_{0<j<i-1} \left(\|D_{n+1}^{[j]}\| \right) \quad \textit{for all values of} \quad i. \tag{5.23}$$

Then for any choice of $\varepsilon > 0$ the inequality (5.18) with y_{n+1} replaced with $y_{n+1}^{[\mu]}$ will be satisfied if

- *the number of iterations carried out in (5.5) is sufficiently large,*
- *the error from the previous time-step, $\|y_n - z_n\|$, is sufficiently small, and*
- *the modified Newton iterative method (5.1) is stopped when $\|\triangle y_{n+1}^{[\mu]}\|$ becomes sufficiently small.*

Proof. Consider (5.15). Apply

- inequality (5.16),
- some basic properties of the norms, and
- ones again inequality (5.16)

to obtain the following bound:

$$
\begin{aligned}
\|y_{n+1}^{[\mu]} - z_{n+1}^{[i]}\| \quad &< \quad (B_{n+1})^i \, \|y_{n+1}^{[\mu]} - z_{n+1}^{[0]}\| \\
&+ \quad \frac{\|C_{n+1}\|}{1 - B_{n+1}} (\|y_n - z_n\|) \\
&+ \quad \frac{D_{n+1}}{1 - B_{n+1}} (\|\triangle y_{n+1}^{[\mu]}\|).
\end{aligned}
\tag{5.24}
$$

The first and the second term in the right-hand-side of (5.24) can be bounded in a similar way as in the proof of Theorem 5.3. More precisely, the following two bounds can be written:

$$
(B_{n+1})^i \, \|y_{n+1}^{[\mu]} - z_{n+1}^{[0]}\| < \frac{\varepsilon}{3} \tag{5.25}
$$

and

$$
\|y_n - z_n\| < \frac{1 - B_{n+1}}{\|C_{n+1}\|} \, \frac{\varepsilon}{3}. \tag{5.26}
$$

Consider now the third term in (5.24). Since the Newton iterative method (5.1) is assumed to be convergent, this term can be bounded (when sufficiently many iterations have been performed) as follows:

$$
\|\triangle y_{n+1}^{[\mu]}\| < \frac{1 - B_{n+1}}{D_{n+1}} \, \frac{\varepsilon}{3}. \tag{5.27}
$$

If the last three inequalities, (5.25), (5.26) and (5.27), are applied in (5.24), then (5.18) can be obtained. This proves Theorem 5.4. □

Remark 5.4. The assumption that $\|\triangle y_{n+1}^{[\mu]}\|$ can be made sufficiently small, see (5.27), is realistic. If the Newton process (5.1) is convergent, then this requirement can be satisfied when the number of iterations, μ, is sufficiently large. If (5.16) is satisfied and if the number of iterations i, carried out in (5.5), is sufficiently large, then (5.25) will also hold. On the other hand, the assumption, imposed in (5.26), that the norm of the error from the previous time-step can be made arbitrarily small is not realistic. Therefore, it is worthwhile to relax this requirement.

5.2.3 Evaluation of the "global" error due to the use of $S_{n+1}^{[0]}$

The bounds found in the previous subsection are based on an assumption that the errors made at the previous time-step can be made arbitrarily small; see (5.22) and (5.26). It is desirable to remove this requirement. This can be done in the following way.

It is useful to introduce, for $k = 1, 2, \ldots, n+1$, the following notation in this subsection:

$$B_k^{[\nu_k]} = \max_{0 \le i \le \nu_k} (\|B_k^{[i]}\|), \tag{5.28}$$

$$E_k = \frac{\|C_k\|}{1 - B_k^{[\nu_k]}}, \tag{5.29}$$

$$G_k = \frac{D_k}{1 - B_k^{[\nu_k]}}, \tag{5.30}$$

and

$$E = \max_{1 \le k \le n+1} (E_k), \tag{5.31}$$

where ν_k $(k = 1, 2, \ldots, n+1)$ is the number of iterations needed to achieve the required accuracy when the iterative process defined by (5.5) is used.

It is again convenient to consider the two cases, the idealized case and the more realistic case, from Subsection 5.2.2. The following theorem holds for the idealized case.

Theorem 5.5. *Assume that the exact solutions* $y_1, y_2, \ldots, y_{n+1}$ *of (4.13) for* $k = 1, 2, \ldots, n+1$ *have been used in the computations. Assume also that the following conditions are satisfied (for* $k = 1, 2, \ldots, n+1$ *):*

$$\left(S_k^{[0]}\right)^{-1} \quad \textit{exists}, \tag{5.32}$$

$$B_k^{[\nu_k]} < 1, \tag{5.33}$$

$$E < 1. \tag{5.34}$$

Then for any choice of an $\varepsilon > 0$ *the inequality*

$$\|y_{n+1} - z_{n+1}^{[\nu_{n+1}]}\| < \varepsilon \tag{5.35}$$

will be satisfied if the numbers of iterations ν_k *with* $k = 1, 2, \ldots, n+1$ *are sufficiently large.*

Proof. Consider (5.24). Replace $y_{n+1}^{[\mu]}$ and $\Delta y_{n+1}^{[\mu]}$ with y_{n+1} and 0 respectively. Replace $z_{n+1}^{[i]}$ and z_n with $z_{n+1}^{[\nu_{n+1}]}$ and $z_n^{[\nu_n]}$. Use (5.28) and (5.29). The result is

$$\begin{aligned}\|y_{n+1} - z_{n+1}^{[\nu_{n+1}]}\| &< \left(B_{n+1}^{[\nu_{n+1}]}\right)^{\nu_{n+1}} \|y_{n+1} - z_{n+1}^{[0]}\| \\ &+ E_{n+1}(\|y_n - z_n^{[\nu_n]}\|).\end{aligned} \tag{5.36}$$

Consider the first term in the right-hand-side in (5.36). Since (5.33) holds, it is clear that this term can be made smaller than any $\delta > 0$ if ν_{n+1} is sufficiently large. This means that the following inequality can be obtained from (5.36).

$$\|y_{n+1} - z_{n+1}^{[\nu_{n+1}]}\| < E_{n+1}(\|y_n - z_n^{[\nu_n]}\|) + \delta. \tag{5.37}$$

Write (5.37) for $k = n+1, n, n-1, \ldots, 2, 1$. Multiply the first of these inequalities with I, the second with E_{n+1}, the third with $E_{n+1}E_n$, $\ldots$ and the last one with $\prod_{k=0}^{n} E_{n+1-k}$. Sum all these inequalities. These transformations lead to

$$\begin{aligned}\|y_{n+1} - z_{n+1}^{[\nu_{n+1}]}\| &< \left(\prod_{k=1}^{n+1} E_k\right) (\|y_0 - z_0\|) \\ &+ \left[\sum_{j=0}^{n+1} \left(\prod_{k=n+1-j}^{n+1} E_k\right)\right] \delta.\end{aligned} \tag{5.38}$$

The last inequality, inequality (5.38), can be rewritten in a simpler form by using (5.31) and ((5.34):

$$\|y_{n+1} - z_{n+1}^{[\nu_{n+1}]}\| < E^{n+1} (\|y_0 - z_0\|) + \frac{1}{1-E}\,\delta. \tag{5.39}$$

Assume now that

- the difference between the two starting approximations is sufficiently small, and
- δ is sufficiently small.

The first of these two requirements is natural; one often starts with $y_0 = z_0$. The second requirement will be satisfied when the number of iterations at each time-step is sufficiently large (which is assumed in the theorem). If both requirements are satisfied, then the following two relationships can be obtained:

$$\|y_0 - z_0\| < \frac{\varepsilon}{2}, \tag{5.40}$$

$$\delta < (1 - E)\, \frac{\varepsilon}{2}, \tag{5.41}$$

The assertion of the theorem, the inequality (5.35), can be obtained from (5.39) by using (5.40) and (5.41). □

Remark 5.5. The requirement for using the exact solutions $y_1, y_2, \ldots, y_{n+1}$ of (4.13), which is imposed in Theorem 5.5, can be replaced by a requirement for convergence of the Newton iteration process defined by (5.1) at every time-step.

Theorem 5.6. *Assume that the Newton iteration process defined by (5.1) is convergent at every iteration step* k *with* $k = 1, 2, \ldots, n+1$. *Assume that inequalities (5.32) - (5.34) are satisfied for all appropriate values of* k*; i.e. for* $k = 1, 2, \ldots, n+1$. *Then for any choice of an* $\varepsilon > 0$ *the inequality*

$$\|y_{n+1}^{[\mu_{n+1}]} - z_{n+1}^{[\nu_{n+1}]}\| < \varepsilon \tag{5.42}$$

will be satisfied if the numbers of iterations μ_k *and* ν_k *with* $k = 1, 2, \ldots, n+1$ *are sufficiently large.*

Proof. Consider (5.36). The exact solutions y_k must now be replaced by approximations $y_k^{[\mu_k]}$, for $k = n, n+1$. An extra term must be added now to the right-hand-side of (5.36); to take into account that (4.13) is not solved exactly (as assumed in Theorem 5.5) and some errors are present when the iterative process (5.1) is used. These changes lead to the following relationship:

$$\begin{aligned} \|y_{n+1}^{[\mu_{n+1}]} - z_{n+1}^{[\nu_{n+1}]}\| &< \left(B_{n+1}^{[\nu_{n+1}]}\right)^{\nu_{n+1}} \|y_{n+1}^{[\mu_{n+1}]} - z_{n+1}^{[0]}\| \\ &+ E_{n+1}(\|y_n^{[\mu_n]} - z_n^{[\nu_n]}\|) \\ &+ G_{n+1}(\|\triangle y_{n+1}^{[\mu_{n+1}]}\|). \end{aligned} \tag{5.43}$$

The first term in the right-hand-side in (5.43) can be made smaller than any $\delta/2 > 0$ if ν_{n+1} is sufficiently large, because (5.33) holds. The third term in the right-hand-side in (5.43) can be made smaller than any $\delta/2 > 0$ if μ_{n+1} is sufficiently large, because $\|\triangle y_{n+1}^{[\mu_{n+1}]}\|$ can be made arbitrarily small when the iterative process (5.1) is convergent. This means that (5.43) can be simplified as follows:

$$\|y_{n+1}^{[\mu_{n+1}]} - z_{n+1}^{[\nu_{n+1}]}\| < E_{n+1}(\|y_n^{[\mu_n]} - z_n^{[\nu_n]}\|) + \delta. \tag{5.44}$$

The inequality (5.44) can be treated in the same way as the inequality (5.37), because these two inequalities are of the same form. This means that the following two inequalities can be obtained in precisely the same way as the corresponding inequalities (5.38) and (5.39) were obtained in the proof of Theorem 5.5:

$$\|y_{n+1}^{[\mu_{n+1}]} - z_{n+1}^{[\nu_{n+1}]}\| < \left(\prod_{k=1}^{n+1} E_k\right) (\|y_0 - z_0\|) + \left[\sum_{j=0}^{n+1}\left(\prod_{k=n+1-j}^{n+1} E_k\right)\right] \delta \tag{5.45}$$

and

$$\|y_{n+1}^{[\mu_{n+1}]} - z_{n+1}^{[\nu_{n+1}]}\| < E^{n+1} (\|y_0 - z_0\|) + \frac{1}{1-E}\, \delta. \tag{5.46}$$

It is clear that one can assume that (5.40) and (5.41) are satisfied (using the same arguments as in the proof of Theorem 5.5). The assertion of Theorem 5.6, the inequality (5.42), can then be obtained from (5.46) by using (5.40) and (5.41). □

The requirement $E < 1$ is essential in the proofs of both Theorem 5.5 and Theorem 5.6. This requirement can be slightly relaxed (by allowing E to become equal to one or even greater than one (but if the latter case takes place, then E must stay very close to one; more precisely it is required that E is of the form $1+\varepsilon$ with $\varepsilon << 1$). If this condition is satisfied, then the following two theorems hold.

Theorem 5.7. *Let all assumptions made in Theorem 5.5 and Theorem 5.6, excepting (5.34), be satisfied. Then the following two inequalities hold for* $E \neq 1$ *and for any* $\delta > 0$*:*

$$\|y_{n+1} - z_{n+1}^{[\nu_{n+1}]}\| < E^{n+1} (\|y_0 - z_0\|) + \frac{1-E^{n+1}}{1-E}\, \delta \tag{5.47}$$

and

$$\|y_{n+1}^{[\mu_{n+1}]} - z_{n+1}^{[\nu_{n+1}]}\| < E^{n+1} (\|y_0 - z_0\|) + \frac{1-E^{n+1}}{1-E}\, \delta. \tag{5.48}$$

Proof. The first assertion of Theorem 5.7 can be obtained from (5.38), while (5.45) must be used to obtain the second assertion. □

Theorem 5.8. *Let all assumptions made in Theorem 5.5 and Theorem 5.6, excepting (5.34), be satisfied. Then the following two inequalities hold for* $E = 1$ *and for any* $\delta > 0$*:*

$$\|y_{n+1} - z_{n+1}^{[\nu_{n+1}]}\| < \|y_0 - z_0\| + n\, \delta \tag{5.49}$$

and

$$\|y_{n+1}^{[\mu_{n+1}]} - z_{n+1}^{[\nu_{n+1}]}\| < \|y_0 - z_0\| + n\, \delta. \tag{5.50}$$

Proof. As in the proof of Theorem 5.7, the first assertion of Theorem 5.8 can be obtained from (5.38), while (5.45) must be used to obtain the second assertion. □

Remark 5.6. The inequalities from the assertions of Theorem 5.7 and Theorem 5.8 are telling us that if

- E is equal to one or greater than one but very close to one,
- the quantities $\|y_0 - z_0\|$ and δ are sufficiently small, and
- n is not very large,

then one should expect to get good results by using the partitioning algorithm.

An example (an ODE system arising in the chemical part of a large-scale air pollution model), where E is greater than one but very close to one, will be given in the next section.

5.3 An application to air pollution problems

In this section it will be shown that the conditions, under which the theorems in the previous sections hold, are satisfied when ODE systems arising in the chemical parts of some large air pollution models are to be treated numerically; as, for example, the large-scale model discussed in Zlatev [282] and used in Ambelas Skjøth et al. [9], Bastrup-Birk et al. [18], Harrison et al. [128], Zlatev et al. [292], [293], [294], [295], [297]. It was mentioned in the previous chapters that one of the chemical schemes used in this model is the CBM IV scheme (described in Gery et al. [118]). There exist several versions of this chemical scheme. The CBM IV chemical scheme implemented in the model from Zlatev [282] contains 35 chemical species. The partitioning algorithm used has been obtained by applying the ideas discussed in Hertel et al. [134], the same partitioning has also been used in Alexandrov et al. [6], Skelboe and Zlatev [227] and Zlatev [283] (see also the previous chapter). It contains one large (13×13) block. The following species are gathered in this block: NO, NO_2, O_3, $O^{(1)}D$, OH, HO_2, NO_3, N_2O_5, HNO_2, PAN, C_2O_3, $PHEN$ and PHO. All other species are forming small (1×1) blocks.

It is clear that the theorems in Section 5.2 will be satisfied for any partitioning when the time-step size chosen, Δt, is sufficiently small. However, it is important, from a practical point of view, to be able to find a partitioning that gives good results also when large time-step sizes are used. It has experimentally been shown in Alexandrov et al. [6] and Skelboe and Zlatev [227] (some results are also given in the previous chepter) that the partitioning algorithm described above can successfully be used in the numerical treatment of ODE systems arising in the chemical parts of some large air pollution models even when the time-step size is rather large. Now, having the results from this chapter, we are able to explain **why** this is the case.

The same six scenarios as those used in the previous chapter will also be considered here. A qualitative characterization of these scenarios is given in Table 4.2.

It should be emphasized again that the six scenarios cover some typical situations that occur often in the numerical treatment of many of the situations studied by air pollution models.

System (4.1) has again, as in the previous chapter, been solved, for any of the six scenarios, over a time interval of 42 hours (starting at six o'clock in the morning and continuing the time-integration until 24 o'clock in the next day). The computations have been carried out with a time-step size $\triangle t = 30$ seconds in Table 5.1 - Table 5.3 and with different values of the time-step size $\triangle t$ in Table 5.4. In the beginning of the time-integration process and at the end of every period of 900 seconds, the matrices A_N , S_N , W_N and $B_N^{[0]}$ have been calculated for $N = 0, 1, \ldots, 168$. While the calculation of the first three matrices is straight-forward, a numerical time-integration subroutine has been used to compute the matrix $T_N^{[0]}$ which is needed in the calculation of $B_N^{[0]}$. After that the eigenvalues of these four matrices have been calculated for each N , $N = 0, 1, \ldots, 168$ and the spectral radii of all matrices have been found. The LAPACK subroutine DGEEV, see Anderson et al. [11], has been used to calculate the eigenvalues of the four matrices.

The intervals, in which the spectral radii of the relevant matrices vary (for $N = 1, 2, \ldots, 168$) are given

- in Table 5.1 (for the matrices A_N and S_N), and
- in Table 5.2 (for the matrices W_N and B_N).

The values of E for the different scenarios that were used in the previous chapter (see Table 4.2) are shown in Table 5.3.

Finally the values of the quantities B , $\rho(C_N)$, where $N = 0, 1, \ldots, 168$, and E , obtained when different time-step sizes are used, can be seen in Table 5.4.

The results shown in Table 5.1 to Table 5.4 indicate that the following conclusions can be drawn for the systems of ODEs arising in the chemical parts of some air pollution models (as, for example, in the models studied in Zlatev [282]):

1. The spectral radii of the first two matrices, A_N and S_N, are practically the same; see Table 5.1. In fact, there is no difference when the results are rounded to the second significant digit; see Table 5.1.

2. The spectral radii of $W_N = A_N - S_N$ are much smaller than those of A_N and S_N; compare the results shown in Table 5.2 with those in Table 5.1.

3. The spectral radii of $B_N^{[0]}$, which are given in the last column in Table 5.2, are considerably less than one (which indicates that convergence will also take place if the time-step size is larger than the time-step size used in this chapter (i.e. if $\triangle t > 30$ seconds). Strictly speaking, it is necessary to calculate $B_N^{[i]}$ and $T_N^{[i]}$ at every iteration i, but it has been verified experimentally that these quantities do not vary too much from one iteration to another.

Table 5.1: The intervals in which the spectral radii vary for the matrices A_N and S_N, $N = 1, 2, \ldots, 168$ (when $\triangle t = 30$ is used).

Scenario	$\rho(A_N)$	$\rho(S_N)$
1	$[2.2 * 10^6, 2.6 * 10^6]$	$[2.2 * 10^6, 2.6 * 10^6]$
2	$[2.2 * 10^6, 2.6 * 10^6]$	$[2.2 * 10^6, 2.6 * 10^6]$
3	$[2.2 * 10^6, 2.6 * 10^6]$	$[2.2 * 10^6, 2.6 * 10^6]$
4	$[2.2 * 10^6, 2.6 * 10^6]$	$[2.2 * 10^6, 2.6 * 10^6]$
5	$[2.2 * 10^6, 2.6 * 10^6]$	$[2.2 * 10^6, 2.6 * 10^6]$
6	$[2.2 * 10^6, 2.6 * 10^6]$	$[2.2 * 10^6, 2.6 * 10^6]$

4. The values of the quantity E are greater than one, but very close to one; see Table 5.3.

5. The values of quantity E becomes closer to one when the time-step size is decreased. From Table 5.4 it is clearly seen why this is so. The values of $\rho(C_k)$ remain nearly the same (equal to 1.0, for all time-step sizes, when rounded to the second significant digit), while the values of B are decreasing because the values of B depend explicitly on the time-step size; see also (5.8).

Table 5.2: The intervals in which the spectral radii vary for the matrices W_N and B_N, $N = 1, 2, \ldots, 168$ (when $\triangle t = 30$ is used).

Scenario	$\rho(W_N)$	$\rho(B_N)$
1	$[3.4 * 10^{-4}, 4.6 * 10^{-2}]$	$[2.4 * 10^{-4}, 3.3 * 10^{-2}]$
2	$[7.7 * 10^{-3}, 2.1 * 10^{-1}]$	$[3.7 * 10^{-4}, 2.1 * 10^{-2}]$
3	$[6.7 * 10^{-2}, 8.2 * 10^{-1}]$	$[5.8 * 10^{-3}, 5.7 * 10^{-2}]$
4	$[1.1 * 10^{-3}, 3.2 * 10^{-1}]$	$[1.3 * 10^{-3}, 4.3 * 10^{-2}]$
5	$[2.1 * 10^{-2}, 2.0 * 10^{-1}]$	$[4.3 * 10^{-3}, 2.5 * 10^{-2}]$
6	$[6.7 * 10^{-2}, 8.7 * 10^{-1}]$	$[5.6 * 10^{-3}, 6.4 * 10^{-2}]$

The theorems in Section 5.2 have been proved for **any** matrix norm. The spectral radii of the matrices have consistently been used in this section. The main relationships between these two concepts are described by the following two statements, which are valid for any square real matrix Q (where $Q \in R^{N_s \times N_s}$ and N_s is an arbitrary positive integer):

- The spectral radius of any matrix $Q \in R^{N_s \times N_s}$ is less than or equal to any norm of Q ($\rho(Q) \leq \|Q\|$); see, for example, Wilkinson [267].

Table 5.3: The values of the quantity E for the different scenarios (when $\triangle t = 30$ is used).

Scenario	E
1	1.034249651
2	1.021791917
3	1.060402003
4	1.046055431
5	1.025220454
6	1.068816212

Table 5.4: Values of the quantities B, $\max(\rho(C_N))$ (where $N = 0, 1, \ldots, 168$) and E for different time-step sizes and for Scenario 5.

Stepsize	B	$\max(\rho(C_N))$	E
30.00	$2.5 * 10^{-2}$	1.0	1.025
10.00	$1.4 * 10^{-2}$	1.0	1.014
1.00	$2.6 * 10^{-3}$	1.0	1.0026
0.10	$3.6 * 10^{-4}$	1.0	1.00036
0.01	$5.3 * 10^{-5}$	1.0	1.000053

- The spectral radius of any matrix $Q \in R^{N_s \times N_s}$ is equal to the infinum taken over all norms of Q in $R^{N_s \times N_s}$ ($\rho(Q) = inf(\|Q\|)$); see, for example, Belitzkii and Ljubich [20]. In other words, for any $\varepsilon > 0$ there exists a norm of matrix $Q \in R^{N_s \times N_s}$ such that $\|Q\| = \rho(Q) + \varepsilon$.

5.4 Applicability to other models

Partitioning is a rather general procedure, which is very successful when certain conditions (discussed in the previous sections of this chapter) are satisfied. Therefore, it is worthwhile to check whether these conditions are satisfied or not when the chemical part of the model under consideration is to be treated numerically. If the conditions are satisfied, then it may be worthwhile to apply the method of partitioning, which has been discussed in this chapter and in Chapter 4. One should expect the conditions studied in this chapter to be satisfied for many of the models discussed in Section 7 of Chapter 1.

5.5 Concluding remarks and plans for future work

Several concluding remarks concerning the results obtained in this chapter and some plans for future research in this direction are needed.

Remark 5.7. It has been assumed in the end of Section 5.1 (and this assumption has been used in Section 5.2) that a sufficiently accurate approximation $z_{n+1}^{[\nu]}$ can be obtained in ν iterations from (5.5). In fact, (5.5) consists of several independent sub-systems. The iterative process in each of these sub-systems will in practice be carried out until the norm of the error of the current iterate becomes less than some prescribed tolerance. This shows clearly that the numbers of iterations needed to achieve the accuracy required will in general be different for the different sub-systems. Therefore, in the practical computations, ν should be viewed as the largest number of iterations needed in the iterative solution of the different sub-systems in (5.5).

Remark 5.8. All results have been proved under the assumption that the modified version of the Newton iterative method defined by (5.1) is used. It can, however, be easily verified that all results hold also in the case where the classical Newton iterative method (4.18) is used and in the case where the other modified version of the Newton iterative method is chosen (i.e. the modified version described in Remark 4.2).

Remark 5.9. Theorem 5.1 to Theorem 5.8 have been proved under the usual assumption, not explicitly stated, that all computations can be carried out in exact arithmetic. If a computer is used, then rounding errors (due to operations with numbers with a finite number of digits) are practically unavoidable. The results will hold if the influence of the rounding errors is sufficiently small. In many practical problems the accuracy requirements are low and the above requirement will be satisfied. Nevertheless, an extension of the results presented in this chapter, in which some rounding error analysis is included, is desirable.

Remark 5.10. It is worthwhile to extend the results, which were proved in Section 5.2 for the Backward Euler Method to other time-integration methods (as, for example, for some Backward Differentiation Formulae of order higher than one and/or for some methods of Runge-Kutta type).

Remark 5.11. The results obtained in Section 5.4 have been used in Section 5.5 to explain why the partitioning used in Alexandrov et al. [6] and Skelboe and Zlatev [227] gives good results. One can use the results from Section 5.4 in an attempt to find automatically a good partitioning. This will be a topic for future research.

Remark 5.12. It is also necessary to compare the partitioning algorithms with other algorithms that are popular and efficient when ODE systems arising in air pollution models are to be treated numerically; see, for example, Chock et al. [46], Jay et al. [145], Odman et al. [187], Sandu et al. [211], [213], [212], Shieh et al. [221], Verwer and van Loon [259], Verwer and Simpson [260], Verwer et al. [258]. The extrapolation time-integration methods used by Deuflhard and his co-workers ([63], [64], [65], [66], [67]) or methods of Runge-Kutta type as those studied in Butcher [37] are promising candidates when large air pollution models are to be used and, thus, comparisons with these methods are also needed.

Remark 5.13. It may be worthwhile to apply some sparse matrix technique in the treatment of (4.1). This topic was discussed in Subsection 4.3.6 for the partitioning procedure applied to the system of ODEs arising from the chemical part of DEM. The same technique is also applicable in the case where partitioning procedures are applied to other scientific and engineering problems.

Remark 5.14. In the application of the partitioning procedure to the air pollution model, which was made in this chapter (and also in Chapter 4), it has been assumed that some splitting procedure has been applied and, after that, the partitioning procedure has been applied only to the chemical sub-model. It is interesting to investigate the possibility of using some partitioning procedures also in the case where no splitting procedure is applied in the treatment of large-scale models on computers. The application of partitioning procedures in the case where no splitting is used might lead to more efficient computational process, which is highly desirable.

Chapter 6

Efficient organization of the matrix computations

After the discretization of the spatial and/or the temporal derivatives one must normally perform some matrix computations during the treatment of the selected numerical algorithms. Typical examples for matrix computations, which often have to be carried out when large-scale models are treated on computers, are:

- matrix-vector multiplications,
- inner products, and
- solving systems of linear algebraic equations (including here the important case where the coefficient matrix of the system is large and sparse).

The treatment of these matrix computations is often very time consuming. This is especially true when systems of linear algebraic equations are to be solved. This is why it is important to select fast methods for performing matrix computations. Several such methods will be discussed in this chapter.

In the beginning of this chapter (the first three sections) it is assumed that some of the splitting procedures that were discussed in Chapter 2 have been applied. The application of an appropriate splitting procedure leads to computational tasks involving structured matrices (banded matrices and in some cases tri-diagonal matrices). Efficient numerical methods for the treatment of such matrices are well described in the literature. Moreover, very efficient standard packages, which can easily be applied when large scale models are handled, exist and can in many cases be downloaded from the Internet. Two well-known examples are

- the package LAPACK (see Anderson et al. [11]), and
- the package ITPACK (see Kincaid et al. [150], [151]).

In four sections of this chapter (from Section 4 to Section 7), we discuss some problems which arise when no splitting technique is used. The direct discretization of the system of PDEs (1.1), i.e. the discretization of the system of PDEs without using splitting procedures, results in large non-linear systems of algebraic equations, which have to be treated at every time-step. The application of the Newton iterative procedure leads to the solution of many systems of linear algebraic equations. The coefficient matrices of these systems are general sparse matrices. It is important to find fast methods for the solution of the systems of linear algebraic equations, i.e. methods that exploit efficiently the sparsity of the coefficient matrices. The main principles on which such methods should be based are described in Section 5 to Section 7. Numerical examples are given there in order to illustrate the efficiency of the proposed algorithms.

The fact that the algorithms proposed and discussed in the first seven sections of this chapter are applicable not only when large-scale air pollution models are treated, but also when many other large-scale mathematical problems are to be handled is discussed in Section 8.

Some general conclusions about the algorithms described in this chapter are given in the last section, Section 9.

6.1 The horizontal advection-diffusion sub-model

If finite differences are chosen in the discretization of the spatial derivatives, then the matrices arising when the horizontal advection-diffusion phenomena are handled are banded matrices (see also Chapter 4). Banded matrices also arise when finite elements are selected instead of finite differences. Matrices that can efficiently be treated by applying Fast Fourier Transforms (FFTs) arise in the case where pseudo-spectral methods are used (see Zlatev [282] or Zlatev et al. [289]). The matrix computations arising in these three cases will be discussed in this section.

6.1.1 Matrix computations when finite differences are used

As stated above, the matrices arising when finite differences are used to discretize the spatial derivatives are banded matrices. It is necessary to distinguish between the case where the time-integration methods used is explicit and the case where the time-integration method is implicit.

- **Application of explicit time-integration methods.** If the finite difference discretization of the spatial derivatives in the horizontal advection-diffusion sub-model is followed by some explicit time-integration method, then the system of PDEs is transformed to a system of ODEs $dg/dt = Hg$, where as stated above H is a banded matrix. Only matrix-vector multiplications involving banded matrices are to be carried out in this case. These operations can be performed in a very straight-forward way. One can also apply standard subroutines from BLAS (Basic Linear Algebra Subprograms); see, for example, Dongarra et al. [78], [79].

- **Application of implicit time-integration methods with a direct solver for the resulting systems of linear algebraic equations.** If the finite difference discretization of the spatial derivatives is followed by some implicit time integration method, then systems of linear algebraic equations, whose coefficient matrices $I - \gamma \triangle t H$ are banded, are to be handled at every time-step. The coefficient matrices depend on values of the wind velocities at the grid-points and, thus, these vary both in space and time. One can apply direct methods for the solution of linear systems of algebraic equations with banded matrices. Subroutines from LAPACK (see Anderson et al. [11]) or LINPACK (see Dongarra et al. [77]) can efficiently be used when this choice is made.

- **Application of implicit time-integration methods with an iterative solver for the resulting systems of linear algebraic equations.** The alternative choice is the use of iterative methods. Subroutines from package ITPACK might be used when the second choice is made (see Kincaid et al. [150], [151]). It should be noted here that some kind of preconditioning might be necessary when iterative methods are selected. It is not always very easy to find a good preconditioner (thus, several trials with different preconditioners might be needed). Many of the iterative methods used in ITPACK can be used together with different preconditioners. Efficient methods for large linear systems are discussed in details in the recently published book of H. A. van der Vorst (see [256]). The advantage of the use of iterative methods is that matrices with wide bandwidths may be efficiently treated when the rate of convergence is fast. However, the rate of convergence for some matrices might also be very slow when good preconditioners are used. This may happen, because the matrices are in general non-symmetric. If the latter case occurs, then it might be more profitable to go back to the direct methods (i.e., if the convergence rate of the iterative process is very slow, then it might be more profitable to switch to direct methods).

6.1.2 Matrix computations when finite elements are used

In this case it is also necessary to distinguish between the two major groups of numerical methods (explicit methods and implicit methods) which might be applied in the time-integration part.

- **Application of explicit time-integration methods.** If the finite element discretization of the spatial derivatives in the horizontal advection-diffusion sub-model is followed by some explicit time-integration method, then one has in general to treat problems, see (3.2), of the type: $P dg/dt = Hg$, where P is a constant matrix depending of the particular finite element discretization used, while H is depending of the wind velocities and, therefore is varying in both time and space. Both matrices, P and H, are banded matrices. Both direct methods and iterative methods can be applied in the solution of $Pdg/dt = Hg$.

(a) **Direct methods.** Let us start with the direct methods. The fact that P is normally constant can be exploited (matrix P can be factorized in the very beginning of the computational process and, after that, the factors can be used during the whole computational process).

(b) **Iterative methods.** It should be noted that P is a symmetric and positive definite matrix. Therefore, many iterative methods will be convergent. Moreover, experimental results show that the convergence rate is fast when P arises from air pollution models. This means that iterative methods can successfully be used in many cases.

The choice of the methods depends on the splitting procedure used and on the grid. Refined grids will often lead to matrices with a large bandwidth. Therefore, the iterative methods might be more profitable than the direct methods in this case. Iterative methods might also be more profitable when some vector computers are used (because these vectorize very well, while the vectorization of the direct methods for banded matrices is normally rather poor).

- **Application of implicit time-integration methods.** If the finite element discretization of the spatial derivatives in the horizontal advection-diffusion sub-model is followed by some implicit time-integration method, then again, as in the previous case, one has in general to solve systems of linear algebraic equations arising after the discretiziion of systems of ODEs of type (3.2), i.e. systems of the type: $Pdg/dt = Hg$, which are discretized this time by applying some implicit method for solving systems of ODEs. The application of an implicit time-integration algorithm leads to the solution, at every time-step, of systems of linear algebraic equations whose coefficient matrices are of the type $P - \gamma \triangle t H$, where γ depends on the particular time-integration algorithm chosen while $\triangle t$ is the time-step size. This case is very similar to the case where the finite differences of the spatial derivatives is followed by implicit time-integration methods. Indeed, matrices $I - \gamma \triangle t H$ are to be used (instead of $P - \gamma \triangle t H$) when finite differences are applied instead of finite elements. This means that in principle the same methods for solving systems of linear algebraic equations as those discussed in the previous subsection can be applied when implicit time-integration algorithms are selected after the spatial discretization carried out by either finite differences or finite elements.

6.1.3 Matrix computations for the pseudo-spectral method

As mentioned in Chapter 3, one can also use a pseudo-spectral discretization during the treatment of the advection-diffusion sub-model. The application of such methods is discussed in detail in Zlatev [277], [278], [282] as well as in Zlatev et al. [289]. The pseudo-spectral method is based on the use of truncated Fourier series. Therefore, it is important in this case to apply some efficient fast Fourier transforms (FFTs). FFTs proposed by Swarztrauber [241] and Temperton [246] (a very good implementation of the Temperton's FFTs can be found in the NAG Library [185]) have been used in a previous version of DEM (see again Zlatev [277], [278],[282] and Zlatev et al. [289]).

The pseudo-spectral discretization was implemented in DEM (by applying explicit time-integration algorithms when the semi-discretized system is treated). It is not very clear whether an efficient implementation can be achieved when implicit time-integration methods are used. The implementation of the pseudo-spectral discretization in DEM (where, let us reiterate, it was followed by applying explicit time-integration algorithms in the discretization of the advection-diffusion sub-model) is very efficient on vector processors like CRAY. It is also very accurate.

Periodic boundary conditions are required (see also Chapter 3) when this algorithm is used, which is a serious drawback, because it is not very easy to satisfy this requirement (however, it should be mentioned here that there exist algorithms, which can be used to handle the problems arising by the requirement for periodicity on the boundaries; see, for example, Lyness [170] and Roache [202], [203]).

The pseudo-spectral method is not used in the newest versions of DEM (these versions are united in a common model, UNI-DEM, which will be discussed in the next chapter).

6.2 Matrices in the vertical exchange sub-model

The numerical treatment of this sub-model was discussed in Chapter 3. Finite elements are used in the discretization of the spatial derivatives. The θ-method is used after that in the time-integration part. This means that we have finite elements followed by an implicit time-integration algorithm, which leads to the solution of systems $Pdg/dt = Hg$. The solution of systems of ODEs of this type was discussed in the previous section.

However, the situation that occurs in the vertical exchange sub-model is much simpler, from computational point of view, than the situation discussed in the previous section when the complexity of the computational process is taken into account, because the application of some splitting procedure leads to many independent problems (one for each vertical grid-line; i.e. 8 064 000 problems are to be treated at every time-step when the model with 35 chemical species is discretized on $480 \times 480 \times 10$ grid). This is why the matrices which arise and have to be treated in this part are tri-diagonal. The treatment of systems with tri-diagonal matrices is very cheap; the computational cost being $O(3N_z)$. This simplifies the implementation of the algorithms. Therefore, the vertical exchange sub-model does not cause great computational difficulties when the vertical exchange is separated from the other physical and chemical processes by using some splitting procedure.

Standard subroutines for tri-diagonal matrices are easily available. Such subroutines can, for example, be found in package LAPACK (see Anderson et al. [11]) or in the older package LINPACK (see Dongarra et al. [77].

6.3 Matrices arising in the chemical sub-model

It has been mentioned several times in the previous chapters that the chemical part of the model is normally the most time consuming module. The computational work in this part of the model consists of the solution of a very large number of **stiff** ODE systems (one such system per grid-point), which can be treated as parallel tasks. One can use the following three methods (see also Chapter 4) in the solution of each of the small ODE systems.

- QSSA algorithms (either the original QSSA algorithm or the improved QSSA algorithm).
- Classical integration methods for solving stiff systems of ODEs (such methods are normally implicit and, therefore, lead to the solution of systems of non-linear algebraic systems). The use of two simple methods (the Backward Euler Method and the Trapezoidal Rule) was discussed in Chapter 4. Four other methods for solving systems of ODEs (the Implicit Mid-point Rule, a Diagonally Implicit Runge-Kutta method of order 2, a Fully Implicit Three-stage Runge-Kutta Method of order six and a Rosenbrock method of order 2) were also used; see Chapter 2.
- Methods based on applying partitioning (the fact that some of the chemical reactions are fast while others are slow is exploited in the construction of partitioned methods). Partitioning procedure as well as some properties of these procedures were discussed both in Chapter 4 and in Chapter 5.

The numerical treatment of the chemical part of a large-scale air pollution models by using the three types of methods listed above was discussed in Chapter 4. Before starting the discussion in this section it is important to emphasize the fact that

- the QSSA algorithms can normally be implemented by using only simple matrix-vector operations, while
- systems of linear algebraic equations are to be solved both when the classical integration methods for solving stiff systems of ODEs are applied and when methods based on applying partitioning are developed and used.

6.3.1 Implementation of the QSSA algorithms

As mentioned above, the QSSA algorithms, both the original QSSA algorithm and the improved QSSA algorithm, can normally be implemented by using matrix-vector products. The matrix is a general sparse matrix and it might be worthwhile to exploit the sparsity during the implementation. It is clear that the number of simple arithmetic operations needed in the treatment of the chemical reactions at each grid point is constant when the QSSA algorithm discussed in Chapter 4 is used. This means that the parallel tasks are perfectly balanced.

The main conclusion is that if only the computational issues are taken into account, the QSSA algorithms are very efficient. Unfortunately, there may arise problems with the accuracy when some concentrations vary very quickly (this has been demonstrated by using some simple examples in Chapter 4).

6.3.2 Implementation of classical time-integration algorithms

If classical time-integration algorithms as well as algorithms based on partitioning are applied, then non-linear systems are to be treated. This means that iterative methods are to be used and, therefore, systems of linear algebraic equations are to be solved at each iteration. There are many systems of linear algebraic equations, but each of these systems is small. The number of equations in each of these systems is equal to the number of chemical species involved in the model. If the model with only 35 chemical species is discretized on a $480 \times 480 \times 10$ grid, then 2 304 000 systems, each of them containing 35 equations, have to be treated at every time-step. This is why it is possible to neglect the fact that the Jacobian is sparse and to use dense matrix technique in the treatment of the matrix operations (for the methods based on partitioning this was illustrated in Subsection 4.3.5, but precisely the same ideas can be used also when any classical time-integration method is applied without partitioning).

The use of sparse matrix technique in this particular case (where the matrices treated are small) can be efficient only if the special rules discussed in Subsection 4.3.6 are used (in fact these rules are used only for the algorithm based on partitioning in Subsection 4.3.6, but they can also be applied in connection with any classical time-integration method).

The use of the classical time-integration algorithms and/or algorithms based on partitioning leads to more accurate computations (see again Chapter 4). However, these methods are normally more expensive (especially when the classical time-integration methods are directly, i.e. without using some partitioning procedure, implemented in the subroutines for the solution of the chemical ODE systems) and, therefore, the efficiency depends strongly on the ability of the method to produce sufficiently accurate results when a **large** time-step size is used.

The use of the classical time-integration algorithms and/or algorithms based on partitioning may give some problems with load balancing of the parallel tasks. The reason for this is the fact that some iterative method has to be applied in the solution of the non-linear systems arising after the space and time discretization. While the number of simple arithmetic operations per a system of linear algebraic equations, i.e. per one iteration in the solution of (4.18), is remaining the same during the whole computational process, the number of iterations per a non-linear system of type (4.17) is varying from one grid-point to another. This problem will further be discussed in the next chapter.

6.3.3 The QSSA algorithms versus the other algorithms

The discussion in the previous subsection indicates clearly that it is difficult to answer the question:

> *Is it better to use the QSSA algorithms or should one prefer the other algorithms (the classical time-integration algorithms and/or algorithms based on partitioning)?*

The essential issue is to find out whether the QSSA algorithms applied with a small time-step size are still better than the other algorithms in spite of the fact that the other algorithms are used with a larger time-step size. At present we have no definite answer to the above question. Much more experiments with different numerical algorithms in different situations are needed in order to obtain a reliable answer. We are continuing the experiments with different numerical methods.

6.4 Treatment of the model without splitting

In the previous sections of this chapter it was assumed that some splitting procedure is used. This simplifies, very considerably, the computations. This is why splitting procedures are, as mentioned in the previous chapters, used in all large-scale air pollution models that are operationally run in different comprehensive studies as those, which will be discussed in Chapter 8 and Chapter 9.

The price which has to be paid is the fact that splitting will normally cause errors and, moreover, it is difficult to control these errors (see also Chapter 2). Therefore, as mentioned several times in the previous chapters, it is important to avoid splitting when this is possible. The size of the computational tasks when no splitting is used will be discussed in this section.

If no splitting is used, then the discretization of the spatial derivatives in (1.1) with either finite differences or finite elements leads to a large stiff system of ODEs (the stiffness being introduced mainly by the presence of the chemical terms, but the vertical exchange also introduces stiffness). The number of equations in the resulting system, obtained when splitting is not used, is equal to the product of grid-points and chemical species. The time-integration must be carried out by implicit time-integration algorithms for stiff systems of ODEs, because stiffness is introduced by some of the physical and chemical processes (as mentioned above the terms containing chemical reactions are the major reason for the stiffness of the resulting system of ODEs, but the terms describing the vertical transport are also introducing stiffness in the semi-discretized model when no splitting procedure is used). This means that non-linear systems of algebraic equations are to be handled at every time-step when no splitting procedure is applied in the discretization of (1.1). The application of the Newton iterative method (or some of its modifications, see Remark 4.1 and Remark 4.2 in Chapter 4) leads to the solution of several large and sparse systems of linear algebraic equations at every time-step. Some attempts to handle these systems with general methods for sparse matrices from Zlatev [281] were not very successful (however, we are still working on the improvement of the implementation of sparse matrix codes: some results that are related to the performance of sparse matrix codes will be given in the following sections).

The results in this direction, which have been obtained by us until now, can be summarized as follows. On the available computers (some SUN computers at the Danish Centre for Scientific Computing), we were able to run problems of the same size as the model with 35 chemical species discretized on $(96 \times 96 \times 10)$ grid. However, the computational times were very large. We were not able to run the model for period longer than a couple of months. The version, in which no splitting procedure is applied, may be (and is) applicable in some studies related to short periods with high pollution levels. However, this version of the model cannot yet be used in comprehensive studies over long time-periods and/or with many different scenarios.

It is not clear, at present, whether it is possible to design some special sparse matrix technique which will reduce substantially the computational time. On the other side, it is worthwhile to continue the efforts to avoid the use of splitting procedures. The computers are becoming faster and faster, which means that one should expect that the treatment of large-scale air pollution models without using splitting will become possible in the near future and the modellers have to be well prepared for this, which means that some efficient codes for handling general sparse matrices have to be developed and attached as soon as possible to the existing large-scale air pollution models. This is also valid for other large-scale mathematical models used in other fields of science and engineering.

As mentioned above, some current results, obtained with codes for general sparse matrices, in which an attempt to exploit the properties of the modern computer architectures is made, will be presented in the next three sections.

6.5 Use of sparse matrix techniques

The matrices, which appear after the discretization of large-scale mathematical models (not necessarily air pollution models) by using either finite differences or finite elements are banded matrices. There are well-described and easily available tools for treating such matrices; such as BLAS (Basic Linear Algebra Subprograms) described in Dongarra et al. [78], [79], LAPACK documented in Anderson et al.

static storage scheme cannot be used when the structure of the non-zero elements in matrix A is dynamically changing during the computational process. As a typical example consider the major formula

$$a_{i,j}^{(s+1)} = a_{i,j}^{(s)} - \frac{a_{i,s}^{(s)} a_{s,j}^{(s)}}{a_{s,s}^{(s)}}, \quad s = 1, 2, \ldots, n-1, \quad i, j = s+1, s+2, \ldots, n, \tag{6.2}$$

with

$$a_{s,s}^{(s)} \neq 0, \quad a_{i,j}^{(1)} = a_{i,j} \in A, \tag{6.3}$$

which is used when a system $Ax + b$ of linear algebraic equations is solved by the well-known Gaussian elimination (the application of the Gaussian elimination leads to the calculation of two triangular matrices, a lower triangular matrix L and an upper triangular matrix U, such that $LU = A + E$, where E is a result of the rounding errors which are practically unavoidable in calculations on computers; see Zlatev [281]). It is clear that if $a_{i,j}^{(s)}$ is equal to zero but both $a_{i,s}^{(s)}$ and $a_{s,j}^{(s)}$ are non-zeros, then a new non-zero element, $a_{i,j}^{(s+1)}$, will be produced in a position, (i, j), where a zero element was located before the performance of the arithmetic operations in (6.2). Such a new non-zero element is called fill-in. The appearance of fill-ins and the need to store both the fill-ins and some information about the fill-ins (row numbers, column numbers, etc.) dynamically in the appropriate arrays show why the static storage scheme for sparse matrices cannot be used in the solution of linear algebraic equations by Gaussian elimination. Much more complicated storage schemes, called dynamic storage schemes for general sparse matrices, have to be used in this case.

Different dynamic storage schemes for sparse matrices can be designed (see Chapter 2 in Zlatev [281]). Schemes based on row ordered lists are popular. The non-zero elements are ordered by rows in a one-dimensional real array, say array ALU. If a non-zero element is located in position K of array ALU, then its column number must be stored in position K of a one-dimensional integer array $CNLU$. The length of these two arrays is $NN \geq 2 * NZ$. The integer NN must be large enough in order to ensure place, "elbow room", for fill-ins, which in general will be produced during the computations and which have dynamically to be inserted in the storage scheme. If row I $(1 \leq I \leq N)$ of matrix A starts in position $K1$ $(1 \leq K1 \leq NN)$ and ends in position $K2$ $(1 \leq K1 < K2 \leq NN)$, then we must have $ROWSTART(I) = K1$ and $ROWEND(I) = K2$, where $ROWSTART$ and $ROWEND$ are two integer arrays of length N.

The row ordered list provides full information about the locations of the non-zero elements of matrix A. It is efficient to use this list when the elements of the matrix are scanned by rows. It is desirable, however, to be able to scan the non-zero elements also by columns in order to make the computational process more efficient. This is done by using a similar structure, a column ordered list. This structure contains three integer arrays, $RNLU$ of length $NN1$ with $NN1 \geq NZ$

as well as $COLSTART$ and $COLEND$ (both of length N). The row numbers of the ordered by columns non-zero elements are stored in array $RNLU$. The row numbers of column J $(1 \leq J \leq N)$ are stored from position $COLSTART(J)$ to position $COLEND(J)$ in array $RNLU$. Note that the non-zero elements of matrix A are not needed in the column ordered list.

Assume that a new non-zero element, fill-in, is created in position (i, j) of matrix A. Then the following actions are carried out in the row ordered list:

- the non-zero elements of row i and their column numbers are copied to the end of arrays ALU and $CNLU$,
- the positions where the non-zero elements of row i were located before making the copy to the end of the row ordered list are declared as free positions (by setting zeros in the all locations between $ROWSTART(i)$ and $ROWEND(i)$ in array $CNLU$),
- the new non-zero element and its column number are stored after the last occupied position in ALU and $CNLU$ (this position is saved because it will be needed when the next fill-in is to be inserted in the row ordered list), and
- the contents of $ROWSTART(i)$ and $ROWEND(i)$ are updated.

Similar transformations are to be carried out in the column ordered list in order to insert the row number i of the fill-in created in position (i, j):

- the row numbers of the non-zero elements of column j copied to the end of arrays $RNLU$,
- the positions where row numbers of the non-zero elements of column j were located before coping them to the end are declared as free positions (by setting zeros in all locations between $COLSTART(j)$ and $COLEND(j)$ in array $RNLU$),
- the row number of the new non-zero element is stored after the last occupied position in $RNLU$ (this position is saved because it will be needed when the row number of the next fill-in is to be inserted in the column ordered list), and
- the contents of $COLSTART(j)$ and $COLEND(j)$ are updated.

It is clear that in general the length of the arrays ALU, $RNLU$ and $CNLU$ will not allow us to make too many copies. Therefore, the structure of the lists (the row ordered list and the column ordered list) should from time to time be compressed (utilizing here the fact that, when rows and columns are moved to the end of the appropriate arrays, some free locations are, as explained above, created in the lists). This action is called "garbage collection".

Garbage collections are expensive. The number of garbage collections can be reduced in the following way (exploiting here the fact that some free locations are created between two rows and/or two columns every time when a copy of a row or a column is made in the end of the appropriate ordered list). If a fill-in is produced, then a copy of its row(column) should be made only if there are no free positions at the beginning or at the end of the row(column). If there are free positions then the fill-in and its column number (the row number of the fill-in) should be put there, updating appropriately the information about row start or row end (column start or column end).

Another way to reduce the number of garbage collections is by using pivoting for preservation of the sparsity of the original matrix. Pivoting for general sparse matrices will be discussed in the next subsection.

Finally, the number of the garbage collections can be reduced if some small elements are dropped (in fact, replaced by zeros). Assume that the updated by (6.2) element $a_{i,j}^{(s+1)}$ satisfies the following inequality:

$$| a_{i,j}^{s+1} | \leq RELTOL \max_{s \leq j \leq N} | a_{i,j}^{(s)} | . \tag{6.4}$$

where $0 \leq RELTOL < 1$ is a user-selected parameter, called the relative drop-tolerance. Then $a_{i,j}^{(s+1)}$ is dropped. In fact, it is removed from the lists by making some simple modifications. Assume that $a_{i,j}^{(s+1)}$ is stored in position K of array ALU. Then one must set

- $ALU(K) = ALU(ROWEND(i))$,
- $CNLU(K) = CNLU(ROWEND(i))$,
- $CNLU(ROWEND(i)) = 0$, and
- $ROWEND(i) = ROWEND(i) - 1$.

Assume that the row number of $a_{i,j}^{(s+1)}$ is stored in position $K1$ of array $RNLU$. Then one must set

- $RNLU(K1) = RNLU(COLEND(j))$,
- $RNLU(COLEND(j) = 0$, and
- $COLEND(j) = COLEND(j) - 1$.

It is seen that by dropping the element $a_{i,j}^{(s+1)}$ not only might a new copy of a row (column) be avoided, but also a free location where fill-ins can be stored might be created. Thus, dropping small elements makes the computational process more efficient, but this process also introduces errors in the triangular factors L and U. Therefore, these factors cannot be used directly to calculate the solution of $Ax = b$ when positive values of the relative drop-tolerance are used. However, they could be used as a preconditioner in an iterative process for solving $Ax = b$. Preconditioning will be discussed in one of the following subsections.

6.5.2 Pivoting for sparse matrices

If dense matrices are to be handled in connection with solving systems of linear algebraic equations by Gaussian elimination, then pivoting (either partial pivoting or complete pivoting) is used in an attempt to preserve the numerical stability of the computational process (see Golub and Van Loan, [122], Stewart [232], Wilkinson [266], [267]). Pivoting, for the preservation of the numerical stability, has to also be used when general sparse matrices are treated. However, the situation here is more complicated because pivoting has also to be used in an attempt to preserve (as much as possible) the sparsity of the original matrix. These two requirements (the preservation of the numerical stability and the preservation of the sparsity) work in opposite directions. Therefore a compromise is needed. Both the experimental results and some theoretical results indicate that the pivotal strategy proposed in Zlatev [272] is a good compromise. This pivoting strategy will be outlined in this subsection (more results about different pivotal strategies for sparse matrices can be found in Duff et al. [81], and Zlatev [281]).

Denote by A_s the sub-matrix that has to be updated by using (6.2). Let $a_{i,j}^{(s)} \neq 0$ be an element in A_s. Assume that $r(i,s)$ is the number of non-zero elements of row i which are in A_s, and that $c(j,s)$ is the number of non-zero elements of column j which are in A_s. The Markowitz cost of element $a_{i,j}^{(s)}$ is defined by

$$M_{ijs} = [r(i,s) - 1]\, [c(j,s) - 1]. \tag{6.5}$$

It is worthwhile to select as a pivotal element the element $a_{i,j}^{(s)}$ with a minimal Markowitz cost when only the preservation of the sparsity is taken into account. If the element with a minimal Markowitz cost is found, then it must be moved (by row and column interchanges) to position (s,s). When this is done, the calculations with formula (6.2) are to be performed.

There are two major difficulties related to the strategy which has been sketched above:

- the search for the non-zero element with a minimal Markowitz cost can be very expensive when the matrix is large, and
- no attempt to preserve the numerical stability of the computations is carried out when pivots are chosen by using only the minimal Markowitz cost.

It was observed in Zlatev [272] that the search for the element with the minimal Markowitz cost could successfully be replaced by a search for an element with the best Markowitz cost in a few "best" rows only (rows with minimal numbers of non-zero elements are called "best" rows here). Experiments in Zlatev [272] indicated that the search in three "best" rows only is normally giving very good results. In an attempt to preserve also the numerical stability an additional constraint is to be imposed: the pivotal element must satisfy the following inequality:

$$\mid a_{i,j}^{(s)} \mid \geq u \max_{s \leq k \leq N} \mid a_{i,j}^{(s)} \mid, \tag{6.6}$$

with some parameter u which satisfies $0 < u \leq 1$. Parameter u is called the stability factor in Zlatev [272] and [281].

It should be mentioned here that the rows must be ordered so that if $i < k$ the number of non-zero elements in row i is less than or equal to the number of non-zero elements in row k and, moreover this order must be kept during the whole factorization process (this ordering is fully explained in Zlatev [281]).

The idea of using (6.5) in the choice of a good pivot was first suggested in Markowitz [180]. Other implementations of this idea and/or other pivoting strategies for general sparse matrices can be found in Davis and Davidson [58], Davis and Yew [59], Duff et al. [81], George and Ng [115], Gilbert and Peierls [119], Pissanetzki [195], Sherman [219], Tewarson [247], Zlatev [281].

6.5.3 Using preconditioning

As mentioned in the previous subsection, dropping small non-zero elements can be useful in the attempts to preserve better the sparsity of the original matrix and, thus, to improve the performance of the selected sparse algorithm. While the performance is often improved when dropping of small non-zero elements is applied during the Gaussian elimination, the accuracy of the calculated triangular matrices L and U will often become very poor. This is why a preconditioning is necessary when this approach is used.

Ten preconditioned methods

Ten well-known numerical methods for solving systems of linear algebraic equations by using a direct method based on the Gaussian elimination and preconditioned methods for sparse matrices are discussed below. The methods are listed in Table 6.1. Some information about the abbreviations, as well as references, where more details about the ten methods can be found, are given in Table 6.1.

The method for calculating directly the solution (DIR) is based on the use of a sparse matrix technique for calculation of an LU factorization of the coefficient matrix of the system of linear algebraic equations without dropping small non-zero elements, as in (Zlatev [281]).

The same method is also used to calculate approximate LU factorizations, which are used as preconditioners in the iterative methods. Let us reiterate here that a special parameter $RELTOL$ (relative drop-tolerance), $0 \leq RELTOL < 1$, is used to calculate an approximate LU factorization. The use of this parameter to drop small non-zero elements during the Gaussian elimination was described in one of the previous subsections.

It should be added here that the relative drop-tolerance can also be used to drop all small non-zero elements in the original matrix before the beginning of the Gaussian elimination. The situation is very similar to that described in the previous subsection. If the absolute value of an element a_{ij} of the original coefficient matrix is smaller than the product of $RELTOL$ and the largest in absolute value in row

i, then a_{ij} is dropped (replaced by zero). This means that a_{ij} is dropped when the following inequality

$$| a_{i,j} | \leq RELTOL \max_{1 \leq j \leq N} | a_{i,j} | \tag{6.7}$$

is satisfied. Note that (6.7) is very similar to (6.4), but while (6.7) is applied to the elements of the original matrix, (6.4) is applied to the elements calculated at some stage $s+1$ of the Gaussian elimination.

If $RELTOL$ is set equal to zero, then the direct solution can be calculated (see also the previous paragraph). However, if $0 < RELTOL < 1$ then the LU factorization obtained by using the selected positive value of the relative drop-tolerance $RELTOL$ must be used as a preconditioner in one of the iterative methods.

If $RELTOL$ becomes larger, then in general the obtained LU factorization is becoming less accurate, but more sparse. This means that the number of iterations needed to obtain the required accuracy of the solution of the system of linear algebraic equations will tend to become greater for larger values of $RELTOL$, but the computational work per iteration will, in general, become smaller. If $RELTOL$ is very large, then the rate of convergence can be very slow or the preconditioned iterative method may even not converge (the calculations must be repeated by using a smaller value of the relative drop-tolerance $RELTOL$ when this happens; it might be necessary sometimes to reduce the value of $RELTOL$ several times). This shows that it is very important to select robust and reliable stopping criteria when preconditioned iterative methods are used. The proper implementation of such stopping criteria will allow us to decide

- whether the required accuracy is achieved or not (and to stop the iterative process when the required accuracy is achieved), and
- whether the iterative process is slowly convergent or does not convergent at all and,
  - to reduce the relative drop-tolerance, and
  - to recalculate the *LU* factorization with the reduced drop-tolerance

  when there are problems with the convergence.

It should be emphasized here that robust and reliable stopping criteria are also needed when pure iterative methods are applied.

Preconditioning is always used when the system of linear algebraic equations is solved by the last seven methods in Table 6.1. The Modified Orthomin Method (Zlatev [281]) is used both as a pure iterative method (PURE) and as a preconditioned iterative method (ORTH). If an approximate LU factorization is used in IR, then this method can be considered as a preconditioned iterative method.

Table 6.1: Solvers for systems of linear algebraic equations with general sparse coefficient matrices, which are used in this section.

No.	Method	Abbreviation	Reference
1	Direct solution	DIR	[272], [281]
2	Iterative Refinement	IR	[274], [281]
3	Pure Modified Orthomin	PURE	[262], [281]
4	Preconditioned Modified Orthomin	ORTH	[262], [281]
5	Conjugate Gradients Squared	CGS	[230]
6	Bi-Conjugate Gradients STAB	BiCGSTAB	[255]
7	Transpose-Free Quasi Min. Residual	TFQMR	[107]
8	Generalized Minimum Residual	GMRES	[207]
9	Eirola-Nevanlinna Method	EN	[86], [263]
10	Block Eirola-Nevanlinna Method	BEN	[270]

It must be emphasized here that if the preconditioned iterative method used either is not convergent or the convergence is not sufficiently fast, then the preconditioner can **automatically** be improved, as described above, by reducing the relative drop-tolerance $RELTOL$ (see more details in Zlatev [281]).

6.5.4 Description of the stopping criteria

Let x_n be the approximation of the exact solution x of the system of linear algebraic equations $Ax = b$ obtained after n iterations of the chosen iterative method. Let $r_n = b - Ax_n$ be the corresponding residual vector. We shall assume that all components of x are of the same order of magnitude and use vector norms in the stopping criteria. The desired accuracy of the approximate solution will be defined by a parameter $ACCUR$.

(A) Traditionally used stopping criteria

One of the following stopping criteria is as a rule selected when an iterative method is used (see, for example, Freund [107], Saad and Schultz [207], Vorst [255]): stop the iterative process if

$$\|x_n - x_{n-1}\| \leq ACCUR, \tag{6.8}$$

$$\|r_n\| \leq ACCUR, \tag{6.9}$$

$$\frac{\|x_n - x_{n-1}\|}{\|x_n\|} \leq ACCUR, \tag{6.10}$$

$$\frac{\|r_n\|}{\|r_0\|} \leq ACCUR. \tag{6.11}$$

(B) Difficulties related to the traditional stopping criteria

Difficulties which may arise when (6.8) is used. It is intuitively clear that if the iterative process is slowly convergent, then (6.8) can be satisfied even if the exact error $\|x_n - x\|$ is much larger than $ACCUR$.

Difficulties that may arise when (6.9) is used. Assume that $x_n = x+\epsilon_n$, where ϵ_n is the exact error after n iterations. Then $r_n = b - Ax_n = b - A(x+\epsilon_n) = -A\epsilon_n$. From this relationship we can conclude that the fact that (6.9) is satisfied does not always mean that the required accuracy is achieved and the fact that (6.9) is not satisfied does not necessarily mean that $\|x_n - x\| > ACCUR$. Indeed, consider the scalar case (the number of equations is equal to one) and assume that $ACCUR = 10^{-4}$, $\epsilon_n = 10^{-4}$ and $A = 1$. Then (6.9) is telling us that the required accuracy has been achieved and this is true. Replace $A = 1$ with $A = 100$. Now (6.9) is telling us that the required accuracy has not been achieved, while the opposite is true. Finally, replace $\epsilon_n = 10^{-4}$ with $\epsilon_n = 10^{-2}$ and $A = 1$ with $A = 10^{-2}$. In this case (6.9) is telling us that the required accuracy has been achieved, which is certainly not true. This scalar example shows that check (6.9) will give good results in the scalar case **only** when the problem is scaled so that $A = 1$.

It is clear that similar problems (as the problems discussed above for the scalar case) are also unavoidable in the multi-dimensional case. However, the difficulties are much greater in this case, because it is not immediately clear what kind of scaling should be used in order to get a reliable estimate by using (6.9). It can easily be shown that (6.9) will give good results if a very good preconditioner C has been selected and the preconditioned system $CAx = Cb$ is solved instead of $Ax = b$. To show that this is true, assume that the preconditioner C is equal to A^{-1}. Let $ACCUR = 10^{-4}$ and $\|\epsilon_n\| = 10^{-4}$. Denote by r_n^* the residual vector $Cb - CAx_n$ of the preconditioned system. Then $\|r_n^*\| = \|CA\epsilon_n\| = \|\epsilon_n\| \leq ACCUR$. Thus, very good preconditioning makes (6.9) reliable. It is very often recommended to use in the stopping criterion the residual vector of the original system also when preconditioning is used. The above analysis indicates that if the preconditioner is very good, then the use of the residual vector of the preconditioned system should be preferred. If the preconditioner is not very accurate, then the situation is not clear. The extension of the analysis of the scalar case to the multi-dimensional case indicates that both the residual vector of the original system and the residual vector of the preconditioned system may cause difficulties in the latter case.

Difficulties which may arise when (6.10) is used. It is clear that if the first approximation x_0 is very bad and if the rate of convergence is very slow, then (6.10) will probably give very bad results.

Difficulties which may arise when (6.11) is used. In some evolutionary problems, where many time steps are used, one normally takes as a first approximation x_0 the accepted solution at the previous time-step. If the process approaches the steady state case, then x_0 chosen in this way can be very accurate. If this happens, then $\|r_0\|$ will become very small and, thus, the check (6.11) can cause severe problems.

General conclusion about the use of the traditional stopping criteria. The above analysis shows that the simple stopping criteria (6.8) - (6.11) may cause difficulties in some situations. We shall illustrate by numerical examples that this actually happens when the problem is **difficult** for the iterative methods.

(C) Derivation of more reliable stopping criteria

Checking the behaviour of the iteration parameters. More reliable stopping criteria can be derived as follows. In many iterative methods the key operation is the calculation of a new approximation $x_{n+1} = x_n + \alpha_n p_n$. If the process converges then we have that $x_n \rightarrow x$ as $i \rightarrow \infty$ which implies $x = x_0 + \sum_{i=0}^{\infty} \alpha_n p_n$, where p_n is some vector calculated during iteration n. The convergence of the infinite sum implies that $\alpha_n p_n \rightarrow 0$ as $n \rightarrow \infty$. The reverse relations do not hold, but it is nevertheless reasonable to expect that if the following two conditions are satisfied:

- $\alpha_n p_n \rightarrow 0$ as $i \rightarrow \infty$

and

- α_n does not vary too much and α_n is of order $O(1)$,

then $x_n \rightarrow x$ as $i \rightarrow \infty$.
The requirement related to α_n indicates that it is reasonable to require that the following conditions are satisfied:

(A1) α_n does not decrease too quickly,

(A2) α_n does not increase too quickly, and

(A3) α_n does not oscillate too much.

If any of the requirements (A1) - (A3) is not satisfied, then the stopping criteria should be made more stringent (which essentially leads to reducing the value of $ACCUR$).

Checking the behaviour of convergence rate. Assume again that the iterative process is convergent. Then the following inequality holds:

$$\|x - x_n\| \leq |\alpha_n| \, \|p_n\| \left(1 + \sum_{j=n+1}^{\infty} \frac{|\alpha_j| \, \|p_j\|}{|\alpha_n| \, \|p_n\|} \right) \tag{6.12}$$

Assume that

$$\frac{|\alpha_j|\ \|p_j\|}{|\alpha_n|\ \|p_n\|} = (RATE)^{j-i} < 1 \quad for \quad \forall\, j > i. \tag{6.13}$$

Then (6.12) reduces to

$$\|x - x_n\| \le |\alpha_n|\ \|p_n\| \left(\sum_{j=0}^{\infty} (RATE)^j \right) = \frac{|\alpha_n|\ \|p_n\|}{1 - RATE}. \tag{6.14}$$

In general, $RATE$ varies from one iteration to another. Denote by $RATE_n$ the value of $RATE$ at iteration n. If $RATE_n$ does not vary too much during several iterations and if all three conditions (A1) to (A3) are satisfied, then the last term in (6.14) can be used in the stopping criteria with the following modification: $RATE$ should be replaced by some appropriate parameter as, for example, $0.25(RATE_{n-3} + RATE_{n-2} + RATE_{n-1} + RATE_n)$; see Gallivan et al. [111] and Zlatev [281]. If the requirement that $RATE_n$ does not vary too much is not satisfied, then some extra requirements (leading again to the use of a reduced value of $ACCUR$) are to be imposed.

If we replace x with x_{n-1} in (6.14), then the stopping criterion can be derived by imposing the requirement that the important quantity $|\alpha_n|\ \|p_n\|$ satisfies:

$$|\alpha_n|\ \|p_n\| \le (1 - RATE)ACCUR \tag{6.15}$$

It is quite clear that this criterion can be considered as an improved version of the stopping criterion defined by (6.8). This criterion is only used when the condition $RATE < 1$ is satisfied. If this condition is not satisfied and $n < n_0$ ($n_0 = 7$ is used in our experiments), then the iterative process is continued. This means that if the behaviour of the iterative process in the beginning is not stable, we allow some extra iterations (hopping that the process will gradually be stabilized). If $RATE \ge 1$ and $n \ge n_0$ then the iterative process is declared as divergent and stopped.

The stopping criterion defined by (6.15) is used in most of the experiments that will be discussed in the following part of this chapter.

(D) Discussion of the rules and some extensions

The particular rules (based on the above principles) for the Modified Orthomin Method are described in Gallivan et al. [111] and Zlatev [281]. Similar rules were developed for each of the other iterative methods in Table 6.1. In some of the iterative methods updating is made not by the formula $x_{n+1} = x_n + \alpha_n\, p_n$, but by a formula which contains some linear combination of vectors instead of the term $\alpha_n\, p_n$. However, the rules sketched above can easily be extended to cover these cases too.

It should be mentioned that relative stopping criteria are used in the experiments presented in this chapter. These criteria are obtained by performing in (6.14) the replacements described above and by dividing both sides of of the obtained in this way equality with $\|x_n\|$. We shall use criteria based on norms of the involved vectors.

One of the additional benefits of using stopping criteria based on (6.15) is the fact that these can easily be modified for component-wise checks. A relative component-wise criterion derived from (6.15) for the component k of the error can be written in the following way:

$$\frac{|x_k - x_{kn}|}{|x_{kn}|} \leq \frac{|\alpha_n|\,|p_{kn}|}{(1 - RATE)\,|x_{kn}|} \leq ACCUR. \tag{6.16}$$

Component-wise stopping criteria are useful in the cases where the components of the solution vector are varying in a wide range (this is the case for some problems arising in atmospheric chemistry, see Chapter 4 and Zlatev [282]). These criteria require some more careful work in the cases where (A1) - (A3) are not satisfied and/or when the values of $RATE_n$ vary considerably from one iteration to another. This is why only relative stopping criteria based on (6.15) will be used in the remaining part of this chapter.

6.5.5 Generating test matrices according to different requirements

In order to ensure systematic investigation of the abilities of the discussed in the previous section stopping criteria to produce reliable results, we need test-matrices that have the following properties:

- matrices of different orders can be created,
- the pattern of the non-zero elements can be varied, and
- the condition number of the matrices can be varied.

A matrix generator which creates matrices that satisfy these three conditions has been developed. The matrices that are created by using this matrix generator depend on four parameters:

- the order N of the matrix,
- a parameter C by which the location of some of the non-zero elements can be varied, and
- parameters δ and γ by which the size of certain elements can be varied.

The first two parameters, N and C, are integers, while δ and γ are real numbers. All four parameters can be freely varied.

The non-zero elements of a matrix created by this matrix generator are obtained by using the following rules:

- all elements on the main diagonal are equal to 4,
- the elements on the diagonal with first positions (1,2) and (1,C+1) are equal to $-1+\delta$,
- the elements on the diagonals with first position (2,1) and (C+1,1) are equal to $-1-\delta$, and
- the elements on the diagonal with first position (1,C+2) are equal to γ.

The matrix created with $N = 16$, $C = 4$, $\delta = 3$ and $\gamma = 5$ is given in Fig. 6.2 in order to show how the above rules are used.

4	**2**	0	0	**2**	**5**	0	0	0	0	0	0	0	0	0	0
-4	**4**	**2**	0	0	**2**	**5**	0	0	0	0	0	0	0	0	0
0	**-4**	**4**	**2**	0	0	**2**	**5**	0	0	0	0	0	0	0	0
0	0	**-4**	**4**	**2**	0	0	**2**	**5**	0	0	0	0	0	0	0
-4	0	0	**-4**	**4**	**2**	0	0	**2**	**5**	0	0	0	0	0	0
0	**-4**	0	0	**-4**	**4**	**2**	0	0	**2**	**5**	0	0	0	0	0
0	0	**-4**	0	0	**-4**	**4**	**2**	0	0	**2**	**5**	0	0	0	0
0	0	0	**-4**	0	0	**-4**	**4**	**2**	0	0	**2**	**5**	0	0	0
0	0	0	0	**-4**	0	0	**-4**	**4**	**2**	0	0	**2**	**5**	0	0
0	0	0	0	0	**-4**	0	0	**-4**	**4**	**2**	0	0	**2**	**5**	0
0	0	0	0	0	0	**-4**	0	0	**-4**	**4**	**2**	0	0	**2**	**5**
0	0	0	0	0	0	0	**-4**	0	0	**-4**	**4**	**2**	0	0	**2**
0	0	0	0	0	0	0	0	**-4**	0	0	**-4**	**4**	**2**	0	0
0	0	0	0	0	0	0	0	0	**-4**	0	0	**-4**	**4**	**2**	0
0	0	0	0	0	0	0	0	0	0	**-4**	0	0	**-4**	**4**	**2**
0	0	0	0	0	0	0	0	0	0	0	**-4**	0	0	**-4**	**4**

Figure 6.2: A matrix created with $N = 16$, $C = 4$, $\delta = 3$ and $\gamma = 5$.

When the non-zero elements of the matrix are created by using the above four rules, the type of the matrix depends on the choice of parameters δ and γ in the following way:

- symmetric and positive definite matrices (similar to the matrices obtained by the discretization of the Laplacian operator by the 5-point rule) are created when $\delta = \gamma = 0$,
- the matrices are symmetric in structure when $\delta \neq 0$ and $\gamma = 0$ (i.e. $a_{ij} \neq 0$ implies $a_{ji} \neq 0$ for $\forall i, j$, where $i, j = 1, 2, \ldots, N$),

- general matrices are produced when $\gamma \neq 0$.

The condition numbers of the created matrices are normally increased when the order is increased and/or the values of the parameters δ and γ are increased. Some condition numbers are given in Table 6.2. It is seen that the condition numbers of the matrices created by the pair $\delta = 0,\ \gamma = 0$ are greater than the condition numbers created by the two other pairs. It will be shown in the next section, however, that the iterative methods perform normally better in the solution of the systems created with the pair $\delta = 0,\ \gamma = 0$. This is, of course, not a surprise, because symmetric and positive definite matrices are generated when the pair $\delta = 0,\ \gamma = 0$ is used.

Table 6.2: Condition numbers of matrices that are generated by different values of parameters N, C, δ and γ.

N	C	$\delta = 0,\ \gamma = 0$	$\delta = 4,\ \gamma = 0$	$\delta = 4,\ \gamma = 4$
100	10	9.3E+1	1.5E+1	3.3E+1
400	20	3.5E+2	2.8E+1	6.3E+1
900	30	7.7E+2	4.2E+1	9.1E+1
1600	40	1.4E+3	5.5E+1	1.2E+2
2500	50	2.1E+3	6.8E+1	1.5E+2
3600	60	3.0E+3	8.1E+1	1.8E+2
4900	70	4.1E+3	9.4E+1	2.0E+2
6400	80	5.3E+3	1.1E+2	2.3E+2
8100	90	6.7E+3	1.2E+2	2.6E+2
10000	100	8.3E+3	1.3E+2	2.9E+2

It is easy to verify the fact that the requirements that were stated in the beginning of this section are satisfied when parameters N, C, δ and γ are used to generate test-matrices.

6.5.6 Testing the reliability of the stopping criteria

Matrices have been created by selecting five values of δ ($\delta = 0,\ 2,\ 4,\ 8,\ 16$). For each value of δ, six values of γ were chosen with $\gamma = 0(1)5$. This means that 30 pairs $(\delta,\ \gamma)$ were used in the experiments. For every pair $(\delta,\ \gamma)$, 20 matrices were created with $C = 10(10)200$ and $N = C^2$. Every matrix was factorized by using 10 values of the relative drop-tolerance $RELTOL$, $RELTOL = 0$ and $RELTOL = 2^{-k}$ where $k = 9(-1)1$. The LU factorizations calculated by the sparse code which was sketched in the previous sections (much more details can be found in Zlatev [281]) were used as preconditioners of the iterative methods. An experiment with matrices produced by a fixed pair (δ, γ) leads to the solution of 1840 systems of linear algebraic equations (200 per each preconditioned iterative method and 20 both for DIR and PURE) in the range from $N = 100$ to $N = 40000$. The right-hand-side b for each system of linear algebraic equations is created with a solution x all components of which are equal to 1.

The conclusions, which will be made in this section and in the following section, are based on results obtained in all runs discussed above (i.e. not only on the results presented in the Table 6.3 - Table 6.6). The ability of the stopping criteria to successfully terminate the iterative process is mainly discussed, but some computational times are also given (see Table 6.6). The computational times become even more important when the matrices are large (i.e. of order greater than 10^6). Large sparse matrices will be handled in the next section.

Accuracy requirements

It is worthwhile to test the stopping criteria both in the case where the accuracy requirement is stringent and when a low accuracy is required. A stringent accuracy requirement is defined by setting $ACCUR = 10^{-10}$, i.e. it is desirable to obtain an approximate solution, the relative error of which satisfies $\|x - x_n\|/\|x\| \leq 10^{-10}$, when iterative methods are used. In a similar way, $ACCUR = 10^{-4}$ is used in the experiments, when a low accuracy is required.

The code tries to evaluate the relative error and to stop the calculations when the evaluated relative error becomes less than $ACCUR$. The run is counted as successful every time when this happens. The maximal number of successful runs for a given pair (δ, γ) is 200 for every preconditioned iterative method (20 matrices, each of them being factorized by using 10 values of $RELTOL$) and 20 for DIR and PURE.

The error evaluated by the code can be less than $ACCUR$, while for the exact error this relation is not satisfied. If this happens, then the code is "lying". It is also necessary to count how many times the code is "lying". The run is accepted as successful when the exact error of the solution obtained by the code is less than $10 * ACCUR$, which means that the accuracy requirement is slightly relaxed when the number of accepted runs is counted. Thus, it is declared that the code is "lying" when it is telling us that the accuracy requirement is satisfied, but the exact error is greater than $10 * ACCUR$.

During the iterative process, the requirements for stopping the iterations are increased when certain conditions are not satisfied (when some appropriate parameters of the method, $RATE$, α and others, are varying too much). Because of these increased requirements, which are necessary in the efforts to make the stopping criteria more reliable and more robust, there is a danger that the code will tell us that the computed solution is not sufficiently accurate (the evaluated error is greater than $ACCUR$), while the exact error satisfies the relaxed requirement $\|x - x_n\|/\|x\| \leq 10 * ACCUR$. Therefore, it is also necessary to count the cases in which the code cannot find out that the accuracy requirement has been satisfied.

Results obtained by using the stringent accuracy test

Some results, which are obtained with the stringent accuracy test $ACCUR = 10^{-10}$, are given in Table 6.3.

One should expect that the iterative methods will perform best when the pair $(\delta = 0,\ \gamma = 0)$ is selected, because in this case the matrices are symmetric and positive definite. It is seen from Table 6.3 that this is so.

It was also expected that the results obtained by using the second pair $(\delta = 4,\ \gamma = 0)$ will be better than the results obtained by using the third pair, $(\delta = 4,\ \gamma = 4)$, because in the former case the matrices are symmetric in structure and the number of non-zero elements is smaller. However, the results for the third pair are often better.

The number of cases where the code is telling us that the accuracy required is achieved, while the exact error is greater than $10 * ACCUR$ is not very high (see the results in the columns under "WR" in Table 6.3). Also the cases where the fact that the accuracy required has been achieved remains undetected are low (see the results in the columns under "UN" in Table 6.3). Nevertheless, the results indicate that some more work on the stopping criteria is desirable.

Table 6.3: Numerical results for matrices generated with three pairs (δ, γ) are solved by $ACCUR = 10^{-10}$. "SU" refers to the number of successful runs. "WR" refers to number of cases where the code is reporting successful runs, but the exact error does not satisfy the relaxed accuracy requirement. "UN" refers to number of cases in which the code is not able to detect the fact that the accuracy requirement is satisfied. If DIR is used, then the run is successful only when $RELTOL = 0$. If PURE is used, then the solution does not depend on the value of $RELTOL$ ($RELTOL$ is only used to calculate a preconditioner).

	$\delta = 0,\ \gamma = 0$			$\delta = 4,\ \gamma = 0$			$\delta = 4,\ \gamma = 4$		
Method	SU	WR	UN	SU	WR	UN	SU	WR	UN
DIR	20	0	0	20	0	0	20	0	0
IR	173	0	1	154	0	2	159	0	0
PURE	16	0	0	19	0	1	17	0	1
ORTH	194	0	0	170	0	0	177	0	0
CGS	174	0	0	163	1	0	144	0	0
BiCGSTAB	200	1	0	177	2	0	176	0	0
TFQMR	175	3	5	167	5	3	177	2	3
GMRES	194	0	2	173	0	8	181	0	0
EN	174	0	6	158	0	0	152	0	1
BEN	188	0	6	159	0	0	182	0	2

Results obtained when the accuracy requirement is low

The same matrices have also been tested with an accuracy requirement $ACCUR = 10^{-4}$. The results are given in Table 6.4.

Comparing the results in Table 6.4 with those in Table 6.3, it is seen that the number of successful runs is increased when the value of $ACCUR$ is decreased. For some methods and for some pairs (δ, γ) the difference is rather considerable. Compare, for example, the results for CGS in Table 6.4 when the pair $(\delta = 4, \gamma = 4)$ is used with the corresponding results in Table 6.3.

Results obtained by using the traditional stopping criteria

It is also necessary to test the performance of the traditionally used stopping criteria, i.e. the stopping criteria that are based on one of the following four tests (without any attempt to evaluate the convergence rate and/or the reliability of any of the important iteration parameters): (i) $\|r_n\| < ACCUR$, (ii) $\|r_n\|/\|r_0\| < ACCUR$, (iii) $\|x_n - x_{n-1}\| < ACCUR$ and (iv) $\|x_n - x_{n-1}\|/\|x_n\| < ACCUR$. Some results, obtained when $ACCUR = 10^{-10}$ are given in Table 6.5. It should immediately be mentioned, that it was extremely difficult to run the experiment when the second of these criteria is used. This was especially true when the preconditioner was a very good approximation of A^{-1}, which is, for example, true when the preconditioner is obtained by using $RELTOL = 0$). Therefore, we replaced this criterion with (ii.a) $\|r_n\|/\max(\|r_0\|, 1.0) < ACCUR$. This means that a pure residual test is carried out when $\|r_0\| \leq 1.0$, while a relative residual test is used in the stopping criteria when $\|r_0\| > 1.0$. The results in Table 6.5 were obtained by using the mixed criterion (ii.a). Moreover, only the results obtained by using stopping criteria based on (i), (ii.a) and (iii) are given in Table 6.5. The results obtained by using a stopping criterion based on (iv) are very similar to those obtained by using (iii), because all components of the solution vectors are equal to one in the experiments.

The comparison of the results given in Table 6.5 with the corresponding results in Table 6.3 (the results in columns 2 to 4) shows that

- the numbers of "successful runs" (i.e. the cases where the code tells us that the problem is successfully solved) is increased when the traditionally used stopping criteria is used,

however, it is unfortunately also true that

- the comparison of the calculated by the code error with the exact one reveals the fact that very often the problem has not been successfully solved.

On the other hand, there are only a few cases where the code cannot detect that the desired accuracy is achieved.

The comparison of the results in columns 2 to 4 in Table 6.3 with the results in Table 6.5 indicates that one must be careful when the stopping criteria are selected

Table 6.4: Numerical results for matrices obtained when three pairs (δ, γ) are solved with $ACCUR = 10^{-4}$. "SU" refers to the number of successful runs. "WR" refers to number of cases where the code is reporting successful runs, but the exact error does not satisfy the relaxed accuracy requirement. "UN" refers to number of cases in which the code is not able to detect the fact that the accuracy requirement is satisfied. If DIR is used, then the run is successful only when $RELTOL = 0$. If PURE is used, then the solution does not depend on the value of $RELTOL$ ($RELTOL$ is only used to calculate a preconditioner).

	$\delta = 0,\ \gamma = 0$			$\delta = 4,\ \gamma = 0$			$\delta = 4,\ \gamma = 4$		
Method	SU	WR	UN	SU	WR	UN	SU	WR	UN
DIR	20	0	0	20	0	0	20	0	0
IR	182	0	4	157	0	0	159	0	0
PURE	17	0	1	20	0	0	18	0	2
ORTH	194	0	2	170	0	0	177	0	0
CGS	200	3	0	174	0	0	189	1	0
BiCGSTAB	200	3	0	177	0	2	175	1	1
TFQMR	188	7	0	171	1	0	180	9	2
GMRES	200	0	0	181	0	0	181	0	0
EN	178	0	22	176	14	0	194	27	0
BEN	192	0	8	160	0	0	182	0	1

for the iterative methods that are to be used in the treatment of large scientific and engineering problems. In order to achieve better reliability and robustness of the iterative methods, it is desirable to apply more advanced stopping criteria in which attempts are made

- to evaluate the convergence rate, and
- to prevent stopping the iterative process when some iteration parameters vary in a wide range.

Stopping criteria based on these principles have been applied in an attempt to improve the performance of the fourth traditionally used stopping criterion (that based on checking $\|x_n - x_{n-1}\| / \|x_n\|$). It is clear that similar principles might be applied in the efforts to improve the performance of the other three stopping criteria.

Comparison of computational times

The results in Table 6.3 indicate that both DIR and PURE seem often to work rather well (when the accuracy is taken into account). It is also important to compare the computational times (especially in the important case where the matrix

Table 6.5: Numerical results obtained when three traditionally used stopping criteria with $ACCUR = 10^{-10}$ are applied in the solution of 20 problems with matrices obtained with the parameters $\delta = 0$ and $\gamma = 0$ and with 10 values of the relative drop tolerance for each matrix. The quantities used in the stopping criteria are given in the first row of the table. "SU" refers to the number of successful runs. "WR" refers to number of cases where the code is reporting successful runs, but the exact error does not satisfy the relaxed accuracy requirement. "UN" refers to number of cases in which the code is not able to detect the fact that the accuracy requirement is satisfied. If DIR is used, then the run is successful only when $RELTOL = 0$. If PURE is used the solution does not depend on the value of $RELTOL$ ($RELTOL$ is only used to calculate a preconditioner).

	$\|r_n\|$			$\|r_n\|/max(\|r_0\|, 1.0)$			$\|x_n - x_{n-1}\|$		
Method	SU	WR	UN	SU	WR	UN	SU	WR	UN
DIR	20	0	0	20	0	0	20	0	0
IR	180	150	0	176	138	0	180	144	0
PURE	17	15	0	16	4	1	16	7	1
ORTH	194	97	0	194	13	2	194	21	0
CGS	178	37	0	178	22	0	178	0	0
BiCGSTAB	200	100	0	200	18	0	200	41	0
TFQMR	192	78	0	191	32	1	200	36	0
GMRES	200	116	0	200	34	0	200	116	0
EN	184	141	0	180	9	0	184	14	0
BEN	198	160	0	194	24	0	196	30	0

is large). The results in Table 6.6 show that great savings in computational time can be achieved when the preconditioned methods are used with a well-chosen drop-tolerance. The computational times are reduced in some cases by a factor more than 100 when preconditioned methods are used instead of the pure iterative method. Sometimes the preconditioned iterative methods converge and give good results while the pure iterative method is not convergent. There are also some problems. If only one system is to be solved, then it is difficult to find the best value of $RELTOL$. However, in practice one has normally to solve a long sequence of problems with similar matrices. In such a case, it will be profitable to spend some extra time in the beginning of the process in order to find a good value of $RELTOL$. After that one should expect to solve the remaining systems efficiently by using the obtained value of the drop-tolerance.

Some observations from the numerical experiments

The results presented above show that the preconditioned methods can be a very useful tool when large scale problems have to be handled (see the results in Table 6.3 to Table 6.6). However, several requirements have to be satisfied in order to achieve reasonable efficiency. First, one should be able to improve the preconditioner when the iterative process is slowly convergent or even not converging. In our case this can be achieved by reducing the value of the relative drop-tolerance $RELTOL$.

It is also highly desirable to achieve the required accuracy. More advanced stopping criteria were developed and used for this purpose. The results indicate that the solution process becomes more reliable when these stopping criteria are properly implemented in the codes.

Table 6.6: Best computational times for $N = 40000$ together with the value of the $RELTOL$ for which they were obtained. "N.C" refers to the case where the method does not converge. The runs have been carried out on a SUN workstation.

	$\delta = 0,\ \gamma = 0$		$\delta = 4,\ \gamma = 0$		$\delta = 4,\ \gamma = 4$	
Method	Time	RELTOL	Time	RELTOL	Time	RELTOL
DIR	45.3	0	46.9	0	409.9	0
IR	45.4	0	2.9	2^{-5}	39.6	2^{-8}
PURE	222.7	-	301.5	-	N. C.	-
ORTH	19.6	2^{-9}	2.9	2^{-5}	36.8	2^{-7}
CGS	12.3	2^{-9}	2.7	2^{-5}	38.0	2^{-8}
BiCGSTAB	11.2	2^{-8}	2.7	2^{-5}	37.5	2^{-8}
TFQMR	13.7	2^{-8}	3.0	2^{-5}	37.5	2^{-8}
GMRES	15.0	2^{-8}	3.0	2^{-6}	37.7	2^{-8}
EN	13.3	2^{-8}	3.1	2^{-5}	37.5	2^{-8}
BEN	13.6	2^{-8}	2.8	2^{-5}	37.2	2^{-8}

Many hundreds of runs have been used in the experiments (some of the results obtained in these experiments are shown in Table 6.3 to Table 6.6). Nevertheless, more experiments might give additional useful information, which can be used to further improve the efficiency. The development of a benchmark of matrices, where matrices of different orders, with different structure and with different properties can be easily generated and used to check systematically the efficiency of the numerical algorithms for solving systems of linear algebraic equations, is a challenging task. Such a benchmark will be very useful both for developers of new algorithms and for users in their search for an appropriate algorithm. It should be mentioned here that other generators for sparse matrices are described in detail in Zlatev [281]; some of them will be used in the next section.

The efficiency of the results obtained by the different algorithms is normally dependent on the particular matrices involved in the problems that are to be treated. The efficiency does not automatically mean a requirement to reduce the CPU time as much as possible. Other parameters may also be important for a particular user. For example, some users might be interested in reduction of the storage requirements. It should be noted here that the use of a large $RELTOL$ will often result in a reduction of the storage needed but the CPU time can be increased considerably because the process becomes slowly convergent and many iterations are needed to achieve the required accuracy.

The above conclusions show clearly that several different algorithms must be included in a good package for solving systems of linear algebraic equations.

6.6 Utilizing the cache memory for large sparse matrices

The problems solved in the previous section were very useful for checking the reliability of the stopping criteria, but all these problems were not really large (indeed, these problems are very small according to the modern standards). In this section we shall discuss the solution of very large problems (the sizes of these problems being comparable with the sizes of real-life problems appearing in air pollution modelling and in some other areas of science and engineering). The main purpose in the remaining part of this chapter will be to demonstrate the fact that it is necessary to exploit efficiently the cache memories of the high-speed computers when such problems are treated. Moreover, we shall show that some very large matrices can be handled on the existing computers **only** when the cache memory is properly utilized.

6.6.1 Major assumptions related to the matrices

We shall again assume that systems of N linear algebraic equations written in the following form:

$$Ax = b, \quad A \in R^{N \times N}, \quad b \in R^N, \quad x \in R^N, \tag{6.17}$$

are to be solved. Matrix A is assumed to be

- very large,
- non-symmetric (i.e. $a_{i,j}$ is in general not equal to $a_{j,i}$), and
- very badly scaled.

All three assumptions are important when problems arising in air pollution modelling are to be handled. It has been seen in the previous chapters of this book that the discretized models contain millions of equations. Therefore, **in this part of Chapter 6, "very large" means that $\mathbf{N > 10^6}$**. Both the advection part and the chemical part of an air pollution models are leading after the discretization to non-symmetric matrices. This is why the second assumption is imposed. The last assumption is made in order to meet the requirements appearing in the chemical part of large-scale air pollution models, where matrix A will depend on the concentrations of the chemical species, which vary in the range from $O(10^{-3})$ to $O(10^{13})$ because these quantities are measured by using the number of molecules in cubic centimeter (see Chapter 4).

It should be emphasized here that the direct use of pure iterative methods is not advisable when the second and the third assumptions are satisfied.

In many large practical problems, matrix A is banded. However, if A is very large, then the bandwidth is also large and the use of direct methods for banded matrices might become, as mentioned in the previous sections of this chapter, prohibitive (in spite of the fact that very efficient software for banded matrices is available; see Anderson et al. [11]).

Matrix A is sparse. However, it is also very difficult to exploit the sparsity directly when matrix A is very large. The major reason for this is the fact that severe cache problems arise, because the structure of the sparse matrices is irregular and it is very difficult to utilize properly the different caches of the computer available. Of course, there are many other difficulties (and these were discussed in the previous sections of this chapter). The appearance of many of fill-ins during the factorization is another major difficulty in the solution of very large systems of linear algebraic equations. Many fill-ins do appear when the matrices are very large even if a good pivoting strategy (as, for example, some of the pivoting strategies introduced in Zlatev [272] and [281]; these strategies are briefly described in one of the previous subsections) is used in the efforts to preserve better the original sparsity pattern of matrix A.

It is well-known that general sparse matrices cannot directly be treated in an efficient way. It is necessary to reorder the non-zero elements of the matrix. An attempt

- to reorder the data (consisting of the non-zero elements of the matrix) into blocks, and then
- to carry out the computations by executing successively the blocks (so that only the non-zero elements of the block under consideration participate in the computational process when the block is executed).

will be discussed here. The major advantage of such an approach is the possibility to obtain better utilization of the cache memories. However, the algorithm can also successfully be used on parallel computers. This property of the algorithm will be demonstrated in the next chapter.

6.6.2 Description of the algorithm

The algorithm used to obtain an approximate LU-factorization of matrix A (that can be used as a preconditioner in the solution of the system of linear algebraic equations by some iterative method), consists of the following five steps.

Step 1 - Obtaining an upper block-triangular matrix

A simple reordering is performed in an attempt to push the non-zeros in the lower triangular part of matrix A as close to the main diagonal as possible. First the columns of A are ordered. Let c_j be the number of non-zeros in column j. Column interchanges are applied to obtain a permutation of A such that

$$j < k \quad \Rightarrow \quad c_j \leq c_k. \tag{6.18}$$

After that the rows are ordered in a similar way. Assume that r_i is the number of non-zeros in row i. Row interchanges are applied to obtain a permutation (of the matrix obtained after the column interchanges) such that

$$i < m \quad \Rightarrow \quad r_i \leq r_m. \tag{6.19}$$

The result of the column and row interchanges performed as described above is a matrix in which the non-zero elements under the main diagonal are located as close as possible to the main diagonals. This fact allows us to represent easily the matrix obtained after these interchanges as an upper block-triangular matrix with rectangular diagonal blocks.

Some other reordering algorithms can also be used. For example, the algorithm LORA from Gallivan et al.[110] can be applied. The advantage of the algorithm sketched above is that it is very cheap; its computational cost is $O(NZ)$, where NZ is the number of non-zeros in A. The computational cost of LORA is $O(NZ \log N)$ which is a considerable increase when N is very large. On the other hand, a more expensive computationally, but also more advanced, algorithms (such as LORA) may lead to a better reordering of the matrix and, thus, to a better performance in the remaining steps of the algorithm.

Step 2 - Dividing the reordered matrix into block-rows

The reordered, as described in the previous subsection, matrix can be divided into several block-rows, the diagonal blocks of which are rectangular matrices. Denote the reordered matrix by B (i.e. $B = PAQ$, where P and Q the permutation matrices that are induced by the row and column interchanges). It is desirable to divide matrix B into q block-rows which can be processed either sequentially or in parallel. A simple example for such a division with $q = 4$ is given in Fig. 6.3.

While the special case $q = 4$ is used both in Fig. 6.3 and in the other figures (in order to facilitate the understanding of the ideas), all remarks given below are for the general case (i.e. for an arbitrary q).

Note that all the blocks located under the diagonal blocks B_{ii}, where $i = 1, 2, \ldots, q-1$ contain only zero elements. This is essential in the attempts to achieve parallel computations when the different block-rows are processed.

It should also be noted that the diagonal blocks are rectangular. The number of rows is greater than or equal to the number of columns for the first $q - 1$ diagonal blocks. The situation is changed for the last diagonal block B_{qq}, i.e. the number of columns is greater than or equal to the number of rows.

$$\begin{vmatrix} B_{11} & B_{12} & B_{13} & B_{14} \\ 0 & B_{22} & B_{23} & B_{24} \\ 0 & 0 & B_{33} & B_{34} \\ 0 & 0 & 0 & B_{44} \end{vmatrix}$$

Figure 6.3: Division of matrix B into rectangular blocks when q = 4 .

Step 3 - Factorizing the block-rows

Gaussian transformations are carried out in order to produce zero elements under the main diagonal of each diagonal block B_{ii}, $i = 1, 2, \ldots, q$. These calculations can be performed in parallel. The result of performing Gaussian transformations for the matrix given in Fig. 6.3. is shown in Fig. 6.4.

While the special case $q = 4$ is used in Fig. 6.4 in order to facilitate the understanding of the ideas, the remarks given below are again for the general case (i.e. for an arbitrary q).

It is seen (compare Fig. 6.4 with Fig. 6.3) that each of the first $q-1$ block-rows of matrix B is divided into two block-rows. Consider the diagonal block-row B_{ii}, $i = 1, 2, \ldots, q-1$. Assume that this diagonal block contains m_i rows and n_i columns with $m_i \geq n_i$. Then, as a results of the Gaussian transformations in block-rows i, the diagonal block B_{ii} is split to an upper triangular matrix $C_{ii} \in R^{n_i \times n_i}$ and a zero matrix of order $(m_i - n_i) \times n_i$.

Consider block B_{qq} and assume that it contains m_q rows and n_q columns with $m_q \leq n_q$. The results of performing Gaussian transformation in block-row B_{qq} will lead to an upper triangular matrix $C_{qq} \in R^{n_q \times n_q}$ and a matrix $E_{qq} \in R^{m_q \times (n_q - mq)}$.

The dimensions of the remaining blocks in Fig. 6.4 are listed below:

- $C_{ij} \in R^{n_i \times n_j}$ with $i = 1, 2, \ldots, q-1$, $j = 2, 3, \ldots q-1$, and $i < j$;
- $D_{ij} \in R^{(m_i - n_i) \times n_j}$ with $i = 1, 2, \ldots, q-1$, $j = 2, 3, \ldots q$ and $i < j$;
- $E_{iq} \in R^{n_i \times (n_q - m_q)}$ with $i = 1, 2, \ldots, q-1$;
- $F_{iq} \in R^{(m_i - n_i) \times (n_q - m_q)}$ with $i = 1, 2, \ldots, q-1$.

The computations related to the blocks C_{ij} $(i, j = 1, 2, \ldots, q$ and $i \leq j)$ as well as the computations related to the blocks E_{iq} $(i = 1, 2, \ldots, q-1)$ are finished at the end of Step 3. The further computations are carried out in the remaining blocks. Therefore, it is desirable that these blocks are small (the reordering procedure from Step 1 is performed in attempt to reduce as much as possible the size of these blocks).

C_{11}	C_{12}	C_{13}	C_{14}	E_{14}
0	D_{12}	D_{13}	D_{14}	F_{14}
0	C_{22}	C_{23}	C_{24}	E_{24}
0	0	D_{23}	D_{24}	F_{24}
0	0	C_{33}	C_{34}	E_{34}
0	0	0	D_{34}	F_{34}
0	0	0	C_{44}	E_{44}

Figure 6.4: The result of the factorization of the block-rows of the matrix given in Fig. 6.3.

Step 4 - Producing zeros in blocks D_{ij}

The computations should be continued by producing zeros in the blocks D_{ij} with $i = 1, 2, \ldots, q-1$, $j = 2, 3, \ldots q$ and $i < j$. It is appropriate to carry out the calculations by "diagonals".

During the calculations along the "first diagonal", the pivotal elements of block C_{22} are used to produce non-zeros in D_{12}, the pivotal elements of C_{33} are used to produce non-zeros in D_{23}, and so on. The number of of blocks $D_{i,i+1}$, $i = 1, 2, \ldots, q-1$, which are to be treated, is $q-1$ and the computations needed to produce zero elements in any of these blocks is independent of the computations in the other blocks. Thus, the computations can be carried out in parallel and the number of parallel tasks is $q-1$..

When the calculations along the "first diagonal" are finished, the calculations along the "second diagonal" can be started. The pivotal elements of block C_{33} are used to produce non-zeros in D_{13}, the pivotal elements of C_{44} are used to produce non-zeros in D_{24}, and so on. The number of blocks, which are to be handled, is now $q-2$, which means that the number of parallel tasks is also $q-2$.

The calculations are carried out in a similar manner along the "remaining diagonals". It is important to emphasize that the number of blocks in which zeros are to be produced is reduced by one when the calculations along the next diagonal are to be carried out. This means that the number of parallel tasks is also reduced by one in the transition from diagonal i to the diagonal $i+1$. This is not a big problem when the blocks D_{ij} are small. In order to improve the performance it is worthwhile to choose the number of block-rows q considerably larger than the number of processors p that will be used. We shall consider parallel computations in the next chapter. However, it is important to emphasize the fact that the choice of large number of blocks q is also good in the efforts to utilize better the cache memory of the available computer. This will be illustrate by many numerical examples.

When the calculations in Step 4 are completed, the matrix shown in Fig. 6.4 will be transformed so that the structure given in Fig. 6.5 will be obtained.

$$\left|\begin{array}{ccccc}
C_{11} & C_{12} & C_{13} & C_{14} & E_{14} \\
0 & 0 & 0 & 0 & F_{14} \\
 & & & & \\
0 & C_{22} & C_{23} & C_{24} & E_{24} \\
0 & 0 & 0 & 0 & F_{24} \\
 & & & & \\
0 & 0 & C_{33} & C_{34} & E_{34} \\
0 & 0 & 0 & 0 & F_{34} \\
 & & & & \\
0 & 0 & 0 & C_{44} & E_{44}
\end{array}\right|$$

Figure 6.5: The result of producing zeros in the blocks D_{ij} of the matrix given in Fig. 6.4.

Step 5 - Reordering and finishing the computations

The matrix, which structure is shown in Fig. 6.5, can be reordered (by row permutations) to the form given in Fig. 6.6. It should be emphasized here that it is not necessary to perform the actual row interchanges, it is sufficient to renumber the rows.

$$\left|\begin{array}{ccccc}
C_{11} & C_{12} & C_{13} & C_{14} & E_{14} \\
0 & C_{22} & C_{23} & C_{24} & E_{24} \\
0 & 0 & C_{33} & C_{34} & E_{34} \\
0 & 0 & 0 & C_{44} & E_{44} \\
 & & & & \\
0 & 0 & 0 & 0 & F_{14} \\
0 & 0 & 0 & 0 & F_{24} \\
0 & 0 & 0 & 0 & F_{34}
\end{array}\right|$$

Figure 6.6: The result of reordering the blocks in the matrix given in Fig. 6.5.

The large block consisting of the blocks C_{ij} and the appropriate zero blocks is forming a square upper triangular matrix. The order of this matrix is

$$\sum_{i=1}^{q-1} n_i + m_q \tag{6.20}$$

The block-column formed by the blocks E_{iq} is a rectangular matrix. It is clear that the first dimension of this block is the same as that of the square matrix formed by the blocks C_{ij}, which is given in (6.20). The second dimension of this matrix is $n_q - m_q$.

Simple calculations show that the matrix formed by the blocks F_{iq} is square. Its order is $n_q - m_q$. This matrix is normally not very large. Furthermore, it should be expected that this matrix is relatively dense, because a lot of computations involving the blocks F_{iq} have already been carried out in the previous steps. Therefore, it is worthwhile to switch to dense matrix technique in Step 5 and to apply subroutines for treatment of dense matrices from, for example, Anderson et al. [11].

6.6.3 Numerical results

The numerical algorithm discussed in this section has been tested by using different matrices produced by the matrix generators described in Zlatev [281]. There are several advantages when such generators are used:

- one can produce arbitrarily many matrices (while the number of test matrices in the available databases of general sparse matrices is limited),
- one can vary the size of the matrices,
- one can vary the sparsity pattern,
- one can vary the number of non-zero elements, and
- one can vary the condition number of the matrices tested.

Note that the matrices used in the previous section were suitable for testing the preconditioned methods and the stopping criteria used in connection with preconditioned methods, because it was possible to generate three types of matrices (symmetric matrices, matrices that are symmetric in structure and non-symmetric matrices). It was important to test the stopping criteria in these situations. However, the number of non-zero elements per row was fixed (and very small) for the sparse matrices used in the previous section. Now, when we should like to test the ability of the proposed algorithm to exploit efficiently the cache memory of the available computers, it is important to be able to vary the average number of non-zero elements per row. This is why the matrix-generator, the major properties of which were outlined above, has been chosen. It is difficult to represent a matrix from this new class with a figure similar to Fig. 6.1, because the smallest matrix is of order 22 and it is necessary to use even larger matrices in order to give an adequate information about the structure of the matrices of this type. However, adequate information about these matrices, together with the algorithm used to generate them, can be found in Zlatev [281].

Results from several experiments will be given in this section to demonstrate the performance of the algorithm. It will be shown that the behaviour of the numerical algorithm when the number q of block-rows is varied for very large matrices is different from the corresponding behaviour for small matrices. It should be noted that matrices of order $O(10^5)$ are called here small. Such matrices are considered as large (and even as very large) in many papers on the solution of sparse systems of linear algebraic equations.

The conclusions drawn in this section and in the next one are based on results obtained in much more experiments, than those used in the preparation of Table 6.7 - Table 6.9.

Some attempts to generate matrices which are to some degree similar to those that arise after the discretization of air pollution models were made by applying the following rules:

- the average number of non-zero elements per row was no less than 10,
- the non-zero elements were widely spread, and
- the matrices chosen for the experiments were badly scaled.

All experiments, which are reported here were carried out on SUN computers at DCSC (the Danish Center for Scientific Computing). Some experiments on IBM computers at the University of Aarhus gave quite similar results.

(A) Varying the number of block-rows for small matrices

Systems containing 512 000 linear algebraic equations are used in this subsection. The number of the non-zero elements is 5 120 110 (i.e. the average number of non-zero elements per row is about 10). The number q, of block-rows, is varied in the range from 4 to 512. The results are shown in Table 6.7 (the computational times are given in seconds). Computational times measured in seconds are also used in the remaining tables (Table 6.8 and Table 6.9). If the dropping parameter is used, then it is always set equal to 0.1 (i.e. small non-zero elements are dropped by using (6.4) with $RELTOL = 0.1$). The notation that is applied in all the tables can be explained as follows:

- **ORD time** is the sum of the time needed for the reordering of the matrix (described in Step 1) and the time needed to divide the matrix into block-rows (Step 2).
- **FACT time** is the time need to obtain the approximate LU-factorization, which is used as a preconditioner in the solution part (the preconditioner is calculated in Step 3 to Step 5).
- **SOLV time** is the time needed to obtain a starting approximation for the iterative procedure (the back substitution step) and, after that, to carry out iterations until a prescribed accuracy (an accuracy requirement of 10^{-7} was actually used in all experiments) is achieved by the preconditioned iterative method chosen (the Modified Orthomin Method from Zlatev [281] was actually used in all runs).
- **TOTAL time** is the total computational time spent in the run.

Several conclusions can be drawn by studying the results in Table 6.7 (similar conclusions can be drawn by using the results of many other runs which were performed).

1. The ORD times and the SOLV times do not depend too much on the parameter q (on the number of block-rows).

2. The FACT times (and, therefore, also the TOTAL times) depend on the choice of q. For this matrix, the best result is obtained with the choice $q = 128$. Using a small number of blocks (see the results in Table 6.7 for $q = 4$) is very expensive. It should be mentioned here that the use of a very large number q of blocks is also inefficient.

3. When the FACT time is the best one (i.e. when $q = 128$), the sum of the ORD time and the SOLV time is comparable with the FACT time.

Table 6.7: Computational times (measured in seconds) obtained in the solution of a system of 512000 linear algebraic equations with different values of parameter q (different numbers of block-rows). The number of non-zero elements in the coefficient matrix is 5120110. $RELTOL = 0.1$ is used.

Block-rows	ORD time	FACT time	SOLV time	TOTAL time
4	7.34	647.52	4.68	659.54
32	7.79	73.16	4.72	85.65
128	7.87	17.58	4.85	30.30
512	7.80	53.70	5.63	67.03

(B) Varying the number of block-rows for large matrices

The number of equations in the system of linear algebraic equations is increased 32 times, i.e. from 512000 to 16384000. The number of non-zero elements in the matrix is increased from 5120110 to 163840110, i.e. the average number of non-zero elements per row is again about 10. This very large system was also solved by using different values of q. Results are shown in Table 6.8.

Similar conclusions, as those drawn in the previous subsection, can be drawn by studying the results in Table 6.8.

1. It is clearly seen that both the ORD times and the SOLV times practically do not depend on the parameter q (on the number of block-rows).

2. The FACT times (and, therefore, also the TOTAL times) depend on the choice of q. For this matrix, the best result is obtained with the choice $q = 1024$. Using a small number of blocks is very expensive.

Table 6.8: Computational times (measured in seconds) obtained in the solution of a system of 16384000 linear algebraic equations with different values of parameter q (different numbers of block-rows). The number of non-zero elements in the coefficient matrix is 163840110. $RELTOL = 0.1$ is used.

Block-rows	ORD time	FACT time	SOLV time	TOTAL time
4	259	679563	176	679999
32	270	86244	178	86691
256	256	10845	183	11301
512	268	5390	179	5837
1024	275	2887	185	3348

3. The FACT time is the largest part of the TOTAL time also with the best choice of q (i.e. when $q = 1024$).

It should again be emphasized that the results obtained in many other runs show the same trends. We shall give an example with even bigger matrix in order to illustrate this statement. The number of equations in the system of linear algebraic equations is increased 4 times, i.e. from 16384000 to 65536000. The number of non-zero elements in the matrix is increased from 163840110 to 655360110, i.e. the average number of non-zero elements per row is again about 10. Results are shown in Table 6.9. It is a computational disaster to run this matrix with $q = 4$. Even the run with 32 block-rows is very difficult when the matrix is so large.

Table 6.9: Computational times (measured in seconds) obtained in the solution of a system of 65536000 linear algebraic equations with different values of parameter q (different numbers of block-rows). The number of non-zero elements in the coefficient matrix is 655360110. $RELTOL = 0.1$ is used.

Block-rows	ORD time	FACT time	SOLV time	TOTAL time
32	1497	1401211	936	1403646
512	1491	87338	891	89721
1024	1550	44664	1002	47216

(C) Some conclusions from the runs

If the matrices become very large, then it might become practically impossible to handle them without using some special devices for utilizing efficiently the cache memory of the available computer. On the other hand, the use of the algorithm proposed here may lead to dramatical decrease of the computational time.

An extra benefit of the proposed algorithm is that it might be applied (nearly directly, by adding appropriate OpenMP directives in a few places in the code) on parallel machines. The performance of the algorithm in the solution of very large systems of linear algebraic equations on parallel machines will be discussed in Chapter 7.

6.7 Comparisons with another sparse code

The numerical results obtained with the sparse code (in which an attempt to utilize better the cache memory is made) indicate that very considerable reductions of the computational time can be achieved the following two conditions are satisfied:

- when the matrices are very large, and
- when the number of block-rows is carefully chosen.

The natural question, which has to be answered, is:

> *could we conclude (by using the numerical results presented in the previous section) that algorithms based on using block-rows is efficient?*

The answer is that the results that were presented in the previous section do not allow us to draw such a conclusion, because the algorithm studied by us might be slower (even when used with the best division of the matrix into block-rows) than other codes for sparse matrices. Therefore, it is necessary to compare the performance of our algorithm with the performance of a good code for general sparse matrices.

We have selected the code **SuperLU**. This code is well-known, has been used by many scientists and can be down-loaded from the Internet (this can be done by searching for "SuperLU" using the search-machine Google). Documentation of the code is also available on the Internet, but basic ideas applied in the development of the code are furthermore well-described in several papers:

- The version of the **SuperLU**, which is best designed for sequential computers, is described in Demmel et al. [61].
- The version of **SuperLU**, which is best designed for shared memory computers, is described in Demmel et al. [62].
- The version of **SuperLU** which is best designed for distributed memory computers, is described in Li and Demmel [165].

The sequential version of **SuperLU** is used in this section (the shared memory version of **SuperLU** will be used in Chapter 7).

Dropping of small non-zero elements is not used in **SuperLU**. Therefore, we had to use smaller matrices in this section (in comparison with the matrices which were used in the previous section).

Some results are presented in Table 6.10. It is clear that the following conclusions can be drawn:

- If the code described in the two previous sections is run without dropping, then it is at least comparable with the performance of **SuperLU**.

- The code described in the previous two sections is performing much better when it is used with the option for dropping small non-zero elements (dropping is carried out by using $RELTIOL = 0.1$ in Table 6.10).

It should be mentioned here that the utilization of the cache memory will require some search for the number of blocks that gives best results. However, it has been shown in the previous section that this option is performing best when the matrices treated are very large. If this is the case, then one has, as a rule, to run many different scenarios. Therefore, it is worthwhile to try (at the beginning of the computations) to find a good value of the number of blocks and then to carry out the whole computational process (i.e. to handle all scenarios) by using this number of blocks.

The choice of a good value of the drop-tolerance $RELTOL$ may also require some search at the beginning of the computations. The results in the previous section and in this section show clearly that a good value of $RELTOL$ can give great savings in both computational time and storage. In fact, some of the problems solved in the previous section can be treated successfully only by using a proper positive value of $RELTOL$.

Table 6.10: Comparison of computational times obtained by SuperLU with results obtained by the code in which the cache memory is exploited by using block-rows. The latter code was run both with dropping and without dropping of small non-zero elements. $RELTOL = 0.1$ **is used to drop small elements. Dropping is switched off with** $RELTOL = 0$. **The numbers of blocks used in our sparse matrix code are given in brackets.**

Order	SuperLU	Without dropping	With dropping
128000	33	21 (16)	8 (32)
256000	67	45 (32)	14 (64)
512000	139	112 (32)	30 (128)
1024000	277	237 (64)	68 (256)
2048000	560	551 (96)	179 (512)

6.8 Applicability to other models

The proper organization of the matrix computations is becoming more and more important when high-speed computers with hierarchical memory (based on several levels of caches) are used in the treatment of large computational tasks. Matrix computations are used in the treatment of nearly all large-scale models. Moreover, as emphasized in the previous sections of this chapter, standard methods for efficient performance of different matrix computations are available and can be used in the efforts to handle efficiently different computational tasks arising when the models from Section 7 in Chapter 1 are to be treated on modern high-speed computers.

It should be emphasized that the methods for handling sparse matrices by an attempt to exploit efficiently the cache memory of the available computer, which were discussed in the previous three sections, are very general. Indeed, no special property of the matrices arising in air pollution modelling is exploited during the construction of these methods. Therefore these methods are applicable when many large-scale mathematical models are treated.

6.9 Concluding remarks

Mainly two types of matrices arise when large-scale air pollution models (or large mathematical models from other fields of science and engineering) are to be handled on computers: banded matrices and general sparse matrices. The organization of the computations, in which such matrices are involved, was discussed in this chapter. The main conclusions can be summarized as follows.

- **Treatment of banded matrices.** Everything seems to be quite straightforward when the problems are reduced to tasks involving matrix computations if these matrices are banded, because standard subroutines for handling such operations are available in many commercial packages as well as packages which are public domains. Moreover, it is normally very easy to implement such subroutines in air pollution codes. However, one should be careful. If the discretization has not been carefully done, then the tasks involving matrix computations may be rather inefficient. This will happen if data participating in the computations is stored in very large arrays which will lead to many cache misses and, thus, to deterioration of the performance. The problem of efficient exploitation of the cache memory is a serious one in spite of the fact that **the main principle is very simple: one should work with small amounts of data (stored in short arrays) for as long as possible**. However, it is very difficult to achieve this in practical computations when the models are very big. It is much easier to deal with this problem when some splitting procedure is used. This is especially true for the chemical sub-model, which is very often the most time consuming part of the computational process. The efficient exploitation of the cache memories during the computations in the chemical sub-model will be discussed in the next chapter.

- **Treatment of general sparse matrices.** The treatment of general sparse matrices is causing many more difficulties. An algorithm that allows us to treat efficiently very large matrices has been described in this chapter. This algorithm can very easily be applied on parallel machines (this will be discussed in Chapter 7). It must be noted here that it is necessary to improve further the sparse algorithms for very large matrices. "Very large" means here that the matrices that are to be handled are of order greater than one million (and very often even much greater than one million).

A lot of linear algebra operations have to be carried out after the discretization of any large-scale mathematical model. These operations are normally very time-consuming. Therefore, it is important to perform them in an efficient way. The fact that the order in which the computations are carried out is essential for linear algebra operations was recognized many years ago. A special package for linear algebra equations, BLAS (Basic Linear Algebra Subprograms) has been developed and is commonly used in the treatment of many large-scale mathematical models. The first version of package BLAS was created in 1979 by Lawson, Hanson, Kincaid and Krogh [158]. This version was used in LINPACK (package for solving systems of linear algebraic equations, see Dongarra et al. [77]. The main tool used in this first version in the efforts to improve the efficiency of the computational process was the organization of the double loops. Several more advanced versions of BLAS were developed (see Dongarra et al. [78], [79]) and used in Anderson et al. [11], Barret et al. [15], Demmel [60], Dongarra et al. [80]. Basic Linear Algebra Subprograms for sparse matrices were also developed; see Duff et al. [82]. We have shown in this chapter that the organization of the computations is becoming much more important when the matrices are **really very large** and that special techniques must be used for very large sparse matrices. Many of the matrices discussed in this chapter can be treated on the available computers **only** when the computations are properly organized.

Chapter 7

Parallel computations

The application of suitable numerical methods to the different parts of an air pollution model has been discussed in the previous chapters. The fact that splitting procedures are applied is essential and has been exploited in Chapter 3 to Chapter 6. The application of any splitting procedure leads to several sub-models. Each of these sub-models consists of many (sometimes up to many millions) independent small tasks. In the chemical part, for example, the chemical reactions at a given grid-point can be treated independently from the treatment of the chemical reactions of all the other grid-points. Thus, it was possible to consider relatively small computational tasks when different numerical methods were discussed and/or tested. In this chapter we shall describe how all the small tasks can be efficiently handled when the whole air pollution model is to be treated.

Assume that the whole model (or, in other words, the system of PDEs by which the model is described mathematically) has to be treated on a given computer. The discretization of the system of PDEs (1.1) results as a rule in very time consuming computational tasks. It is very difficult to handle these tasks, even when modern high-speed computers are available. Indeed, if an air pollution model is to be applied on a large space domain by using fine grids, then its discretization will always lead to huge computational problems (see Subsection 1.2.7).

There is an additional great difficulty, which is very often underestimated (or even neglected) when large application packages are moved from sequential computers to modern parallel machines. The high-speed computers normally have a very complicated architecture and, therefore, the task of producing an efficient code for the particular high-speed computer that is available is both extremely hard and very laborious.

In this chapter we shall consider the solution of the following computational tasks, which arise when large-scale air pollution models are handled on modern high-speed computers (it should also be emphasized here that most of the results and conclusions remain true also when large-scale mathematical models arising in other fields of science and engineering are to be treated on parallel computers):

1. The need to optimize the computations when only one processor is used.
2. The need to implement standard tools for parallelization.

Both tasks are important. If the code is giving good speed-ups, but the performance on one processor is bad, then it is not necessarily true that one has achieved good results. If the code is tuned too much to the computer used, then it might be very difficult to switch to another computer when this becomes necessary.

OpenMP ([265]) and MPI (Gropp et al. [124]), are becoming more and more standard. Both OpenMP and MPI tools were used in the efforts to facilitate the transportability of the code from one computer to another.

We shall start by considering the implementation of DEM on a particular computer, an IBM SMP computer, which is done only in order to make the understanding of the main ideas easier. We shall also demonstrate that the code is portable by showing results obtained when it was run on several other computers.

The IBM SMP computer is a relatively advanced parallel computer architecture which allows the user to achieve parallelism on two levels. The architecture of the IBM SMP computer will be discussed in the next section. It should be mentioned here that the OpenMP mode can be applied within the processors of each node of the IBM SMP computer, while the MPI mode should be used across the nodes. It is nearly clear that one can easily switch to the use of only one of these two modes, i.e., one can apply either only the OpenMP mode (but the OpenMP mode can only be used when a shared memory computer is available or within a given node of the IBM SMP computer) or only the MPI mode (MPI was originally designed for distributed memory computers, but it was after that discovered that this tool can successfully be used also when shared memory computers are available).

After the description of the main principles used during the parallelization of the code on the IBM SMP computer, we shall introduce a unified version of DEM and shall illustrate that this new and more advanced code is also portable by running it on several computers.

Scalability of a parallel code is an important property. We shall demonstrate by several examples that our parallel code is scalable.

Parallel codes for sparse matrices might be very important in the effort to avoid the use of splitting procedure. In Section 7.9 it will be shown that the sparse algorithms discussed in Chapter 6 can be run in parallel.

The applicability of the algorithms used in this chapter for the treatment of large-scale mathematical models arising in several other fields of science and engineering on parallel computers will be discussed in Section 7.10.

Finally, some conclusions and remarks will be given in the last section of this chapter, Section 7.11.

In all sections before Section 7.9, (the performance of the air pollution code on parallel computers is discussed in these sections), it is assumed, as in the previous chapters, that some of the splitting procedures presented in Chapter 2 is used (and the discussion is carried out for the different sub-models obtained when the selected splitting procedure is implemented in the code).

The results in Section 7.9 are very general. It is not assumed that any splitting procedure is used in this section. These results are not directly related to the treatment of air pollution codes, but they can easily be applied in any air pollution code (and also in models arising in some other areas).

7.1 The IBM SMP architecture

It is very difficult to achieve high performance on a traditional, either shared memory or distributed memory, parallel architecture when a large application code is run. The code should be adjusted to the properties of the particular parallel machine that is available. Some parallelization tools can be used in the attempts to facilitate this task. Examples for such tools are the Message Passing Interface (MPI, Gropp et al. [124]), OpenMP ([265]) and the Parallel Virtual Machine (PVM, Geist et al. [114]). All these standard tools for parallelization are very efficient when a large application code, which works efficiently on a given parallel computer, is to be ported to another parallel computer. However, the preparation of the first parallel version of a large application code is normally a very difficult job. During the preparation of such a version, one should carry out a lot of work in order to identify the parallel tasks and to reorganize some of the computations (in order to separate the parallel tasks, to achieve a better loading balance, to utilize better the cache memory, etc.).

On the IBM SMP computer (which has a more complicated structure that allows us to achieve two levels of parallelization) the task of improving the performance of a large application code becomes even more difficult. A short description of the architecture of the IBM SMP computer is given below in order to explain why the task of achieving a high computational speed (when large applications are to be run on this computer) is more difficult compared with the case where the computer is either only a shared memory computer or only a distributed memory computer.

The IBM SMP architecture is very similar to the architecture of the CEDAR computer, which has been developed at the Center for Supercomputing Research and Development (CSRD) at the University of Illinois at Urbana-Champaign in the 80ies; see, for example, Kuck et al. [154].

The IBM SMP computer consists of several nodes (the particular computer used in our experiments has only two nodes, but this still allows us to perform many meaningful tests). Each node contains eight processors, which share memory. This means that each node works in a shared memory mode. Each processor of the CEDAR computer also had a vectorization capabilities, which is not the case for the processors of the IBM SMP computer (however, each processor of the IBM SMP computer is much more powerful than the processors of the CEDAR computer).

Each node has its own memory. This means that several nodes work together in a distributed memory mode and, thus, message passing is needed in the communications between the nodes.

It is clear from this short description of the major features of the IBM SMP computer that efforts on two levels are needed in order to obtain an efficient parallel code:

- application of parallelization rules within each node, by which the shared memory mode has to be utilized, and
- application of parallelization rules across the nodes, by which the features of message passing machines have to be exploited.

This information about the IBM SMP computer, which is by no means complete (much more details can be found in the SMP/E Internet Library [228]), is quite sufficient to explain why the task of achieving high performance becomes more difficult. There are at least two reasons for this:

- one should exploit the features of both the shared memory machines and the distributed memory machines, and
- one should be very careful in the efforts to avoid conflicts between the devices used in the shared memory code and the devices used in the distributed memory code.

While the task of achieving high performance becomes more difficult when such a complicated architecture as the IBM SMP computer is used, it is also true that if the work is carefully done, then very good results can be achieved. This will be demonstrated in the next sections.

7.2 Running the code on one processor

It is intuitively clear that the first task, which has to be solved, is to optimize the code when it is run on one processor only (i.e., when it is run in a sequential mode). This is why we first tried to increase the performance of the code when it is run on one processor.

The most important task, which has to be solved when a big application code is to be run on one processor, is to exploit better the cache memory of the processor used. The efforts in this direction and the results obtained when different tests were carried out, will be discussed in this section.

It should be mentioned here that the results derived in this section are valid not only for the IBM SMP computer, but also for any computer on which cache memory is available.

7.2.1 Main modules of the two-dimensional code

The code for the two-dimensional version of the Danish Eulerian Model (DEM), the version with 35 chemical species discretized on a 96×96 grid, is used in most of the experiments, which will be discussed in this section. This code is based on the splitting procedure given in Subsection 2.2.2 applied for the two-dimensional case with the addition of pieces of codes for performing input operations, output operations and some initializations. This choice is made only in order to facilitate the exposition of the results. All statements made in this and in the following sections could easily be extended for the other versions of DEM and also for all options of UNI-DEM (the latter model will be discussed in Section 7.7).

The code representing the two-dimensional version of DEM consists of the following modules:

- **Advection module.** This module treats the horizontal transport. The diffusion subroutines have been added to this module. This means that the advection module is the code that treats the sub-model (2.12) from Chapter 2.
- **Chemical module.** This module handles the chemical reactions involved in the model. The deposition subroutines have been added to this module. This means that the chemical module is the code that treats the sub-model (2.13) from Chapter 2.
- **Initialization module.** All operations needed to initialize arrays and to perform the first readings of input data are united in this module.
- **Input operations module.** Input data are read in this module when such data are available at the time step under consideration. If no input data are available at the time step under consideration, then simple interpolation rules are used, both in space and in time, to produce the needed data. Some more details about the input data which is needed in air pollution models are given in Chapter 1.
- **Output operations module.** This module prepares output data. The total number of output files prepared in this version of the model is eight. The total amount of the output data is about 75 MBytes when the model is run over a time period of one month. For some of the concentrations (ozone and nitrogen dioxide) hourly mean values are calculated. For other concentrations and some depositions, daily mean values are prepared. For all concentrations and depositions studied by the model, monthly mean values are calculated in one of the files. Some additional information about the output data prepared when large-scale air pollution models are run in different scientific studies can be found in Chapter 1.

The initialization module is called only once (in the beginning of the computations), while all other modules are called at every time-step.

If a three-dimensional version is to be applied instead of the two-dimensional version discussed above, then only one extra module for treating the vertical transport is to be added. The same rules as the rules discussed in this section are to be used in the preparation of this extra module.

7.2.2 Running the code without special preparations

The results, which were obtained by the code without any attempt to optimize it for runs on the IBM SMP computer, are given in Table 7.1. The subroutines of the code were compiled by using the following options of the IBM SMP FORTRAN compiler:

xlf_r -O4 -qextchk -qhot -qipa -qstrict -qarch=auto

Thus, although we did not tried to optimize the code for runs on the IBM SMP computer, we compiled it with a very high level of optimization.

Table 7.1: Computational times (measured in seconds) obtained when the first version of the code was run on one processor of the IBM SMP computer. The parts of the computational time spent in the modules (compared with the total time for the run) are given in percent in the last column of this table.

Module	Comp. time	Percent
Chemistry	16147	83.09
Advection	3013	15.51
Initialization	2	0.00
Input operations	50	0.26
Output operations	220	1.13
Total time	**19432**	**100.00**

7.2.3 Structuring the code for the chemical part

The results presented in Table 7.1 show clearly that the chemistry module is the most time-consuming part of the code. Therefore, a lot of efforts were spent in the attempts to improve the performance of this part of the code.

We started with some traditionally used actions (such as unrolling loops, reversing the order of performance the inner and the outer loops in double loops, etc.). The improvements due to these actions were minimal, because many of the modern compilers automatically do such optimizations.

The main improvement was achieved by introducing some blocking when the chemical reactions are performed. This has been achieved by dividing the data into several chunks and performing the computations in the chunks sequentially.

It should be mentioned here that the use of chunks is in some sense similar to the use of blocks in the algorithm for solving very large systems of linear algebraic equations with sparse coefficient matrices (this algorithm has been introduced in Chapter 6; the parallel properties of this algorithm will be described in Section 7.9). The algorithm based on using chunks in the chemical sub-model can be described in the following way. Assume that a module that performs the chemical reactions at a given grid-point (such a module is normally called a **box model**) has been developed. Assume also that the number of grid-points is M. Then the chemical part of the code can be performed by using the code shown in Fig. 7.1.

```
C
         DO I=1,M
             Call the box model
         END DO
C
```

Figure 7.1: Pseudo-Fortran code for executing the chemical part under the assumption that a box-subroutine is used. M **is the number of the grid-points used in the discretization of the space domain of the air pollution model used.**

The important issue is that the body of this loop ("Call the box model") is in fact an inner loop in the above piece of code, because when we execute the code "Call the box model" at step I of the loop, we have to perform successively the chemical reactions for all involved compounds: $1, 2, \ldots, NSPECIES$. This means that the code, which is given in Fig. 7.1, can, in principle at least, be equivalently rewritten as shown in Fig. 7.2.

```
C
         DO I=1,M
            DO J=1,NSPECIES
                Perform all chemical reactions
                that involve the chemical species J
            END DO
         END DO
C
```

Figure 7.2: Rewriting the code given in Fig. 7.1 as a double loop. M **is again the number of the grid-points used in the discretization of the space domain of the air pollution model used, while** $NSPECIES$ **is the number of chemical species that are studied by the model.**

It should be emphasized here that the inner loop normally consists of a very complicated code, where several subroutines are to be called. In fact, in most of the codes describing atmospheric chemical schemes a lot of work has to be done in order to obtain the simple structure of the chemical module given above. The other important fact is that the two codes given in Fig. 7.1 and Fig. 7.2 are extremely inefficient on vector machines. Therefore this code should be replaced by a vectorized version which can be obtained by reversing the order of the loops (this is easy when the double loop is already written in the above form; the difficult task is, let us reiterate this, to rewrite the chemical module as shown above). Reversing the order of the inner and the outer loops in the code given in Fig. 7.2 leads to the code given in Fig. 7.3.

It should be noted that the third code performs well not only on vector machines but also on many other computers (because the stride in the inner loop of the involved arrays is one). However, this code cannot efficiently exploit the cache memory of the computer used. The reason for this is that there are several big two-dimensional arrays which are referenced by rows in the inner loop. Consider, for example, the most important array, the array containing the concentrations of the chemical species. This array is denoted by $C(M, NSPECIES)$ in the code. In step I, $I = 1, 2, \ldots M$, of the inner loop $C(I, J)$ is updated, but the new value of the chemical species J depends on some of the other species K, $K = 1, 2, \ldots, J-1, J+1, \ldots, NSPECIES$. Thus, when we are performing the I'th step of the second code, we have to refer to some addresses in row I of array $C(M, NSPECIES)$. There are eight such arrays. It is intuitively clear that it is worthwhile to divide these arrays into chunks and to carry out the computations by chunks.

```
C
      DO J=1,NSPECIES
         DO I=1,M
            Perform all chemical reactions
            that involve the chemical species J
         END DO
      END DO
C
```

Figure 7.3: Reversing the order in which the two loops in the code given in Fig. 7.2 are executed. M **and** $NSPECIES$ **are as in Fig. 7.2.**

Assume that we want to use $NCHUNKS$ chunks. If M is a multiple of $NCHUNKS$, then the size of every chunks is $NSIZE = M/NCHUNKS$, and the third code given above, i.e., the code given in Fig. 7.3, can be modified as shown in Fig. 7.4.

The computational work, which is carried out in the first three codes (i.e., the codes that are shown in Fig. 7.1 to Fig. 7.3), is the same (only the order in which the operations are performed is different). Both the operations that are performed in the beginning and in the end of the first loop in the fourth code, the code shown in Fig. 7.4, are extra. A straight-forward procedure will be to copy the current chunks of all eight arrays in the corresponding small arrays. However, this is not necessary, because some of the arrays are only used as helping arrays in the chemical module. In fact, copies from five arrays are needed in the beginning of the first loop. The remaining three arrays can be defined locally, as small arrays of length (NSIZE, NSPECIES), in the subroutine that performs the body of the inner double loop of fourth code. This leads to a reduction of the storage used, because three large working arrays of length (M, NSPECIES) are replaced by small arrays of length (NSIZE, NSPECIES). The reduction is very considerable, because NSIZE is normally much smaller than M. Let us illustrate this statement by one example. If the three-dimensional version of the model is discretized on a 480 $\times$ 480 $\times$ 10 grid, then M is equal to 2304000, while $NSIZE = 48$ can successfully be used as in the examples presented in this chapter.

The situation at the end of the first loop is similar; it is necessary to copy back to the appropriate sections of the large arrays only the contents of three small arrays. The number of copies made at the end of the first loop has been reduced from five to three because some information (as, for example, the emissions) is needed in the chemical module (and has to be copied from the large arrays to the small ones), but it is not modified in the chemical module (and, thus, there is no need to copy it back to the large arrays in the end of the first loop).

```
C
      DO ICHUNK=1,NCHUNKS
         Copy chunk ICHUNK from some of the eight
         large arrays into small two-dimensional
         arrays with leading dimension NSIZE
         and second dimension NSPECIES
C
         DO J=1,NSPECIES
            DO I=1,NSIZE
               Perform all chemical reactions
               that involve the chemical species J
            END DO
         END DO
C
         Copy some of the small two-dimensional
         arrays with leading dimension NSIZE
         into chunk ICHUNK of the corresponding
         large arrays
      END DO
C
```

Figure 7.4: Introducing chunks in the code shown in Fig. 7.3. M **and** $NSPECIES$ **are as in Fig. 7.2.** $NCHUNKS$ **is the selected number of chunks, while** $NSIZE$ **is the leading dimension of the small arrays in which data is copied. It is required that the condition** $NCHUNKS * NSIZE = M$ **is satisfied.**

One can ask the question:

> *"Is the improvement obtained by using the piece of code that is shown in Fig. 7.4 optimal?"*

The answer is probably not. However, it will be shown in the next subsection that very good results can be achieved by using this device. Note too that the code given, which is given in Fig. 7.4, can be viewed as a **template** by the use of which an efficient exploitation of the caches of the available computer can be achieved when different numerical methods are used. Indeed, different numerical methods can be used in the body of the double loop in the middle of the code that is presented in Fig. 7.4.

The only task, which remains to be solved, is to find an optimal (or, at least, a good) value of parameter NSIZE for the available computer. However, it should be relatively easy to find a good value by performing several experiments. The choice of NSIZE will be further commented on in the next section.

7.2.4 Illustration of the effect of using chunks

The straightforward use of a box model in the first piece of code (the code given in Fig. 7.1) from the previous subsection (and running it over all grid-points) or the slight improvement given by the second piece of code (i.e., the piece of code shown on Fig. 7.2) corresponds to the use of chunks of minimal size (the size of each chunk being 1). However, the use of this version on a vector processor is, as mentioned above, a disaster. On the other hand, the vectorized version, represented above by the third piece of code (shown in Fig. 7.3), corresponds to the use of chunks with a maximal size (in fact, only one chunk, in our particular case of size 9216, being used). It is very easy to implement the former device (the codes shown on Fig. 7.1 and Fig. 7.2). The latter device (the device illustrated on Fig. 7.3) is very efficient on vector processors. However, on parallel machines with complicated architectures (which impose in a much stronger way the requirement of utilizing caches on several levels) it is best to use chunks with medium sizes (i.e., with $NSIZE$ in the range from 16 to about 100 in Table 7.4), while the vectorized version performs very poor in this case.

Some results, which demonstrate the performance of the code when chunks of different sizes are used, are given in Table 7.2. These runs have been performed on three typical computers:

- a vector processor (Fujitsu),
- a parallel computer with shared memory (SGI ORIGIN 2000), and
- a parallel computer, which can be used in both shared memory mode and distributed memory mode (the IBM SMP computer).

All runs, the results of which are given in Table 7.2, were carried out by using one processor only. It is seen that

- the particular computer that is available has to be taken into account when the size of chunks is to be selected, and
- the proper selection of the size of the chunks will normally lead to considerable savings in computer time and storage (the reduction of storage was explained in the previous section).

As expected, the maximal chunks are the best choice when the vector processor, Fujitsu, is used, while the minimal chunks should not be selected for this computer.

The maximal chunks should not be chosen when the SGI ORIGIN 2000 computer is to be used, while the difference between minimal chunks (chunks of size 1) and medium chunks (chunks of size 48) is not very large (the selection of chunks of medium size being, nevertheless, preferable).

The maximal chunks should not be chosen when the IBM SMP computer is available. The use of minimal chunks is also a bad choice in this case. It is best to select chunks with a medium size.

The numerical results that are given in Table 7.2 illustrate, once again, an important fact:

- the choice of an efficient numerical algorithm might not be sufficient in the efforts to achieve an efficient computational process when modern computers are used.

It is sometimes even **more important** to order the computations properly. Moreover, it is seen from Table 7.2 that this action (achieving a proper order of the computational operations) is computer dependent.

Table 7.2: Computational times (measured in seconds) obtained with different chunks on three computers (using one processor only).

Size of the chunks	Fujitsu	SGI ORIGIN 2000	IBM SMP
1	76964	14847	10313
48	2611	12114	5225
9216	494	18549	19432

We shall proceed in the following way. In the next two sections of this chapter it will be shown:

- how the selection of chunks influence the parallel runs on the eight processors within a given node of the IBM SMP computer, and
- how to solve the task of achieving efficient parallel runs across the nodes of this computer.

7.3 Parallel runs on one node of the IBM SMP computer

As mentioned above, each node of the available IBM SMP computer contains eight processors with shared memory. The results given in Table 7.1 indicate that it is quite sufficient to achieve efficient parallel computations only in the advection module and in the chemistry module in order to improve the performance of the air pollution model when several processors on one node are to be used, because about 98% of the total computational work is spent in these two modules. This means that one has, first and foremost, to identify parallel tasks in the advection and chemistry modules. Moreover, it is desirable for the parallel tasks to be both large and of equal size. There is no problem to find such parallel tasks in the advection module, because each chemical species can be treated separately. Thus, the number of parallel tasks in the advection module is equal to the number of chemical species, and the code used is of the following type:

```
C
        DO J=1,NSPECIES
            Perform the advection-diffusion operations
            for the J'th chemical species
        END DO
C
```

Figure 7.5: The main loop in the advection module. ***NSPECIES*** **is the number of species involved in the model. This loop can be parallelized in OpenMP environment by using the directive "parallel do" in the beginning of the loop.**

The OpenMP directive *"parallel do"* should be inserted in the beginning of this code. If this directive is used, then the loop in the piece of code that is given above will be run in parallel on any shared memory computer when OpenMP options of the compiler are specified. Furthermore, the parallel tasks in the body of the loop in the above piece of code are of equal size. Therefore, there will be a perfect loading balance when the number of chemical species is a multiple of the number of processors. However, this is not the case for our air pollution model; the number of chemical species is 35. Nevertheless, very good speed-ups can be achieved in the advection module; see the computational times and the speed-ups for 2, 4 and 8 processors under the column "Advection" in Table 7.3 and Table 7.4 respectively.

Table 7.3: Computational times (measured in seconds) obtained by using 1, 2, 4 and 8 processors in one node as well as by using all 16 processors in two nodes of the IBM SMP computer.

Processors	Advection	Chemistry	Total
1	933	4185	5225
2	478	1878	2427
4	244	1099	1405
8	144	521	799
16	62	272	424

Table 7.4: Speed ups obtained by using 2, 4 and 8 processors in one node as well as by using all 16 processors in two nodes of the IBM SMP computer.

Processors	Advection	Chemistry	Total
2	1.95 (98%)	2.23 (112%)	2.15 (108%)
4	3.82 (96%)	3.81 (95%)	3.72 (93%)
8	6.48 (81%)	8.01 (100%)	6.54 (82%)
16	15.01 (94%)	15.39 (96%)	12.32 (72%)

There is no problem to achieve efficient parallel computations in the chemistry module. The performance of the chemical reactions at a given grid-point is a parallel task. Thus, the number of parallel tasks is equal to the number of grid-points, and concurrent computations are achieved by inserting a directive *"parallel do"* before the loop in which the chemical reactions are performed. This is the straightforward way of exploiting the parallelism. In the previous section it was shown that chunks can successfully be used in an attempt to better exploit the cache memory. The parallel code shown in Fig. 7.6 can be used when the number of processors is eight (assuming here that the number of chunks is a multiple of eight).

The term *"chemical module"*, which is used in Fig. 7.6, refers to some subroutine which contains the body of the outer loop in the fourth code given in the previous section. The fourth code from the previous section, the code given in Fig. 7.4, can also be called directly (with a *"parallel do"* directive before the first loop). The experiments show that the code with parallel sections, which is given Fig. 7.6, performs slightly better.

It is necessary to point out that the parallelization of the advection module does not depend on the numerical method used. The loops in Fig. 7.5 will be parallelized when any numerical method is selected. This means that the code shown in this figure can be considered as **template**, which can be used in connection of any numerical method for designing a parallel code for the advection module. This illustrates once again why the use of splitting techniques is very efficient when large-scale mathematical models are to be handled on modern computers.

```
C
C$OMP parallel sections
        DO ICHUNK=1,NCHUNKS/8
           Call the chemical module for chunk ICHUNK
        END DO
C$OMP section
        DO ICHUNK=NCHUNKS/8+1,2*NCHUNKS/8
           Call the chemical module for chunk ICHUNK
        END DO
C$OMP section
        DO ICHUNK=2*NCHUNKS/8+1,3*NCHUNKS/8
           Call the chemical module for chunk ICHUNK
        END DO
C$OMP section
        DO ICHUNK=3*NCHUNKS/8+1,4*NCHUNKS/8
           Call the chemical module for chunk ICHUNK
        END DO
C$OMP section
        DO ICHUNK=4*NCHUNKS/8+1,5*NCHUNKS/8
           Call the chemical module for chunk ICHUNK
        END DO
C$OMP section
        DO ICHUNK=5*NCHUNKS/8+1,6*NCHUNKS/8
           Call the chemical module for chunk ICHUNK
        END DO
C$OMP section
        DO ICHUNK=6*NCHUNKS/8+1,7*NCHUNKS/8
           Call the chemical module for chunk ICHUNK
        END DO
C$OMP section
        DO ICHUNK=7*NCHUNKS/8+1,8*NCHUNKS/8
           Call the chemical module for chunk ICHUNK
        END DO
C$OMP end parallel sections
C
```

Figure 7.6: Using parallel sections in the chemical module. $NCHUNKS$ **is the number of chunks. "Chemical module" refers to the body of the outer loop in Fig. 7.4.**

Also the code for performing the chemical sub-model parallel computations, which is shown in Fig. 7.6, can be considered as a **template** for parallel computations. Parallel computations will be achieved for any numerical method applied in the chemistry module. However, in this case the loading balance depends on the selected numerical algorithm (this was not the case for the code for the advection module which is shown in Fig. 7.5). If the improved QSSA algorithm as applied in Subsection 4.3.2 is used, then the loading balance is perfect (because the number of arithmetic operations per processor is the same). The situation changes when the classical time integration methods (Subsection 4.3.3 and Subsection 4.3.4) or methods based on partitioning (Subsection 4.3.5 and Subsection 4.3.6) are used. In these two cases non-linear systems are to be solved by some modified Newton iterative procedure (see Remark 4.1 and Remark 4.2). The number of iterations needed to obtain the required accuracy may change from one grid-point to another. This explains why the loading balance is not perfect when these methods are used. The improved QSSA algorithm is used in this chapter.

Some results can be seen under the column "Chemistry" in Table 7.3 and Table 7.4. The speed-ups for the chemical module are usually better than those achieved in the advection module.

Of course, we are mainly interested in the results for the whole air pollution model. These results can be seen under the column "Total" in Table 7.3 and Table 7.4.

7.4 Parallel runs of the code across the nodes

A version of the parallel code for the IBM SP2 computer (IBM SP2 is a distributed memory parallel computer), which is described in Georgiev and Zlatev [117], has been modified for the runs on two IBM SMP nodes. MPI (Gropp et al. [124]), is used in this code.

The main ideas on which the particular version used in this chapter is based can been sketched as follows. Assume that the concentrations are stored in array $C(96, 96, 35)$ and that two nodes are to be used. Array C is divided into two sub-arrays $C_1(1 : 49, 96, 35)$, and $C_2(48 : 96, 96, 35)$. For every chemical compound s, $s = 1, 2, \ldots, 35$, the sub-arrays C_1 and C_2 contain 49 rows. The advection process (united as in the previous sections with the diffusion process) can be carried out concurrently for each sub-array (assuming here that the wind velocities and the arrays containing the matrices induced by the application of the finite elements are appropriately divided into sub-arrays). It is necessary to update (both in the advection part and in the chemical part) the concentrations of rows 1-48 for C_1 and rows 49-96 of C_2. The concentrations of row 49 are only used as an inner boundary condition in the calculations involving C_1, while the concentrations of row 48 are only used as inner boundary conditions during the calculations involving C_2. Since explicit integration methods, described in Zlatev [276] and [282] (see also Chapter 3 and Chapter 6), are used to treat the systems of ODEs arising after the finite element discretization, the values of the concentrations obtained during the last application of the chemical procedure can be used in the inner boundary conditions. Both the advection part and the chemical part can be treated concurrently for every sub-domain (defined by the two sub-arrays) as in the sequential case provided that the inner boundaries are treated by taking the last updates of the concentrations in the chemical procedure. This means that the following communications must be done at each time-step after the chemical part: the contents of row 48 must be communicated from the first node to the second one, while the contents of row 49 must be communicated from the second node to the first one. Thus, communications take place only once per time step and, moreover, only one row per chemical species is to be communicated by each node. The same procedure, as that described in the previous section, is used within each node. It is assumed here that

- a preprocessor module is used to distribute the data to scratch files in the local memories of the two nodes in the very beginning of the computational process, and

- a post-processor module is used to collect the output data from the scratch files of the two nodes to the output files needed.

More details about the preprocessor and the post-processor can be found in Alexandrov et al. [5], Dimov et al. [70], [71], Georgiev and Zlatev [116], [117] and Owczarz and Zlatev [188], [189], [190].

The situation is very similar if more nodes are used; the main difference being that for the inner sub-domains it is necessary to communicate to neighboring processors two rows per chemical species (again this has to be done only once per time-step).

Results obtained by using the parallel algorithm sketched above are given in the last lines of of the IBM SMP computerTable 7.3 and Table 7.4.

7.5 Scalability of the code

It is important *to preserve* the efficiency of the code when the size of some of the involved arrays is increased (for example, as a result of refining the grid, increasing of the number of chemical compounds, the transition from the two-dimensional version to a three-dimensional version, etc.). This property is often referred to as a scalability of the code. While such a property is highly desirable (the requirements to the air pollution codes are permanently increasing and, therefore, refining the grids is becoming more and more desirable), it is by no means clear in advance whether the code has such a property or not when the modern complicated computer architectures are in use.

Some experiments were performed in an attempt to check the scalability of the parallel devices discussed in the previous sections of this chapter. A (288×288) grid was used in the experiments together with the (96×96) grid considered in

Table 7.5: Computational times obtained by using a refined (288×288) grid and a coarse (96×96) grid. The ratios of the times for the refined and coarse grids are given in the last column. The two codes were run on 16 processors of the IBM SMP computer, the size of the chunks was 48.

Process	(288×288)	(96×96)	Ratio
Advection	1523	63	24.6
Chemistry	2883	288	10.0
Total	6209	424	14.6

the previous sections. This corresponds to a transition from cells of size ($50\ km \times 50\ km$) to cells of size ($16.67\ km \times 16.67\ km$). The number of chemical species was kept 35. This means that in this version of the code the number of grid-points was increased by a factor of 9.

In the advection part, we had also to decrease the time stepsize by a factor of 3 in order to preserve the stability of the computations. Thus, the number of arithmetic operations (or, in other words, the amount of computational work) is roughly speaking increased by a factor of 27 in the advection part.

There was no need to decrease the time stepsize in the chemical part. This means that the number of arithmetic operations (or, in other words, the amount of computational work) is increased, roughly speaking, by a factor of 9 in the chemical part.

This short analysis indicates that if the code is scalable, then the computational times should be increased by factors approximately equal to 27 and 9 in the advection part and the chemical part respectively. The results given in Table 7.5 show clearly that this is precisely what happens (see the ratios between the computational times for the refined grid and the coarse grid in Table 7.5). The conclusion from this experiment is that the code is scalable. This conclusion was further supported

- by some runs with the more precise version obtained by using a (480×480) grid, and
- by performing some runs with the three-dimensional version obtained by using a ($96 \times 96 \times 10$) grid.

7.6 When is it most desirable to improve the performance?

The computers are becoming faster and faster and their memory discs are becoming bigger and bigger. Many of the problems which only several years ago had to be solved on big supercomputers can now be treated on workstations or even on big PCs. Therefore it is reasonable to ask the question:

> *"Why and when is it necessary to improve the performance?"*

This question is further justified by the fact that the new supercomputer architectures are becoming more and more complicated and, thus, it is more and more difficult to achieve efficiency when big application codes are run on the new and modern supercomputers. It is true, however, that if the work is properly done, then the code is becoming very efficient.

Let us reiterate here that the results from the first run of the code on the IBM SMP computer were very poor (see Table 7.1). The total computational time for this non-optimized code was 19432 seconds. We should like to compare this time with the best computational times obtained on the IBM SMP computer when the two versions of the code (defined on a coarse and refined grids respectively) are run. The results, computational times and ratios of improvements when compared with the non-optimized version, are given in Table 7.6.

Table 7.6: The best results obtained by using a refined (288×288) grid and a coarse (96×96) grid on 16 processors with chunks of size 48. The ratios of the total time (19432) given in Table 7.1 and the times in the second column of the present table are given in the third column.

Process	Comp. time	Ratio
(96×96)	424	45.8
(288×288)	6209	3.1

Most of the runs, results from which were presented until now in this chapter, were performed by using meteorological data covering a period of one month (July 1994). However, the model is run on much longer time periods when comprehensive studies are to be carried out. Runs with meteorological data covering a time period of ten years (1989-1998) have been carried out with the Danish Eulerian Model (see Zlatev et al. [296], [298] and Chapter 9 of this book). Recently, 24 scenarios with these data have been completed to study the influence of the biogenic emissions on the formation of ozone in different part of Europe (see Geernaert and Zlatev [112], [113] and Chapter 8). This is a very comprehensive task; 2880 times bigger than the simple experimental task used in the previous sections. More than two years CPU time will be necessary to run the 24 scenarios with meteorological data covering a timer-period of ten years when the non-optimized version of the code is used. While it is perhaps theoretically possible to do this, it is obvious that it is quite impossible to complete such a comprehensive task in practice. At the same time, this task was resolved, very successfully, with the optimized version in several weeks. Some results about the influence of the biogenic emissions on the high ozone levels in Europe were reported at one of the GLOREAM annual workshops, which was held in September 2000 in Cottbus, Germany; as mentioned above more details for this comprehensive study can be found in Geernaert and Zlatev [112] and [113] and Chapter 8.

The results in Table 7.5 and Table 7.6 show that even the refined, on a (288×288) grid, version can be used in the solution of some realistic tasks (but not yet in the solution of the task to run 24 scenarios over a period of ten years).

Finally, even finer resolution is sometimes desirable. Some runs have recently been performed by using a version of the code discretized on a (480×480) grid. Here, also highly optimized versions of large-scale air pollution models are required.

The main conclusion from the discussion given in this section is that, although there are more and more problems which can successfully be handled (without any attempt to optimize the computational process) on workstations and PCs, it is still necessary, when the problems are very large, to run **highly optimized codes on big modern supercomputers**. The set of problems, which could be treated on workstations and PCs will become bigger and bigger in the future, because the computer power of the workstations and PCs will continue to increase. However, the requirements for greater accuracy (i.e., the need to calculate more accurate and, thus, more reliable results) and for obtaining more detailed information are also steadily increasing. This means that the models are becoming bigger and bigger. Of course, this is a well-known fact to the specialists. As an illustration of this fact see again the quotation from the paper of Jaffe [144] in Remark 1.3 in the end of Chapter 1. In this paper Jaffe stated (about 20 years ago) that the computers will always be too slow for the scientists who are dealing with very large applications. He was right, is still right and will certainly continue to be right in the future: there will always exist very advanced and very important, to modern society, scientific problems which cannot be solved even if the fastest available computers are used.

7.7 Unification of the different versions of the model

Recently the Danish Eulerian Model, DEM, was upgraded in a new more powerful model, the Unified Danish Eulerian Model (UNI-DEM). All existing versions of DEM as well as several new versions were united in this new model and all these versions could now be specified as options in the new model **UNI-DEM**. The most important features of UNI-DEM are listed below.

- The new model can be used both as a two-dimensional model and as a three-dimensional model. Ten layers are used in present in the three-dimensional option.
- Different resolutions in the horizontal planes can be specified. Three horizontal grids can be chosen at present: 96×96, 288×288 and 480×480. This leads

to the use of 50 $km \times 50\ km$, 16.67 $km \times 16.67\ km$ and 10 $km \times 10\ km$ cells respectively.

- Different chemical schemes can be selected (chemical schemes containing 35, 56 and 168 chemical compounds are available at present).
- Dividing into chunks can automatically be applied in an attempt to exploit better the cache memory. The user can specify the number of chunks that are to be applied in the job. The use of small chunks is also resulting in reduction of the storage requirements.
- Parallelization based on the use of MPI is available. The OpenMP versions are not yet included in UNI-DEM.

It is important to emphasize here that the different options can be specified by selecting proper values of a few parameters in a very short input file containing no more than ten records. For each parameter used in UNI-DEM (which has to be initialized in the input file) a list of the valid values are given. For example, it is at present allowed to use values 96, 288 and 480 for parameter N_x. It is stated that only square grids are allowed (i.e., N_y must be set equal to N_x). Similar instructions are given for the other key parameters. Thus, it is not necessary either

- to recompile the code, or
- to use another code

if the options are to be changed. The whole work can be done by resetting the values of a few parameters in the input file. Great flexibility is achieved in this way when the code is to be used for different purposes in different comprehensive air pollution studies.

UNI-DEM was run on several SUN computers at the Danish Centre for Scientific Computing (DCSC) as well as on the IBM computer at the University of Århus. The chemical scheme with 35 chemical species was used in all runs, the results of which will be presented in the remaining part of this chapter. The size of the computational tasks for the different options which were used in the runs are given in Table 7.7. The number of time-steps that are needed to complete a job covering meteorological data for one year (the actual year being 1997 and the model was in fact run on time interval of 370 days, because 5 days are needed to start-up the model) are given in Table 7.8. All runs were performed by using 8 processors. The computational times, measured in CPU hours, are given in Table 7.9.

The runs performed on the DCSC's SUN computers, the results of which have been shown in Table 7.7 to Table 7.9, indicate that the following conclusions can be drawn.

- It is possible to use different options of UNI-DEM on the SUN computers. However, more powerful processors and more processors will be needed if

Table 7.7: Numbers of equations when six options of UNI-DEM are used with 35 chemical compounds.

Horizontal spatial resolution	2-D version	3-D version
(96×96)	322 560	3 225 600
(288×288)	2 903 040	29 030 400
(480×480)	8 064 000	80 640 000

Table 7.8: Numbers of time-steps when six options of UNI-DEM are used with 35 chemical compounds.

Horizontal spatial resolution	2-D version	3-D version
(96×96)	35 520	35 520
(288×288)	106 560	106 560
(480×480)	313 120	313 120

Table 7.9: Computational times in CPU hours when six options of UNI-DEM are run on eight SUN processors.

Horizontal space resolution	2-D version	3-D version
(96×96)	1.41	12.39
(288×288)	19.56	152.72
(480×480)	98.72	856.75

the three-dimensional options with fine horizontal resolution are to be used in comprehensive air pollution studies that require runs with many different scenarios.

- The use of more processors is highly desirable (especially for the expensive, with regard to computational time, jobs). The use of 32 processors (instead of eight processors) for the most expensive job (the three-dimensional option of UNI-DEM discretized on a 480×480 grid in the horizontal planes) leads to a reduction of the computational time from 856.75 CPU hours (see Table 7.9) to 192.02 CPU hours (i.e., the speed-up is super-linear, 4.46).

- It is practically impossible to run the most time consuming jobs on one processor. Indeed, it will be necessary to use about 286 days in order to complete the most time expensive run (the three-dimensional option of UNI-DEM discretized on a 480×480 grid). It is clear that such a job will never be completed in practice, because the computer cannot run continuously for 286 days (without re-booting, with no upgrading, with no crashes, etc.).

- The results shown in Table 7.9 can be considered as another illustration of the scalability of the code.

7.8 OpenMP implementation versus MPI implementation

The SUN computers used to calculate the results in previous section, are shared memory machines. Therefore, it is natural to expect that the OpenMP versions of the code will be more efficient than the MPI versions. However, the MPI versions are in fact more efficient. This can be explained as follows. It is important to emphasize here that the introduction of p sub-domains (p being the number of processors which are to be used in the job) leads to a reduction of the main arrays by a factor of p. Consider as an illustration the major arrays used in the chemical sub-model when a two-dimensional option of UNI-DEM is run. The dimensions of these arrays are reduced from $(N_x \times N_y, N_s)$ to $(N_x \times N_y/p, N_s)$. It is clear that this is equivalent to the use of p chunks; see Section 7.2. Chunks of length $N_x \times N_y/p$ are still very large. Therefore, the division into chunks described in Section 7.2 has also to be used (in each sub-domain related to the chemical sub-model) when the MPI versions are used. However, the reduction of the arrays leads to reductions of the copies that are to be made in the beginning and in the end of the algorithm based on chunks from Section 7.2 (see Fig. 7.4). Thus, the reduction of the arrays leads to a better utilization of cache memory.

The automatic reduction in the sizes of the involved arrays and the resulting better utilization of the cache memory, make the MPI versions attractive, even when shared memory machines are available. The MPI versions of UNI-DEM are often performing better than the corresponding OpenMP versions, when the SUN computers at DCSC (Danish Centre for Scientific Computing) are used.

Some results are given in Table 7.10 in order to illustrate the fact that the reduction of the leading dimension of arrays when the MPI versions are used results not only in a considerable reduction of the storage requirements, but also in reduction of the computational times when very large problems are to be handled (the use of refined grid requires the use of a smaller time stepsize in the advection module, while the time stepsize needed in the chemical module remains the same as that used when the coarse grids are used).

It should be noted that for these runs the advection-diffusion part is becoming more expensive than the chemical part.

Table 7.10: Running DEM discretized on a $480 \times 480 \times 1$ grid on 8 processors by using the MPI version and the OpenMP version. The computational times are given in seconds. The time period for these two runs was one year.

Process	MPI version	OpenMP version
ADV	822291	1663812
CHEM	393158	596920
COMM	255785	-
TOTAL	1782752	2614983

7.9 Parallel computations for general sparse matrices

The algorithm for treating general sparse matrices on modern computers with multi-hierarchical memories, which was described in Chapter 6, is very useful in the attempts to exploit more efficiently the cache memory of the computer available. However, it is much more important that this algorithm can very easily be used for parallel computations in the OpenMP mode. This is achieved by a simple insertion of *"parallel do"* directives in several places of the code.

First and foremost, the most time consuming operations related to the factorization of the matrix (these operations are carried out in Step 3 and Step 4 of the algorithm described in Chapter 6) have to be parallelized by using the directive *parallel do"*. The use of this directive in the parallelization of the computations in Step3 and Step 4 is very successful; some results are presented in Table 7.11 and Table 7.12.

OpenMP directives were used in several other places of the code, but without a great success (see the speed-ups obtained in the ordering and iterative parts of the code, which are shown in Table 7.11 and in Table 7.12).

Finally, the code was compiled by selecting high optimization options (which means that the compiler tries to automatically parallelize some loops), but this was not very successful either; see again the numerical results, which are presented in Table 7.11 and Table 7.12.

The statements made above indicate that one should not rely on automatic parallelization and/or on the use of simple OpenMP directives when general sparse matrices are to be handled. The development of special algorithms,

- which are designed for parallel computations, and
- by which an efficient exploitation of the cache memory is achieved,

is much more important in this case.

7.9.1 Some numerical examples

A very large systems produced by the matrix generator used in the previous chapter, a system with 32768000 equations and with 327680110 non-zero elements in its coefficient matrix, was run on up to eight processors. Results are shown in Table 7.11 and in Table 7.12. The same notation as that applied in the tables presented in Section 6.6 is used in the tables given in this section (i.e. ORD time, FACT time, SOLV time and TOTAL time refer to the reordering time, the factorization time, the solution time and the total time, respectively).

The speed-ups have also been calculated by taking the ratios of the computational times obtained in the runs on one processor and the computational times obtained on p processors, $p = 2, 4, 8$. The speed-ups are shown in Table 7.12.

It is seen, from the results shown in Table 7.12, that good speed-ups are achieved in the factorization part. The speed-ups for the total times are also good (because the factorization times are much larger that the times spent for the remaining parts of the computational work).

The speed-ups for the ORD times and the SOLV times are rather poor. We have not developed special techniques for the parallelization of these two parts (excepting the fact that some directives were inserted before several loops in these parts; however, the effect of this attempt to obtain parallelization is, as mentioned above, minimal). It is not very clear at present how to improve the parallel performance in these two parts. Fortunately, the computational time spent in these two parts is much smaller than the computational time spent in the factorization part when the matrices treated are very large. Nevertheless, some efforts have to be carried out in order to improve the parallelization in the ordering part and in the solution part when large systems of linear algebraic equations the coefficient matrices of which are general sparse matrices.

It is interesting to investigate whether the performance will become poorer or not when the order of the matrices becomes larger. An experiment with a matrix of order 65536000 and with 655360110 non-zero elements have been run in order to answer this question. The results from these runs are given in Table 7.13 (computational times) and in Table 7.14 (speed-ups).

The following conclusions can be drawn by using the results shown in the Table 7.13 and Table 7.14:

- the behaviour of the ordering part and the solution part of the code remains the same as for the smaller matrix of order 32768000 (i.e., the computational times are slightly decreasing when more processors are used, while the speed-up remains slightly greater than one), and

- the computational time for the factorization part and the total computational time are reduced by factor approximately equal to two when twice more processors are used (which is a very desirable property of the code).

Table 7.11: Computational times (measured in seconds) obtained in the solution of a system of 32768000 linear algebraic equations when different numbers of processors are used. The number of non-zero elements in the coefficient matrix is 327680110. These runs were performed by using $RELTOL = 0.1$. The number of blocks used in these runs is $q = 512$.

Processors	ORD time	FACT time	SOLV time	TOTAL time
1	584	17710	353	18647
2	541	8911	273	9725
4	558	4521	247	5327
8	513	2346	223	3083

Table 7.12: Speed-ups obtained in the solution of a system of 32768000 linear algebraic equations when different numbers of processors are used (the efficiency achieved is given in brackets). The number of non-zero elements in the coefficient matrix is 327680110. These runs were performed by using $RELTOL = 0.1$ The number of blocks used in these runs is $q = 512$. The computational times for the same runs are given in Table 7.11.

Processors	ORD time	FACT time	SOLV time	TOTAL time
2	1.08 (54%)	1.99 (99%)	1.29 (64%)	1.92 (96%)
4	1.05 (26%)	3.92 (98%)	1.43 (36%)	3.50 (87%)
8	1.14 (14%)	7.55 (94%)	1.58 (20%)	6.05 (76%)

The main conclusion is that the code is producing very good speed-ups when very large matrices are run on the SUN computers of the Danish Centre for Scientific Computing (DCSC).

7.9.2 Comparisons with another code for sparse matrix computations

As in Chapter 6, we shall compare the results obtained by our algorithm with results obtained by **SuperLU** (see Demmel et al. [61], [62] and Li and Demmel [165].

As stated in Chapter 6, dropping of small non-zero elements is not used in **SuperLU**. Therefore, we had to use small matrices in this section (as in the comparisons performed in Chapter 6). In fact, the largest matrix which could be run by **SuperLU** when eight processors are used is of order 1024000. It should be mentioned here that the Fortran interface in the **SuperLU** package has been used in all runs, results of which are given in Chapter 6 and in this chapter. It should also be mentioned that larger matrices (of order up to 4096000) can be handled by using our codc without dropping small non-zero elements. The ability of the code

Table 7.13: Computational times (measured in seconds) obtained in the solution of a system of 65536000 linear algebraic equations when different numbers of processors are used. The number of non-zero elements in the coefficient matrix is 655360110. These runs were performed by using $RELTOL = 0.1$. The number of blocks used in these runs is $q = 512$.

Processors	ORD time	FACT time	SOLV time	TOTAL time
1	1085	70964	676	72724
2	1008	35532	540	37081
4	979	17871	473	19323
8	959	9050	441	10452

Table 7.14: Speed-ups obtained in the solution of a system of 65536000 linear algebraic equations when different numbers of processors are used (the efficiency achieved is given in brackets). The number of non-zero elements in the coefficient matrix is 655360110. These runs were performed by using $RELTOL = 0.1$ The number of blocks used in these runs is $q = 512$. The computational times for the same runs are given in Table 7.13.

Processors	ORD time	FACT time	SOLV time	TOTAL time
2	1.08 (54%)	1.99 (99%)	1.25 (63%)	1.96 (98%)
4	1.11 (28%)	3.97 (99%)	1.43 (36%)	3.76 (94%)
8	1.13 (14%)	7.84 (98%)	1.53 (19%)	6.96 (87%)

to solve much larger systems might be increased very considerably when small non-zero elements are dropped during the computations. This has been demonstrated in Chapter 6 and also in the previous section of this chapter.

Results obtained when a system of 1024000 linear algebraic equations is solved are given in Table 7.15 and Table 7.16. The number of non-zero elements is 10240110; i.e., again the average number of non-zero elements per row is 10.

Computational times obtained in the parallel runs are given in Table 7.15. Speed-ups are given in Table 7.16.

It is seen (see the second column in Table 7.16) that the direct solution version of our code is not performing well when the matrices are not very large. The reason for this is that perhaps the large blocks are not very well balanced in this case. The situation becomes slightly better when the preconditioned version is used (see the third column of Table 7.16). However, the speed-ups obtained by **SuperLU** are also in this case better (see the first column of Table 7.16).

On the other hand, the computational times for the direct version of our code

Table 7.15: Comparison of computational times obtained by SuperLU with results obtained by the code in which the cache memory is exploited by using block-rows. The latter code was run both with dropping and without dropping of small non-zero elements. $RELTOL = 0.1$ is used to drop small elements (the number of blocks is $q = 512$ in this case). Dropping is switched off by setting $RELTOL = 0$ (the number of blocks is $q = 80$ in this case; this choice is probably not optimal).

Processors	SuperLU	Without dropping	With dropping
1	277	217	92
2	162	158	62
4	115	143	45
8	101	140	41

Table 7.16: Speed-ups obtained by SuperLU with results obtained by the code in which the cache memory is exploited by using block-rows. The latter code was run both with dropping and without dropping of small non-zero elements. RELTOL = 0.1 is used to drop small elements (the number of blocks is q = 512 in this case). Dropping is switched off with RELTOL = 0 (the number of blocks is q = 80 in this case; this choice is probably not optimal) The efficiency achieved in the parallel runs is shown in brackets.

Processors	SuperLU	Without dropping	With dropping
2	1.71 (86%)	1.37 (69%)	1.48 (74%)
4	2.41 (60%)	1.52 (38%)	2.04 (51%)
8	2.74 (34%)	1.55 (19%)	2.24 (28%)

are comparable with the computational times for **SuperLU** (compare the results given in the second and the first columns in Table 7.15), while the preconditioned version is performing clearly better for this matrix (compare the results given in the third and the first columns in Table 7.15).

7.10 Applicability to other large-scale models

Some very large models can be treated only when parallel computations are efficiently implemented. On the other hand, the parallelization of a large model is still not an easy task. Therefore, one should try

- to use only standard parallelization tools (this will allow the user to move, in a relatively easy way, the code from one computer to another), and

- to prepare some templates for parallel computations, which can be used when different numerical methods are applied in the discretization of the model under consideration.

Only standard tools for parallelization (such as OpenMP and MPI) are used in the code. It is also worthwhile to reiterate here that the **templates**, which were presented in this chapter are quite general and standard devices.

This short discussions shows clearly that the parallel devices studied in this chapter can successfully be used in the parallelization of other models (and, first and foremost, the models discussed in Section 7 of Chapter 1 not only because these models lead to systems of PDEs that are similar to (1.1), but also because only standard tools and devices were discussed in the previous sections of this chapter.

7.11 Concluding remarks and plans for future work

The optimization of a large air pollution code for runs on several parallel computers have been described in the previous sections of this chapter. The results illustrate the well-known fact that the computations should be carefully organized in order to better exploit the great potential power of the modern high-speed computers (compare the results in Table 7.1 with the results in the other tables). If this is not done then the results can be very bad (see again the results obtained without using chunks on one processor in Table 7.1). On the other hand, a successful reorganization of the computations can improve the performance considerably in most of the cases.

The speed-ups, which were obtained when large sparse matrices are run on several parallel computers, are quite satisfactory for all runs. In fact, they are in the most of the cases close to linear. In some cases the speed-ups are even super-linear. However, the optimal for the available computers speed of computations is still not sufficient to treat the large air pollution models when

- the grid is refined,
- three-dimensional versions of the models are to be used, and
- long simulation processes containing many hundreds of scenarios are to be performed.

Much faster computers (perhaps IBM SMP computers with many more nodes or SUN computers with more processors and more powerful processors) are needed in the situations listed above.

The choice of numerical methods is not very important for achieving efficient parallel computations; see Section 7.2 to Section 7.5. From the contents of these four sections it is clear that all devices used in this chapter can also be applied when any of the numerical methods discussed in Chapter 3 to Chapter 6 is used. The structure of the parallel sections will remain the same when other numerical methods are used.

This is why nearly optimal speed-ups will be achieved for many other numerical algorithms. The size of chunks for which best results can be achieved will in general depend both on the methods selected and on the particular computer used. It is easy to find the best size by performing several experiments where the size of the chunks is varied (similar to experiments carried out in Section 7.2 to Section 7.5).

The ideas used in this chapter are pretty general and could also be efficient on other computers too. Thus, it will be rather easy to port this code to other parallel computers on which MPI and OpenMP are available. PVM tools, see Geist et al. [114], can also be used on message passing computers. In our experiments the MPI versions of the code performed better than the PVM versions.

Three major tasks have to be solved in the attempts to run efficiently a large application code (not necessarily an air pollution code):

- one has to run the code on a fast computer,
- one has to select fast numerical methods, and
- one has to organize very carefully the computational process.

The last of these three tasks is very often underestimated. In this chapter we have shown (by fixing both the computer used and the numerical methods selected) that a success in the efforts to solve the third task in a more efficient manner leads to big savings in computational time. Indeed, the total time for the solution of the problem without any optimization was 19432 seconds (see Table 7.1), while the best achievements (obtained without changing the computer, the compiler options or the numerical methods) was 5225 seconds on one processor (see Table 7.2) and 424 seconds on 16 processors (see Table 7.3). Thus, only by improving the solution of the third task, i.e., by reorganizing the computations, is it possible to achieve results which can be achieved by the non-optimized code only if many more processors are available; the number of processors should (roughly speaking) be greater by more than a factor of two or even three if we want to obtain the same computational times as when the optimized version is run. The crucial role of the proper organization of the computations in the efforts to improve the performance of a big application code on parallel architectures is the most important conclusion from the results presented in Section 7.3 to Section 7.6.

The construction and the implementation of codes for the treatment of very large general sparse matrices, which is mainly relevant when one attempts to avoid the application of splitting procedures, is a very difficult problem. One algorithm, by which an attempt

- to exploit efficiently the cache memory, and
- to parallelize the computations has been described in Chapter 6.

Some results from runs on parallel computers were given in this chapter. The results indicate that in spite of the good speed-ups achieved it is necessary to improve further the algorithm (or to design other algorithms) in order to be able to run successfully large-scale air pollution models (or models arising in other fields of science and engineering) on parallel computers. Furthermore, the codes for sparse matrices that are described in this book are based on the use of OpenMP tools. It has been shown in this chapter (see Section 8) that in some cases MPI tools are more efficient than OpenMP tools. Therefore, the development of a code for general sparse matrices that is based on MPI tools is desirable. For the code described in Section 6, this task (the development of a code based MPI technology) should not be very difficult, because the data is already divided into sub-domains in the attempt to better exploit the cache memory.

One of the great challenges in the near future will be the utilization of computer grids in the treatment of very large applications. It was pointed out, in several places in this book, that there are many problems which we want to solved, even problems that must be solved, but it is impossible to handle these problems on the presently available computers. Such problems must probably be handled on computer grids. Computer grids open the way to resolving a series of challenging tasks. As an example let us mention the task of running several loosely connected large-scale models in order to treat a set of complex problems involving

- weather forecasts on different regional scales (starting with results obtained on a global scale),
- weather forecasts on an urban scale (perhaps in parallel for several urban areas),
- air pollution forecasts on different regional scales,
- air pollution forecasts on an urban scale (perhaps in parallel for several urban areas),
- treatment of the output results in order to prepare them for the people who will use them (utilizing data mining algorithms and high-speed visualization tools),
- sending the relevant data to appropriate media (TV stations, radio stations, Internet sites, GMSs, etc.).

It is clear that computer grids will be very powerful tools in the solution of the very challenging task related to the possibility to treat efficiently the set of problems described above. Such sets of problems are at present solved only by imposing many simplifying (and very often not physical) assumptions. At the same time, it is also clear that a lot of difficulties must be overcomed in the efforts to run such complex tasks efficiently on a computer grid. The greatest difficulties are the tasks of

- achieving reliable and robust transition from one scale to another,

- communicating relevant data from one part of the computational grid to another, and
- preparing the **final** results, which should be easily understandable by the recipients.

Several additional comments related to computer grids and to the utilization of such grids in very large application problems related to air pollution modelling will be given in Chapter 11.

Chapter 8

Studying high pollution levels

Many questions must be answered when the damaging effects caused by high pollution levels are studied. The following three questions are very important and reliable answers to these questions are highly desirable.

- Are the present pollution levels under the prescribed critical levels? Critical levels are such pollution levels which should not be exceed (because the exceedance of these levels will cause damages on plants, animals, eco-systems and/or human health). The critical levels are established on the basis of the present knowledge. Many critical levels are introduced in the European Union (EU); see, for example, the EU Ozone Directive, [96].

- What is the trend of the development of the pollution levels in different parts of the world?

- If some pollution levels are over the critical levels (or if the trends indicate that they will be over the critical levels in the near future), then what are the most appropriate measures that can (must) be taken in the efforts to keep our environment safe?

The Unified Danish Eulerian Model, UNI-DEM, has been used in many comprehensive studies related to air pollution levels in Europe as well as in several European countries (for example, Denmark, Hungary and Bulgaria) in an attempt to get some answers to the three questions stated above. The most important of these studies are listed below.

- Studying the impact of future climate changes on the pollution levels in Denmark and in Europe (see Dimov et al. [72]).

- Studying the influence of the biogenic (natural) emissions on high ozone levels in Europe (see Geernaert and Zlatev [112], [113]).

- Economical estimates of losses of crops in Denmark and in Bulgaria that are caused by high ozone levels (see Dimov et al. [73] and Zlatev et al. [296]).
- Long-term variations of the pollution levels in Denmark, Hungary, Bulgaria and Europe (see Ambelas Skjøth et al. [9], Bastrup-Birk et al. [18] Dimov et al. [73], Havasi and Zlatev [131] and Zlatev et al. [296]).
- Studying the relationship between the human-made emissions and the pollution levels in Denmark, Hungary, Bulgaria and Europe (see again Ambelas Skjøth et al. [9], Bastrup-Birk et al. [18] Dimov et al. [73], Havasi and Zlatev [131] and Zlatev et al. [296]).
- Studying the frequency of appearance of pollution levels that exceed the established by EU critical levels and, thus, might lead to damages on plants, animals and human health (see Geernaert and Zlatev [112], [113] and Havasi and Zlatev [131]).
- Studying the contribution of emission sources from a given area in Europe to another area (for example, studying the contribution from the European emission sources outside a given country to the pollution levels in this country) (see Ambelas Skjøth et al. [9], Bastrup-Birk et al. [18], Dimov et al. [73], Havasi and Zlatev [131], Zlatev et al. [296], Zlatev and Syrakov, [299], [300]).

It is impossible to give full descriptions of all these studies here. However, some selected results will be presented in this chapter. Then the important topic of the impact of future climate changes on the air pollution levels in Denmark and Europe will be discussed in some more detail in the next chapter.

8.1 Exceedance of some critical levels for ozone

High ozone concentrations can cause damage to human health. Therefore some critical levels for ozone have been established in the European Union ([96]). Some of these critical levels are legislated in the EU Ozone Directive (see again [96]). Assume that c_{max} is the maximum of the 8-hour averages of the ozone concentrations in a given day at a given site. If $c_{max} > 60$ *ppb*, then the day under consideration is declared as a "bad" day for the site under consideration. "Bad" days have damaging effects on some groups of human beings (for example people who suffer from asthmatic diseases). Therefore, the number of such days should be reduced as much as possible. Two aims are stated in the EU Ozone Directive:

- **Target aim.** The number of "bad" days in the European Union should not exceed 25 after year 2010.
- **Long-term aim.** No "bad" day should occur in the European Union (the year after which the long-term aim has to be satisfied is not specified in the EU Ozone Directive).

Results showing the distribution of the numbers of "bad" days in Europe have been obtained using UNI-DEM for different years see Zlatev et al. [298]). The situation in 1995 is shown on the upper left-hand-side plot in Fig. 8.1. It is much easier to see different details within the area under consideration if colour plots are used instead of plots in black-white. A colour version of Fig 8.1 is given in Appendix A (see Fig. 8.1.col there and note that colour versions of the other two figures in this chapter are also given in Appendix A). The comments given below refer mainly to the colour version, Fig. 8.1.col, of Fig. 8.1. The use of different colours in Fig. 8.1.col can be explained as follows:

- The areas in Europe where the long-term aim of the EU Ozone Directive is satisfied are given in light green.
- The regions, where the target aim is satisfied are coloured in dark green.
- In the areas coloured in yellow the critical level (the target aim) is exceeded, but by no more than 50%.
- The exceedances of the target aim of the critical level are between 50% and 100% in the areas coloured orange.
- Finally, areas in which the critical level (the target aim) is exceed by greater than 200% are given in red.

8.2 How to reduce the number of "bad" days?

It is clear (see again the upper left-hand-side plot in Fig. 8.1 or the corresponding plot in Fig. 8.1.col) that the target aim for 2010 is not satisfied in some parts of Europe. Therefore, some measures are to be taken in an attempt to satisfy the target aim in all countries of the European Union. Plans to reduce the human-made emissions are one of the means which can be used in the efforts to reduce to acceptable levels the pollution levels. These reductions, the so-called Scenario 2010 (Amann et al. [8]), have been applied in a second run of UNI-DEM. The results are shown in the upper right-hand-side plot of Fig. 8.1 (and in the corresponding plot in Fig.8.1.col). It is seen that the reduced human-made emissions, Scenario 2010, do lead to considerable reductions of the ozone levels. However, it is also seen that the major aim, to reduce the number of "bad" days to less than 25 in the whole of Europe, is still not achieved.

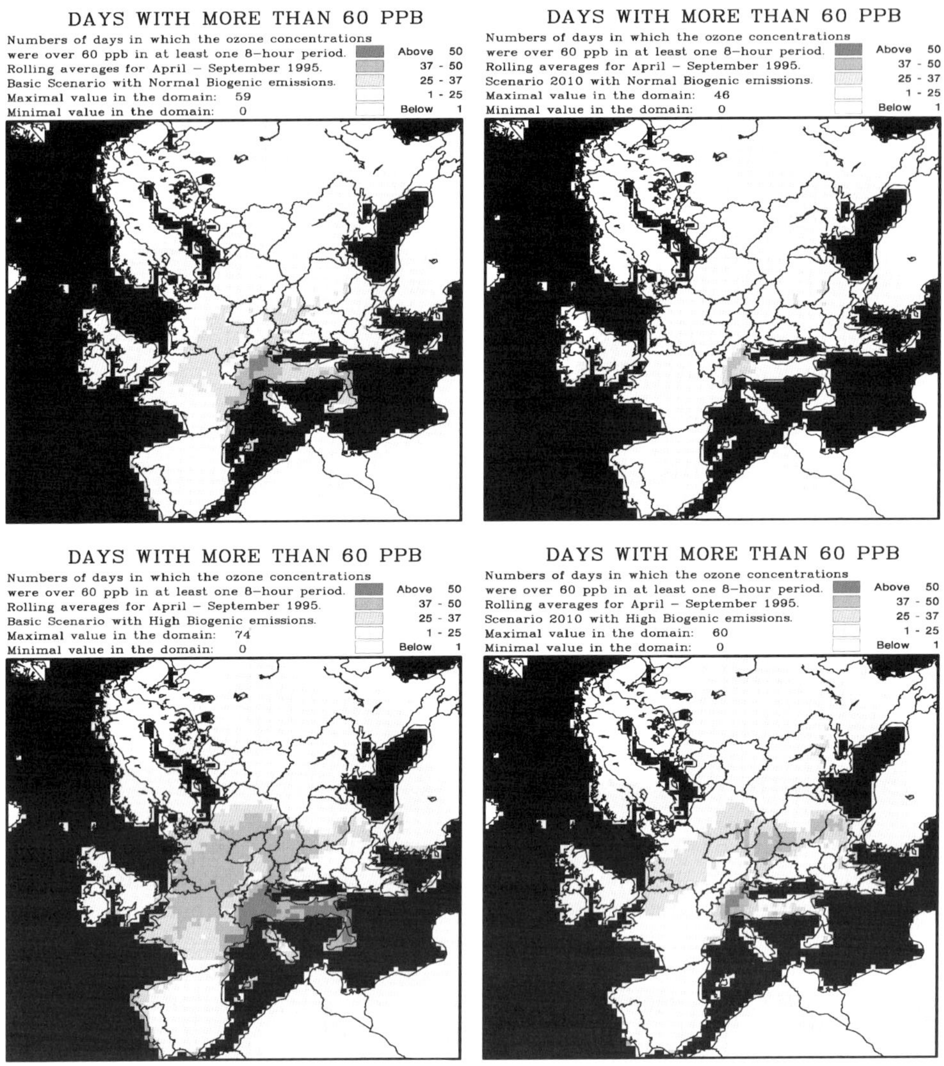

Figure 8.1: Distribution of "bad" days in Europe: (a) results for 1995 human-made emissions with normal biogenic emissions are given on the upper left-hand-side plot, (b) results for the 2010 human-made emissions with normal biogenic emissions are given on the upper right-hand-side plot, (c) results for 1995 human-made emissions with high biogenic emissions are given on the lower left-hand side plot and (d) results for 2010 human-made emissions with high biogenic emissions are given on the lower right-hand side plot. The areas in black are representing the water areas in the model domain (i.e. the black colour is not referring to "days with more than 60 ppb"). A colour version of this figure is given in Appendix A (see Fig. 8.1.col there).

8.3 Influence of the biogenic VOC emissions on the ozone pollution levels

The results shown in the two upper plots of Fig. 8.1 (and in the corresponding plots in Fig.8.1.col) are obtained by biogenic emission inventories calculated by an algorithm based on some ideas from Lübkert and Schöpp [169] and Simpson et al. [225]. The biogenic emissions calculated by this algorithm (they are called "Normal Biogenic Emissions" in Fig. 8.1 and in Fig. 8.1.col) are much smaller than the human-made emissions and, thus, have not a great influence on the ozone pollution levels. However, the uncertainties are very big (see Section 4 in Simpson et al. [225]).

It is worthwhile to test the influence of the biogenic emissions on the ozone levels. To do this we used some suggestions made in Anastasi et al. [10] in order to produce another scenario for the biogenic emissions. According to this scenario, the biogenic emissions from forest trees are roughly speaking increased by a factor of five, while the biogenic emissions from crops were increased by a factor of approximately 20. The scenario for biogenic emissions obtained in this way is called "High Biogenic Emissions" in the two lower plots of Fig. 8.1 (and in the corresponding plots in Fig. 8.1.col).

The two scenarios for human-made emissions (the emissions for 1995 and the expected emissions for 2010) were run with the scenario "High Biogenic Emissions". The results are shown in the two lower plots in Fig. 8.1 (and in the corresponding plots in Fig. 8.1.col).

The results shown on the lower two plots indicate clearly that the influence of the biogenic VOC emissions is considerable and, thus, should not be neglected.

It should be emphasized here that the results presented in this section are only a small part of the results which were obtained by running the 24 scenarios mentioned above. More results from this comprehensive study of the influence of the natural (biogenic) emissions on the high ozone levels in Europe can be found in Geernaert and Zlatev [112], [113].

More and more scientists (see, for example, Bouchet et al., [25], [26]) claim that the biogenic emissions used at present are greatly underestimated. This fact and the results presented in Fig. 8.1 (and in Fig. 8.1.col) indicate that the efforts to control the ozone levels by reduction of the human-made emissions as in Scenario 2010 may not be sufficient to achieve the target aim in the EU Ozone Directive.

The main conclusion is that much more research is needed in this area in the efforts to obtain more reliable and more robust results.

8.4 Studying the transport of pollutants to a given country

It is important to know how big a portion of the pollution in a given country is caused by emission sources that are located outside the country. Several studies related to this topic were carried out. Results related to the contribution of the European sources, which are outside Hungary, to pollution levels in Hungary were presented in Havasi and Zlatev [131].

Some results showing the contribution of the European sources, which are outside Bulgaria, to pollution levels in Bulgaria are given in Fig. 8.2 and, in colour, in Fig. 8.2.col in Appendix A. More details about the influence of the European emission sources on the pollution levels in Bulgaria can be found in Zlatev and Syrakov, [299], [300]).

The SO_2 and NO_x emissions in Europe for 1997 are given in the upper two plots of Fig. 8.2 (and in the corresponding plots in Fig. 8.2.col). It is seen that the SO_2 emissions in Bulgaria are rather high (comparable with the highest emissions in Europe). It is also seen that the largest value of the SO_2 emissions in Europe is rather high (1555 Ktonnes). It should be mentioned here, however, that this happens only in one grid-cell in Sicily where the volcano Etna is located. The SO_2 emissions in the all other grid-cells are at least one order of magnitude smaller.

The NO_x emissions in Bulgaria are considerably smaller than the highest NO_x emissions in Europe.

The contribution of the European sources on the Bulgarian area is shown in the lower two plots in Fig. 8.2 (and in the corresponding plots in Fig. 8.2.col). The results are obtained by running UNI-DEM twice:

- the first time with all European emissions (including the Bulgarian emissions), and
- the second time with all European emissions excluding the Bulgarian emissions.

Consider now a given grid-cell and assume that the result from the first run for the considered chemical species is B, while the corresponding result from the second run is A. The ratios A/B multiplied by 100 are given in the lower two plots of Fig. 8.2 (and in the corresponding plots in Fig. 8.2.col). This means that the contribution of the European sources outside Bulgaria in the Bulgarian area is calculated in percent and presented in the lower two plots of Fig. 8.2 (and in the corresponding plots in Fig. 8.2.col).

It is seen (compare the lower two plots in Fig. 8.2 (or the corresponding plots in Fig. 8.2.col)) that the influence of the European sources on the sulphur dioxide levels in Bulgaria is less than the influence of the European sources on the nitrogen dioxide levels. This should be expected, because, as mentioned above, the SO_2 emissions in Bulgaria are rather high, while the NO_x emissions in Bulgaria are considerably smaller. Thus, the influence of the local SO_2 emissions on the sulphur dioxide levels in Bulgaria should be greater than the influence of the local NO_x emissions on the nitrogen dioxide. This can also be considered as an indication that the model is producing qualitatively correct results (the topic of the qualitative validation of the results, which are calculated by the model, was discussed in Chapter 1).

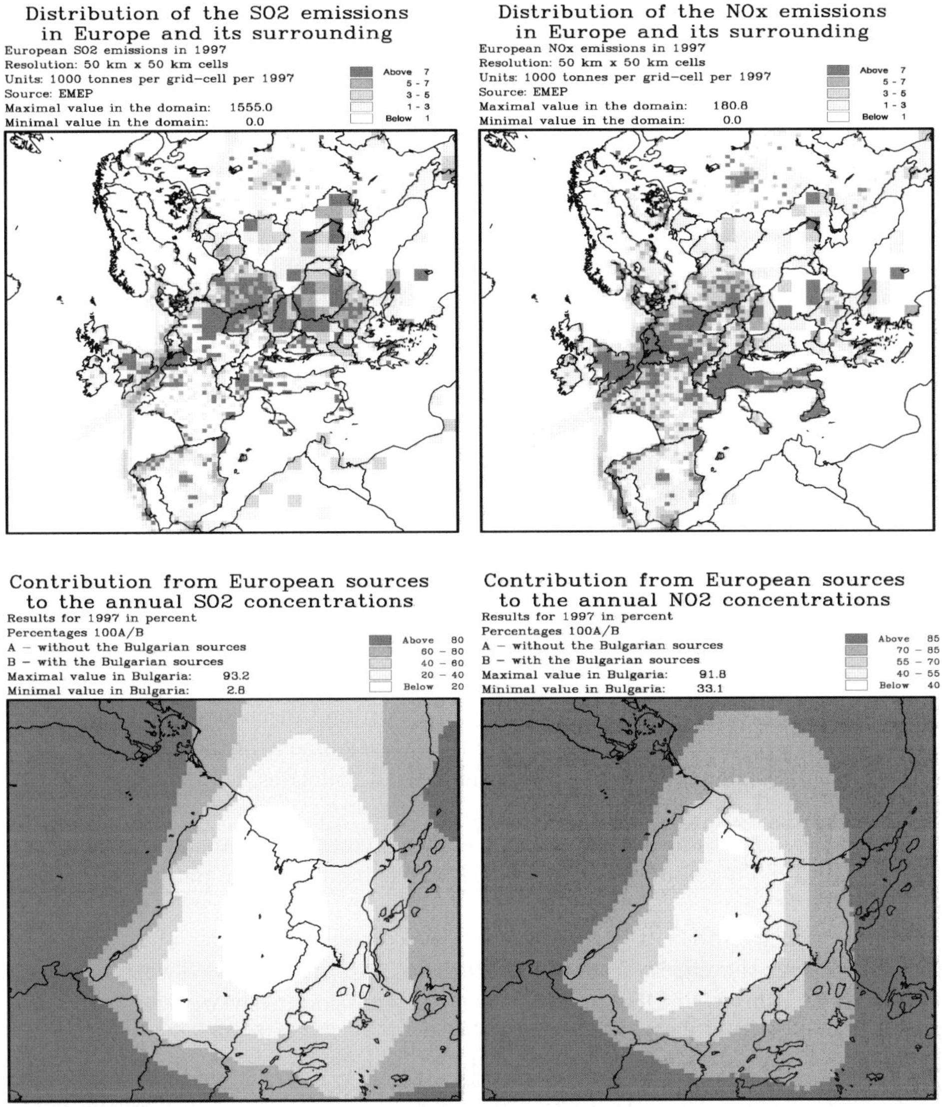

Figure 8.2: Influence of European emission sources to pollution levels in Bulgaria: (a) European SO_2 emission sources, (b) European NO_x emission sources, (c) contribution of the European SO_2 emission sources to the SO_2 concentrations in Bulgaria (in percent), (d) contribution of the European NO_x emission sources to the NO_2 concentrations in Bulgaria (in percent). A colour version of this figure is given in Appendix A (see Fig. 8.2.col there).

Such a study can also be carried out for any other country in Europe. Moreover, one can consider a part of a given country (which is relevant for large European countries (as, for example, Russia, Germany and France). Finally, one can also consider effects involving a group of countries (as, for example, Scandinavia, the Balkan countries, etc.). Some results about the effects of European countries on the high ozone levels in the Balkan countries will be presented in the next section.

8.5 Prediction of the pollution levels for 2010

Expected inventories of the European emissions for 2010 were prepared as a result of a combined work of many European specialists in this fields; among others specialists from IIASA (Laxenburg, Austria) and EMEP (European Monitoring and Evaluation Programme). The obtained emission inventories are often called Scenario 2010 (see also Section 8.2). Some realistic requirements for reductions of the European human-made emissions (compared to the level for 1990) were used in the preparation of Scenario 2010.

The major purpose in the efforts related to the development of Scenario 2010 is to reduce the pollution levels in Europe and (as a consequence of this reduction) to avoid, if possible, the exceeding of some critical levels, which may have different damaging effects.

UNI-DEM was used in different studies related to the expected pollution levels in 2010. Results related to the numbers of "bad" days in Europe were given in Fig. 8.1 (and in the corresponding plots in Fig. 8.1.col) and discussed in Section 8.2. Some results, which show the differences in the development in the different parts of Europe with some more details about the Balkan countries, will be presented in this section.

High ozone levels might have damaging effects (including damaging effects on human health). Therefore one is interested in keeping the ozone daily maxima at a low level.

In the lower two plots of Fig. 8.3 (and in the corresponding plots of Fig. 8.3.col in Appendix A) the effects from using Scenario 2010 are shown. Two runs were performed by using UNI-DEM; one by using the emission inventories for 2010 and another by using the emission inventories for 1997. Then for each grid-cell the ratios of the two runs were calculated and multiplied by hundred (in order to get the changes in percent). Results for the whole of Europe are shown in the left-hand-side lower plot of Fig. 8.3 (and in the corresponding plot in Fig. 8.3.col). Results for the Balkan countries are given in the right-hand-side lower plot of Fig. 8.3 (and in the corresponding plot in Fig. 8.3.col).

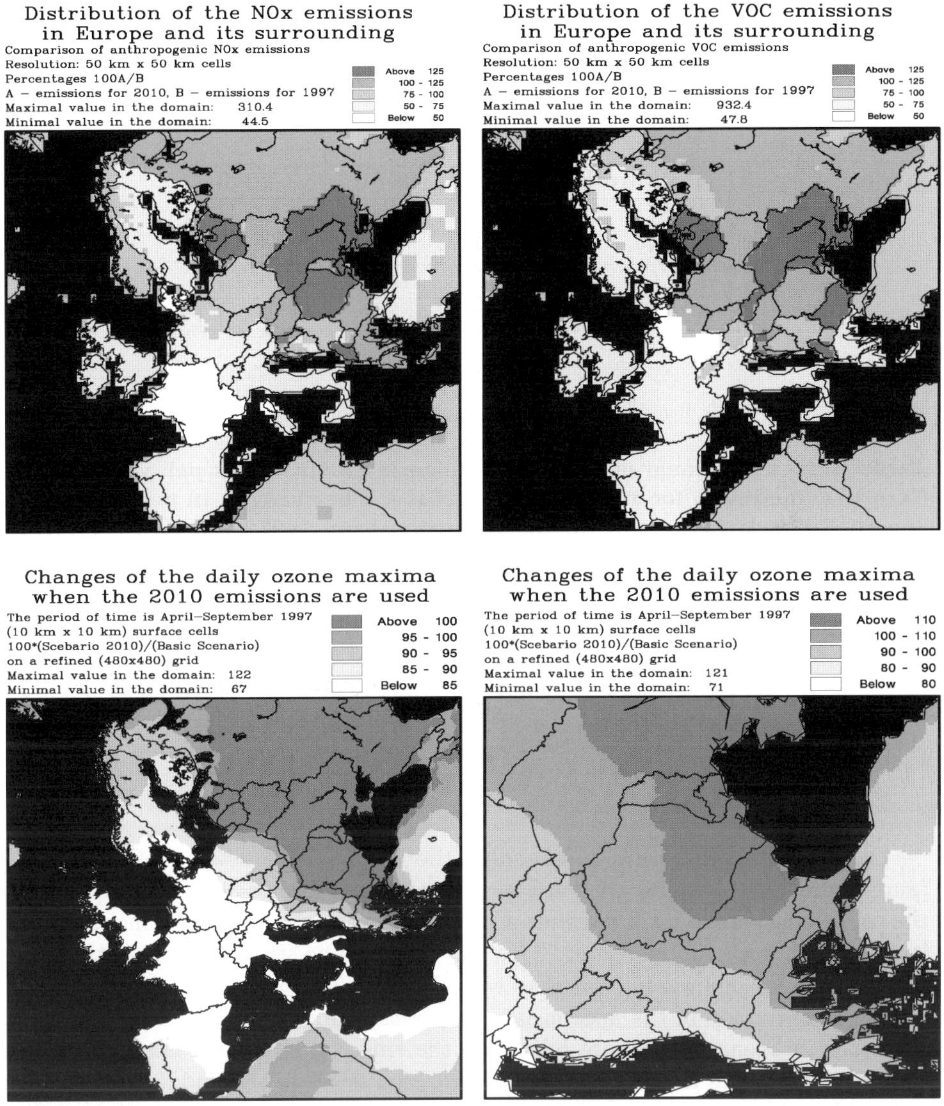

Figure 8.3: Changes of 1997 levels of the emissions and ozone concentrations according to Scenario 2010: (a) changes of the European NO_x emissions, (b) changes of the European VOC emissions, (c) changes of the ozone concentrations in different parts of Europe, (d) changes of the ozone concentrations in the Balkan countries. The areas: in black are representing the water areas in the model domain (i.e. the black colour is not referring to "emissions" or to "daily ozone maxima"). A colour version of this figure is given in Appendix A (see Fig. 8.3.col there).

Table 8.1: The annual emissions in Germany (1000 tonnes per year as SO_2, NO_2, VOC and NH_3 for 1990, 1997 and 2010). The changes, in percent, of the annual German emissions in the transition from 1997 to 2010 are given in brackets.

Pollutant	1990	1997	2010
SO_2	6165	1359	678 (50%)
NO_x	2709	1846	1192 (65%)
VOC	3225	1780	1161 (65%)
NH_3	765	634	573 (90%)

Table 8.2: The annual emissions in Bulgaria (1000 tonnes per year as SO_2, NO_2, VOC and NH_3 for 1990, 1997 and 2010). The changes, in percent, of the annual Bulgarian emissions in the transition from 1997 to 2010 are given in brackets.

Pollutant	1990	1997	2010
SO_2	2008	1365	884 (65%)
NO_x	361	225	303 (135%)
VOC	217	120	210 (175%)
NH_3	144	77	128 (166%)

Consider first the results presented in the left-hand-side lower plot of Fig. 8.3 (and in the corresponding plot in Fig. 8.3.col). It is immediately seen that while the ozone concentrations in Western Europe are reduced, the ozone concentrations in Eastern Europe are increased. This fact seems to be strange, because the main objective which one wants to achieve with Scenario 2010 is to reduce the pollution levels (by reduction of the appropriate emissions). However, the reductions of the emissions in Scenario 2010 are related to the emission levels in 1990. Due to the economical crisis in Eastern Europe the emission levels in the middle of the 1990s were reduced more than the prediction for 2010 levels. For the emissions related to production of ozone, i.e. the NO_x and the VOC emissions, this is illustrated in the two upper plots of Fig. 8.3 (and in the corresponding plots in Fig. 8.3.col). This effect is further demonstrated, this time quantitatively, in Table 8.1 and in Table 8.2. Two typical countries were selected:

- Germany as a representative country in West Europe, and
- Bulgaria as a representative country in Eastern Europe.

The annual emissions for these countries are given in Table 8.1 and Table 8.2 for years 1990, 1997 and 2010.

In these two tables, the emissions for 1990 and 1997 are again taken from the EMEP report written by Vestreng and Støren [261]. The predictions for 2010 are obtained by using the factors given in Amann et al. [8].

The results given in Table 8.1 and Table 8.2 show clearly that in Germany the predicted emissions for 2010 are smaller than the corresponding emissions for for both 1990 and 1997. In Bulgaria, the situation is not the same. The predicted emissions for 2010 are still smaller than the corresponding emissions in 1990, but they are in general greater than the corresponding emissions for 1997. The same relationship between the predicted emissions for 2010 and the actual emissions in the period 1990 to 1999 as that for Germany takes place for most of the other Western countries, while for most of the countries in Eastern Europe the corresponding relationship is similar to that shown in Table 8.2 for Bulgaria.

Let us return now to the expected ozone levels in the countries in the Balkan peninsula. These levels are shown on the right-hand-side lower plot in Fig. 8.3 (and in the corresponding plot in Fig. 8.3.col). It is seen that in general the ozone levels in the Balkan peninsula are expected to become increase by 2010 (compared with 1997).

In some of the countries in the Western part of the Balkan peninsula, however, one should expect some reductions (in spite of the fact that not all countries there will reduce the appropriate emissions). The reductions are due to the reduced amount of ozone transported from Western Europe (where the reductions of the NO_x and the VOC emissions are considerable).

8.6 Some conclusions

It has been shown in this chapter (by using a few examples) that the application of advanced mathematical models in comprehensive environmental studies is very helpful in the efforts to understand better the relationships between high pollution levels (which might cause different damages) and some important parameters that are involved in the models (as, for example, emissions). Furthermore, the results of the application of advanced mathematical models can be used to find out whether the critical levels of harmful pollutants will still be exceeded or not when certain measures (as, for example, emission reductions) will be taken in an attempt to avoid damaging effects from high pollution levels.

Note that most of the comprehensive environmental studies similar to those discussed in this chapter are still carried out with the coarse resolution options ($50\ km \times 50\ km$ cells in the horizontal planes) and with the simplest chemical scheme (which involves 35 chemical species). It is highly desirable to carry out, in the future, similar studies using fine resolution options and chemical schemes involving more chemical species. Moreover, it is also desirable to run UNI-DEM on longer time-intervals with more emissions and/or meteorological scenarios. These objectives can be achieved only when

- the numerical methods are considerably improved,
- the speed of the parallel codes is increased, and
- more powerful computers become available.

Hopefully, it will be possible in the near future to resolve successfully all these problems and, thus, to increase the class of comprehensive environmental studies that can be handled by UNI-DEM.

It should also be mentioned that the successful solution of the three problems above will allow the scientists to apply more advanced mechanisms in the description of the physical and chemical processes involved in the model (some advanced mechanisms are not used at present only because it is not possible to treat the models on the existing now computers when these mechanisms are implemented).

The definitions of some physical and chemical mechanisms should be made more robust for computational studies. This is especially true for the AOT40 values. Some results obtained in this direction were recently published; see, for example, Simpson et al. [226] and Tuovinen [251].

Chapter 9

Impact of future climate changes on high pollution levels

Changes in climate variability as well as changes in extreme weather and climate events in the 20th century, especially those which took place during the last two to three decades of the 20th century, have been discussed in many recent scientific publications. Attempts to project the results of such studies in the future have been made under different assumptions. In this chapter, we have chosen one of the well known scenarios predicting changes of the climate in world during the last 30 years of the 21st century. This scenario is used, together with several general predictions related to the future climate, to produce three climatic scenarios. The derived climatic scenarios are used to calculate predictions for future pollution levels in Denmark and in Europe by applying the Unified Danish Eulerian Model (UNI-DEM), on a space domain containing the whole of Europe.

It is first shown that UNI-DEM is able to produce reliable results. After that the results that are obtained by the climatic scenarios are compared with results obtained by using some more traditional scenarios in an attempt to evaluate the size of the changes in the pollution levels that are due to warming processes in the future climate.

It is explained why calculations over a long time period are essential for such a study. A time-period of ten years was actually used. The results indicate that, although the annual means of the concentrations of many pollutants are rather insensitive to the predicted climate changes, some quantities, which are related to high ozone levels, are increased rather considerably. This is an important fact, because, these quantities, when they exceed certain critical levels (different problems related to exceeded critical levels were discussed in the previous chapter), can cause damaging effects on humans, animals and plants (some of problems related to damaging effects were also discussed in the previous chapter).

9.1 Climate changes and air pollution levels

There have been many programs designed to predict the future state of the world's climate, and to explore the effects of such changes on a variety of policy sectors, e.g., food, water, energy, socioeconomic development, and regional security (see, e.g., Houghton et al. [140]). Environmental degradation is also of concern in future climate scenarios, and much effort has been dedicated to understand changing pressures on an already stressed system. We focus our attention in this chapter on air pollution in the future climate.

Within many regions on planet earth air pollution policy has been regionalized in order to achieve two major aims:

- to control and reduce transboundary pollution, and
- to meet policy objectives to limit air pollution impacts on human health and sensitive ecosystems.

The effort to achieve these two aims has been a daunting task. Within Europe, the Convention on Long Range Transport of Air Pollution (CLRTAP), see [254], has been dedicated to establishing a legal framework for reducing air pollution and assuring the safety of humans and ecosystems within prescribed limits. Limit values for a variety of pollutants have been established, where a finite number of exceedance days per year are allowable. However, in reaching compliance to the air quality directives, very few studies have considered the possibility that climate change may induce a controlling factor in the rate of reaching compliance. In this chapter, we will attempt to answer the question:

> *Will climate change add to the rate of reaching compliance to Europe's air quality policy objectives, or will it make the process of reaching compliance more difficult?*

Without the possibility of accessing future data, we will attempt to answer this question by carrying out a modelling study, based on scenarios of future climate states that are published in the IPCC (Intergovernmental Panel on Climate Change) report prepared by Houghton et al. [140].

We shall introduce three scenarios for climatic changes and shall study the impact of the predicted climatic changes on different pollution levels in Europe by using the results obtained by these scenarios as input data in UNI-DEM. Different options of UNI-DEM are, as stated in the previous chapters, described in detail in Zlatev [282]. Several studies related to exceeded critical levels have already been carried out by using this model; see, for example, Zlatev et al. [296] and the references given there. As mentioned in the previous chapter, the model results have systematically been compared with measurements taken both over land (Ambelas Skjøth et al. [9], Bastrup-Birk et al. [18] and Zlatev et al. [292], [293], [297]) and over sea (Harrison et al. [128]). It is nevertheless necessary to answer two important questions:

- Is the air pollution model chosen, UNI-DEM, producing reliable results?

and

- What is the relationship between changes of pollution levels that are caused by variations of emissions and changes of pollution levels that are caused by variations of meteorological parameters?

In order to answer the first question, we shall compare the results obtained by using UNI-DEM over a time-period of ten years, from 1989 until 1998, with the corresponding measurements. It should be mentioned here that the input data, both the meteorological data and the emission data, that are used in these runs of UNI-DEM are obtained from EMEP (European Monitoring and Evaluation Programme); see Sandnes Lenschow and Tsyro [208] and Vestreng and Støren [261].

In order to answer the second question, we have run UNI-DEM by using two traditional scenarios. In the first of them, we kept the emissions during the whole period, 1989 to 1998, the same as those in 1989; while the same meteorology (the meteorology for 1989) was used in the second traditional scenario for every year.

Not only are the results showing that the pollution levels depend essentially on the variations of both the emissions and the meteorological parameters, but it becomes clear that a long time period should be used in the study of the impact of climatic changes on the pollution levels. This is why we are consistently using a time period of ten years in the whole chapter, i.e. also when the influence of the climatic changes on the concentrations of different air pollutants is studied.

9.2 Definition of six scenarios

Six scenarios, which will be used in the following sections of this chapter, are described are described in this section:

- The first scenario is the basic one (it will be used both in the validation tests and in comparisons with results obtained by the other five scenarios).

- The next two scenarios (or, at least, scenarios that are similar to the next two scenarios) are traditionally used in air pollution studies.

- The last three scenarios are based on some expectations for climatic changes in the last thirty years of the 21st century that are discussed in Houghton et al. [140]. These three scenarios will be called climatic scenarios.

9.2.1 Scenario 1 - Basic Scenario

The Basic Scenario is run for the time period 1989 to 1998. The EMEP emission inventories, Vestreng and Støren [261], for these ten years are used together with meteorological data, Sandnes Lenschow and Tsyro [208], which are also supplied from EMEP. If the model is run for a given year, year N_{YEAR}, $N_{YEAR} \in [1989, 1998]$, then both the emission inventories for year N_{YEAR} and the meteorological data for year N_{YEAR} are used in the run.

The period 1989 to 1998 is an important period when air pollution levels in Europe are studied, because the European emissions were, in general, reduced during this period. For some countries the reductions are rather considerable. This is why long-term calculations for this period are very valuable. The results from such computations can be used in the solution of two major tasks:

- to study the impact of the emission reductions to the pollution levels in some parts of Europe, and
- to validate the model results by comparing them with measurements taken at representative stations in the EMEP network.

The second task is very important. If the model results reflect correctly the emission reductions during the period 1989 to 1998 (by showing reductions of the pollution levels, which do agree with measurements) then the application of the model in runs with other scenarios becomes more reliable. Some results obtained in the treatment of the second task will be presented in Section 9.3.

9.2.2 Scenario 2 - keeping the emissions unchanged

The same period of ten years is used in the second scenario. The meteorological data for year N_{YEAR} are used when the model is run for year N_{YEAR}, $N_{YEAR} \in [1989, 1998]$. The **same** emission inventories, those for year 1989, are used for all years in the interval $[1989, 1998]$. Thus, the changes in the pollution levels are only caused by the variation in the meteorological parameters when this scenario is run (because only the meteorological data are varied). This explains the major purpose with the second scenario:

we should like to see to what extent the meteorological data influence the pollution levels.

The comparison of the results obtained when this scenario is run with the results from the Basic Scenario, i.e. Scenario 1, will reveal the relationship between meteorological data and pollution levels.

9.2.3 Scenario 3 - keeping the meteorology unchanged

In the third scenario the emissions are varied from one year to another as in Scenario 1, while the same meteorological data (the meteorological data for year 1989) are used for all years in the interval [1989, 1998]. The changes in the pollution levels are now caused only by the changes in the emissions (because only the emission inventories are varied in the runs). Scenarios where only the emissions are varied are commonly used in different comprehensive environmental studies; see, for example, Amann et al. [8].

9.2.4 Scenario Climate 1

Several climatic scenarios, called SRES (Special Report on Emissions Scenarios), are discussed in detail in Houghton et al. [140]. We chose to follow the predicted climatic changes in the SRES A2 scenario. In fact the SRES A2 is presented in Houghton et al. [140] on p. 532 as a storyline and scenario family which describe a very heterogeneous world. We digitalized the information related to the annual temperature changes in Europe for the period 2071-2100. The predicted, for this period, changes of the temperature in the different parts of Europe and its surroundings are shown in Fig. 9.1.

In Scenario Climate 1, we used the temperature data for 1989 to 1998 to simulate changes in a ten hypothetical years according to the prescribed in the SRES A2 Scenario increases of the temperatures. This means that for the given hypothetical year, corresponding to year N_{YEAR}, we add some amount α to the temperature at each grid-square of the space domain and at every time step. Consider any grid-square and assume that this grid-square is located in a region in Fig. 9.1 where the prescribed by the SRES A2 scenario annual change is in the interval $[\beta, \gamma]$. The temperature at the grid-square under consideration at a time-step n, $n = 1, 2, \ldots$, is increased by an amount τ_n where τ_n is a randomly generated numbers in the interval $[0, \gamma - \beta]$ (which means that the amount α for the increase of the temperature in the considered grid-square at time-step n is equal to τ_n). The mathematical expectation of the annual mean of the increase of the temperature at each grid-square is given by the quantity $(\gamma - \beta)/2$. It should be mentioned that **only the temperatures** are varied in this scenario. Scenario Climate 1 was run for ten hypothetical years, corresponding to the ten years that are used in the Basic Scenario.

9.2.5 Scenario Climate 2

It is emphasized many times in the comprehensive report written by Houghton et al. [140] that the extreme cases will become even stronger in the future climate. In the second climate scenario an attempt to take into account this expectation has been carried out. More precisely, the following recommendations made in Table 9.6 on p. 575 in Houghton et al. [140] were taken into account when the Scenario Climate 2 was developed:

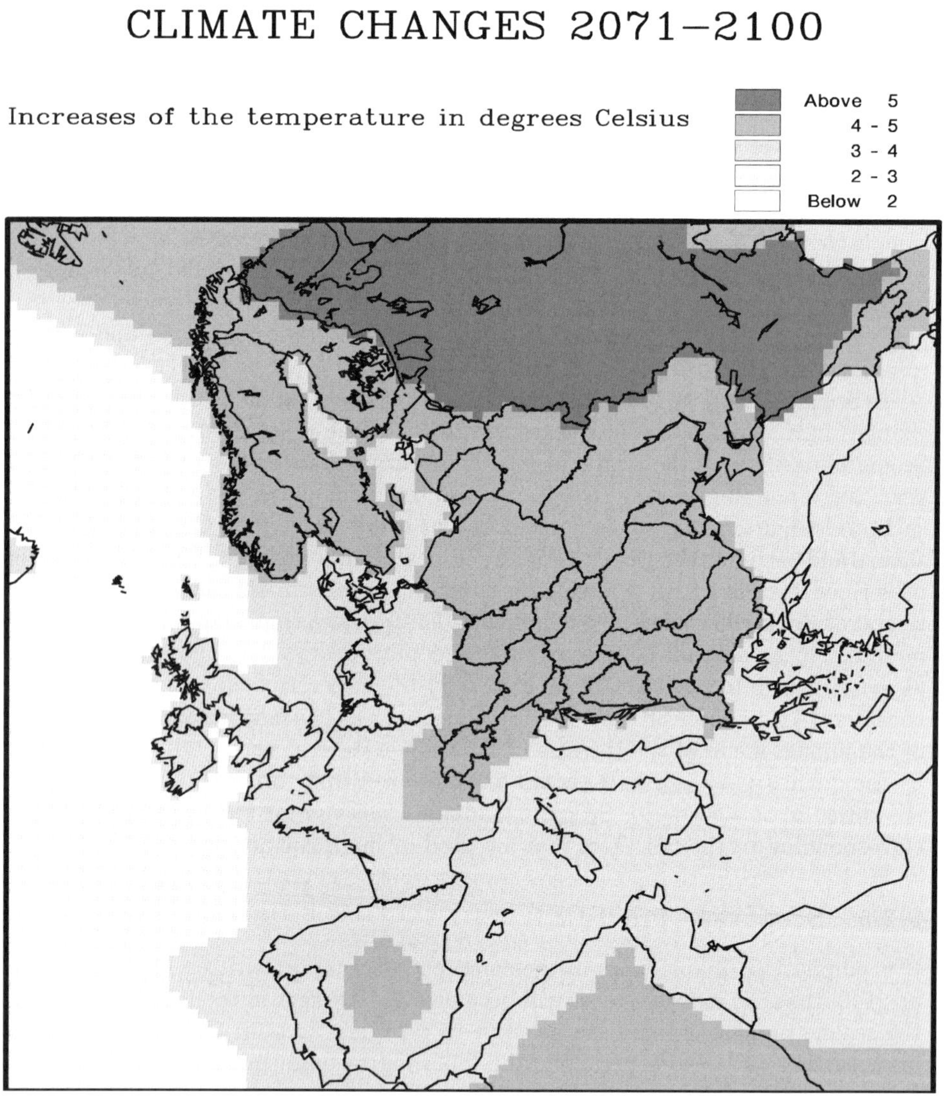

Figure 9.1: Predicted (by Scenario SRES A2, from Houghton et al. [140]) changes of the temperatures in different parts of Europe.

- There will be higher maximum temperatures and more hot days in the land areas.
- There will be higher minimum temperatures, fewer cold days and fewer frost days in nearly all land areas.
- The diurnal temperature range will be reduced over land areas.

It is not very easy to satisfy simultaneously all these three conditions. In this scenario we increased the temperatures during the night with a factor larger than the factor by which the daytime temperatures were increased. In this way, the second and the third requirements are satisfied. The first requirement is satisfied in the following way. During the summer periods the daytime temperatures are increased by a larger amount in hot days. It should be pointed out here that all these changes are carried out only over land. Furthermore, all changes described above are interposed over the changes made in the first climatic scenario. This means that, while the changes of the temperature are more or less smooth in the first climatic scenario, now some additional changes are imposed in order to meet the above three requirements.

In this scenario we also varied the data about cloud cover during the summer periods of the ten hypothetical years. Since there will be more hot days, it is natural to reduce the clouds cover during these periods. A factor of two has been used in the reduction of the cloud cover during the summer periods.

9.2.6 Scenario Climate 3

Two important recommendations are made in Table 9.6 on p. 575 in Houghton et al. [140] in connection with the amount of precipitation events in the future climate:

- There will be more intense precipitation events.
- There will be increased summer drying and associated risk of drought.

In order to satisfy these recommendations we increased the precipitation events during winter (both over land and over water). During summer, the precipitation events in the continental parts of Europe were reduced.

Similar changes in the humidity data were made. The humidity was increased during winter (both over land and over water). The humidity in the continental parts of Europe during summer was reduced.

The changes in precipitation events and humidity imply some changes of cloud cover. The cloud cover during winter was increased, while the same cloud cover as those used in the second climatic scenario were applied in the third climatic scenario.

All these changer were superimposed over the changes made in the second climatic scenario (Scenario Climate 2). The climatic scenario obtained in this way will be referred to as Scenario Climate 3.

9.2.7 Running the six scenarios

Each of the six scenarios was run on several powerful supercomputers by using the (96x96x10) option of UNI-DEM over a time-period of ten years. This is a huge computational task that was handled by using an efficient parallel code, which was discussed in detail in Chapter 7 (some more details can be found in Brown et al. [32], Georgiev and Zlatev [116], [117], Owczarz and Zlatev [188], [189], [190] and Zlatev [278], [280], [282]). The computers, which were actually used, were SUN computers at the Danish Centre for Scientific Computing. Up to 32 processors were applied in these runs. The calculated data were transmitted to workstations at the National Environmental Research Institute, where different graphical tools (see the discussion in Chapter 1) were used to visualize the digital information.

It must be emphasized here that it was possible to accomplish successfully this comprehensive study only because (i) powerful suppercomputer were available and (ii) an efficient powerful code was developed.

9.3 Validation of the results

The first scenario, the Basic Scenario, is used in this section to validate the results, which are calculated by UNI-DEM, by comparing them with measurements taken at Danish sites. The sites are Tange, Keldsnor and Anholt for all measured species excepting ozone. Ozone concentrations are measured on hourly basis in Ulfborg and Frederiksborg. It should be mentioned here that all these stations belong to to the EMEP network of measurement stations. The locations of these five sites are shown in Fig. 9.2.

The curves representing the temporal variations of the annual means of the concentrations (measured and calculated by the model) for the major species (SO_2, $SO_4^=$, NO_2, $HNO_3 + NO_3^-$, $NH_3 + NH_4^+$ and O_3) as well as for four quantities related to high ozone concentrations (AOT40 values for crops and forest trees, numbers of days in which the averaged 8-hour ozone concentrations exceed at least once the limit of 60 ppb and averaged daily ozone maxima, see European Commission, [96], [95] or Zlatev et al. [296]), are given in Fig. 9.3. Curves representing the temporal variation of the averaged concentrations (over the 38 Danish 50 $km \times 50$ km grid-squares) are also given in the plots of Fig. 9.3.

The comparison is not very easy. The stations do not measure regularly. Sometimes, measurements for many years are missing (see the upper plot in Fig. 9.3.c). Even if the station measures for all years (from 1989 until 1998) measurements for many days are sometimes missing. This causes great problems when AOT40 values calculated by the model have to be compared with the corresponding quantities obtained by using measurements. The AOT40 values are normally calculated by using the formula

$$\text{AOT40} = \sum_{i=1}^{N_{hours}} \max(c_i - 40, 0), \tag{9.1}$$

Figure 9.2: Locations of the Danish measurement stations which are included in the EMEP network.

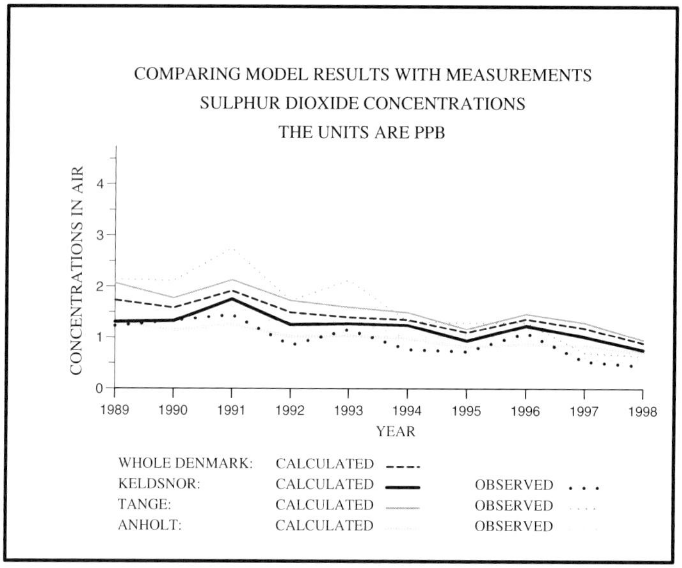

Figure 9.3.a: Comparison of annual means of SO_2 concentrations.

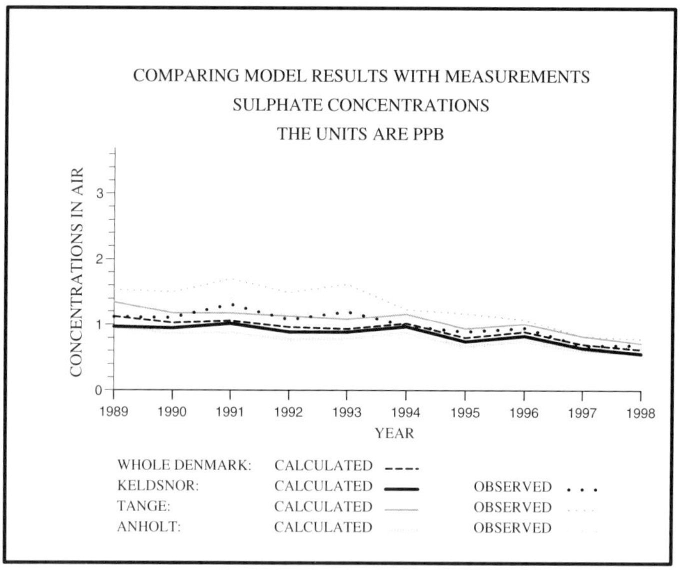

Figure 9.3.b: Comparison of annual means of $SO_4^=$ concentrations.

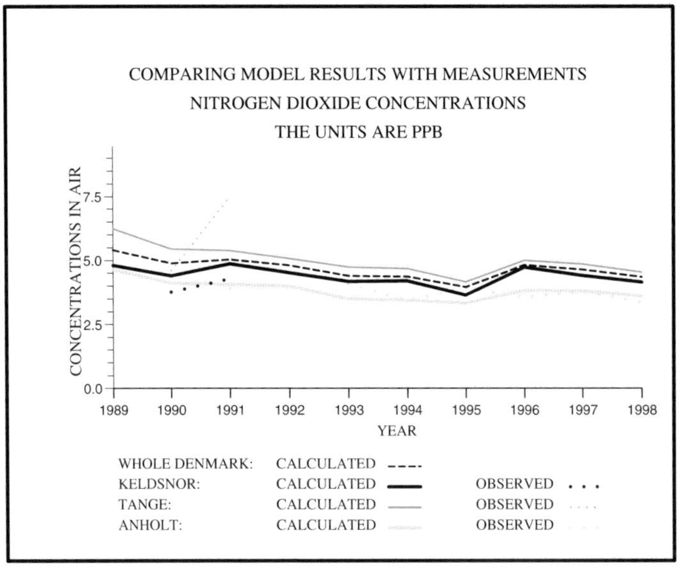

Figure 9.3.c: Comparison of annual means of NO_2 concentrations.

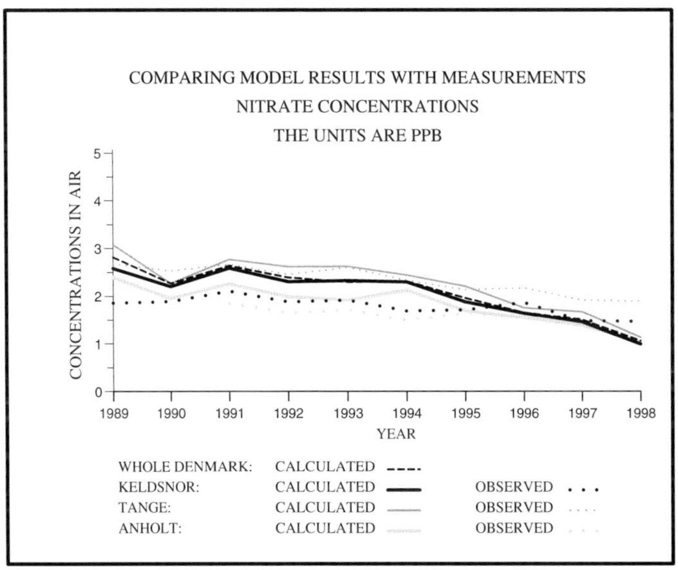

Figure 9.3.d: Comparison of annual means of $HNO_3 + NO_3^-$ concentrations.

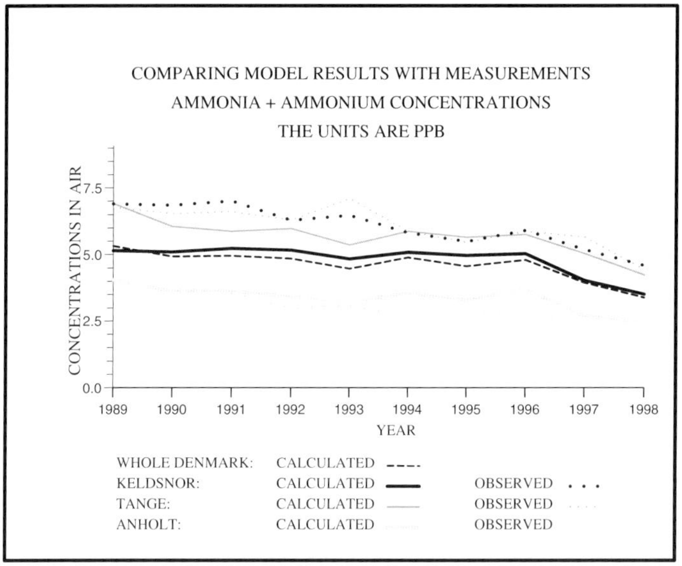

Figure 9.3.e: Comparison of annual means of $NH_3 + NH_4^+$ concentrations.

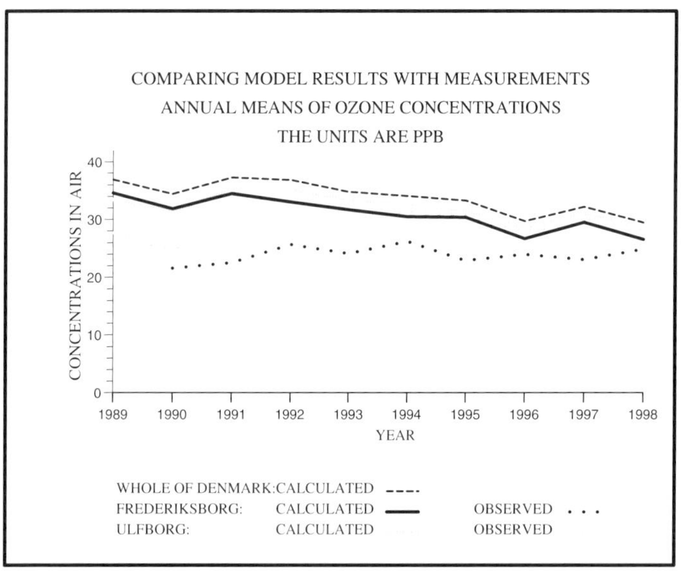

Figure 9.3.f: Comparison of annual means of O_3 concentrations.

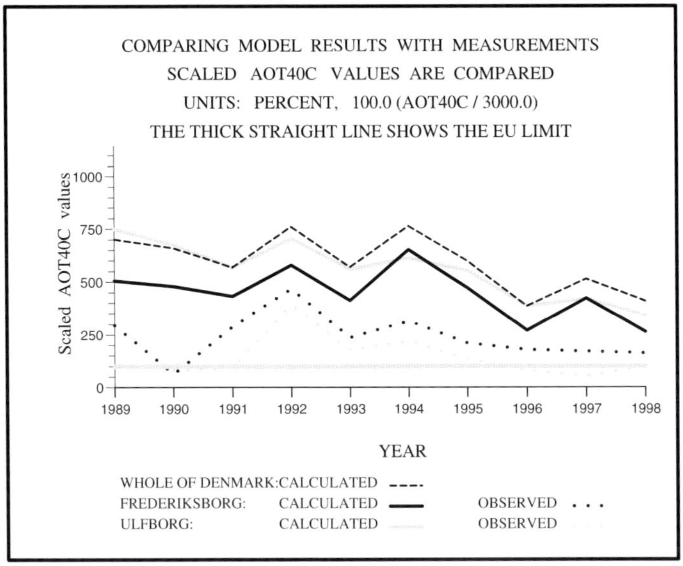

Figure 9.3.g: Comparison of AOT40C (AOT40 for crops) values. The units are ppb.hours.

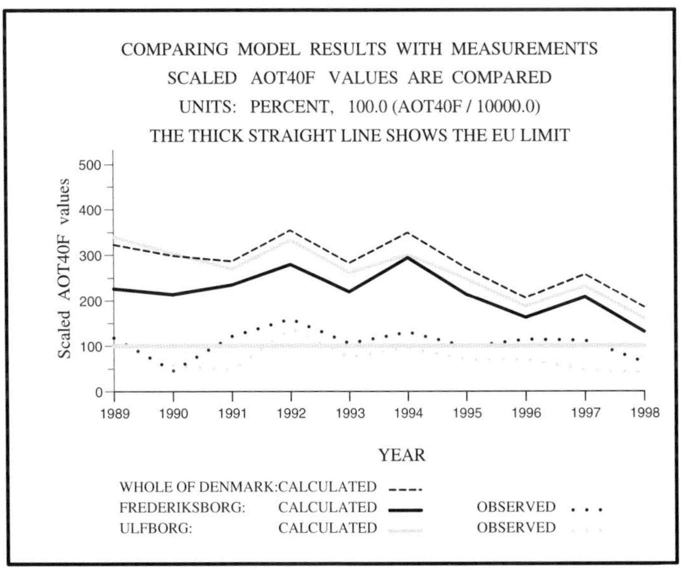

Figure 9.3.h: Comparison of AOT40F (AOT40 for forest trees) values. The units are ppb.hours.

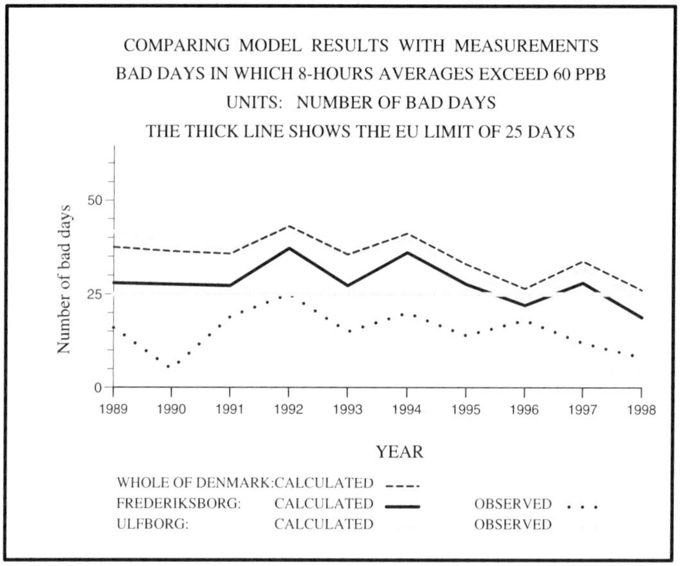

Figure 9.3.i: Comparison of numbers of days in which the 8-hour averaged ozone concentrations exceed at least once the limit of 60 ppb. The units are days.

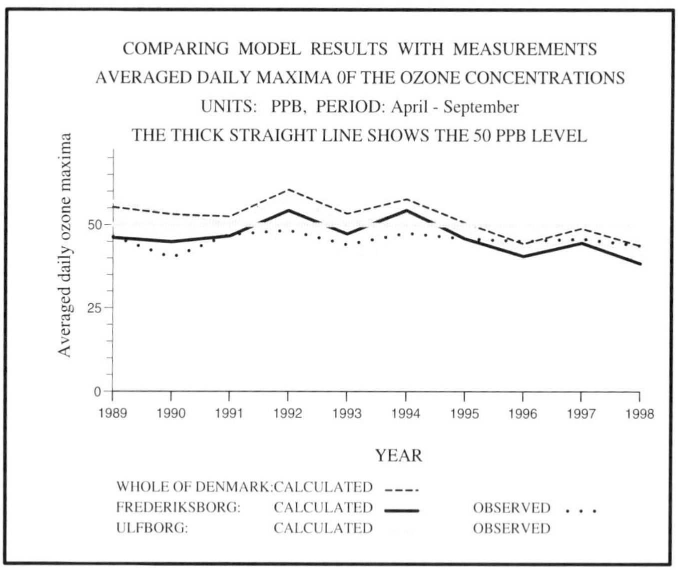

Figure 9.3.j: Comparison of average daily maxima of ozone concentrations. The units are ppb.

where N_{hours} is the number of day-time hours in the period under consideration (for crops this period contains the months May, June and July, while the period from April to September is used for forest trees), c_i is the calculated by some model or measured at some station ozone concentration.

With a model we can calculate c_i for **all** i, $i = 1, 2, \ldots, N_{hours}$. However, when many measurements at the stations under consideration are missing, then one should expect the AOT40 values that are measured at the EMEP stations to be considerably less than the calculated by the model AOT40 values. From Fig. 9.3.g and Fig. 9.3.h it is seen that precisely this happens in our comparisons, because many measurements are missing (see also Table 1 in Zlatev et al. [296]).

Missing measurements cause similar problems also when numbers of days in which the 8-hour averaged ozone exceed at least once the limit of 60 ppb.

It should also be emphasized that some of the quantities that are related to high ozone concentrations (AOT40 values for crops and forest trees and numbers of days in which the averaged 8-hour ozone concentrations exceed at least once the limit of 60 ppb) are very sensitive to small errors (see again Zlatev et al. [296]).

The missing measurements do not cause great problems when the averaged daily ozone maxima are compared. In this case we can use the number of measurements at a given station in order to obtain the averaged value at the measurement station at consideration. This is why the results of the comparison of calculated by the model and measured averaged daily ozone maxima gives much better results than the results of the corresponding comparisons for AOT40 values and for averaged 8-hour ozone concentrations that exceed at least once the limit of 60 ppb (compare Fig. 9.3.j with Fig. 9.3.h - Fig. 9.3.i).

It is clearly seen (Fig. 9.3) that the reductions in the European emissions in the period 1989 to 1998 do result in some reductions of the concentrations (both the measured and the calculated concentrations) of most of the pollutants in Denmark. An exception is the variation of the ozone concentrations. While the calculated results show a slight trend of reduction, the measurements at the two Danish stations show a slight trend of increasing. Note, however, that the model results are closer to the measurements at the end of the period.

9.4 Variations of emissions and of meteorological parameters

Comparing the results obtained with the second and the third scenarios with the results obtained with the Basic scenario (the first scenario) will allow us to evaluate

- the sensitivity of the concentrations to the variations of the meteorological parameters, and
- the sensitivity of the concentrations to the variations of the emissions.

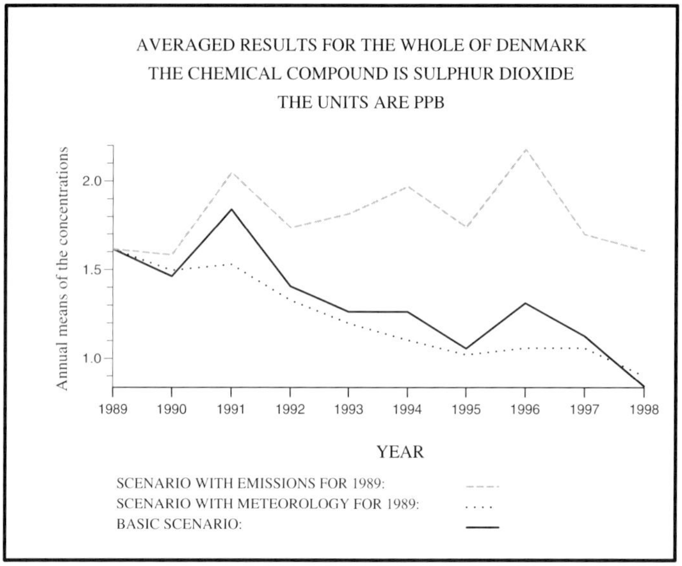

Figure 9.4.a: Averaged (over the Danish cells) annual means of SO_2 concentrations obtained by three scenarios.

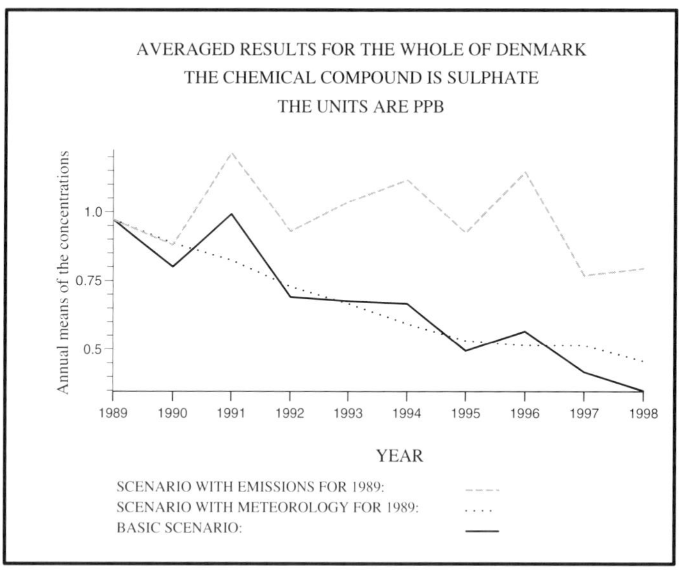

Figure 9.4.b: Averaged (over the Danish cells) annual means of $SO_4^=$ concentrations obtained by three scenarios.

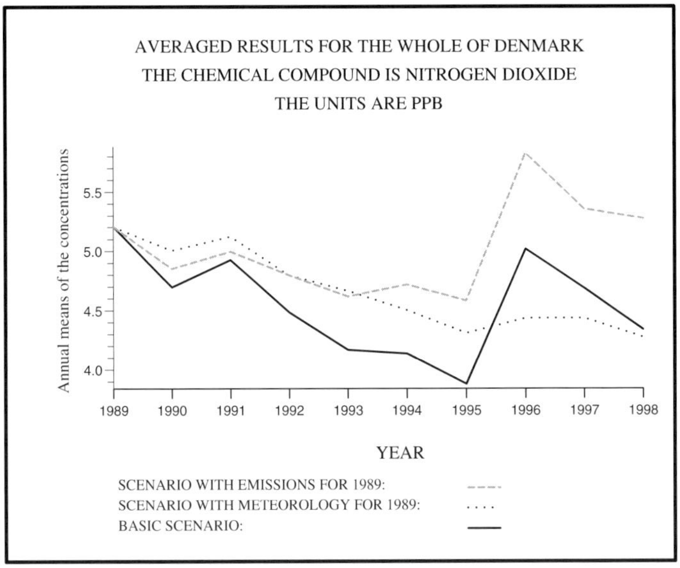

Figure 9.4.c: Averaged (over the Danish cells) annual means of NO_2 concentrations obtained by three scenarios.

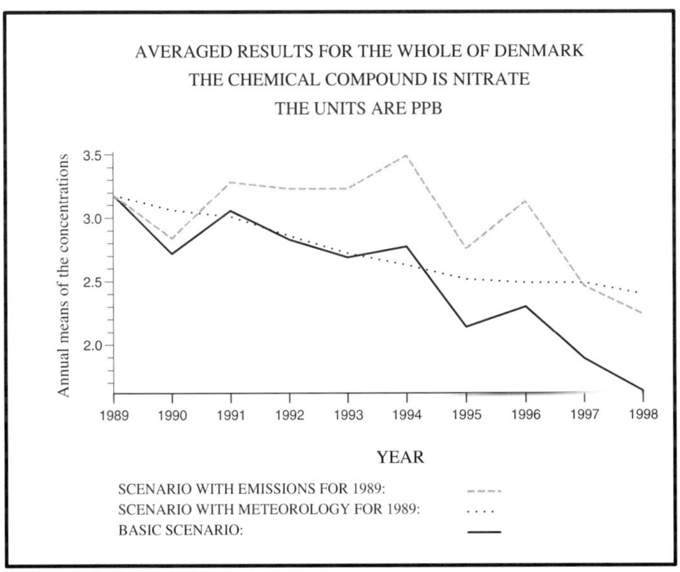

Figure 9.4.d: Averaged (over the Danish cells) annual means of $HNO_3 + NO_3^-$ concentrations obtained by three scenarios.

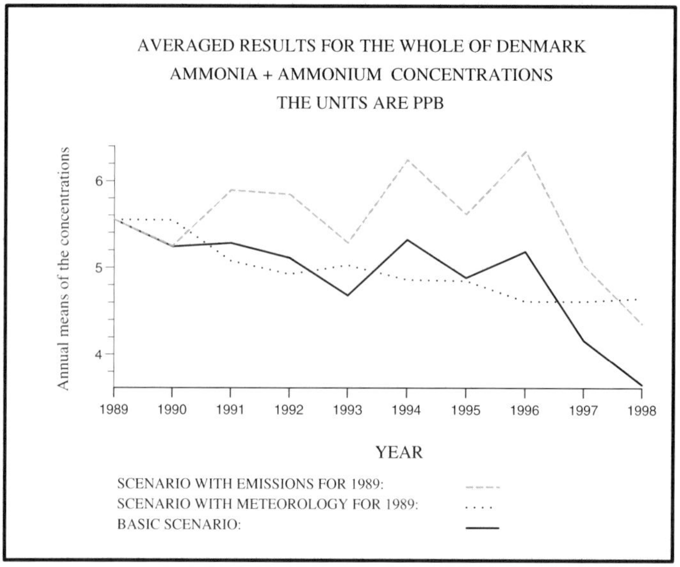

Figure 9.4.e: Averaged (over the Danish cells) annual means of $NH_3 + NH_4^+$ concentrations obtained by three scenarios.

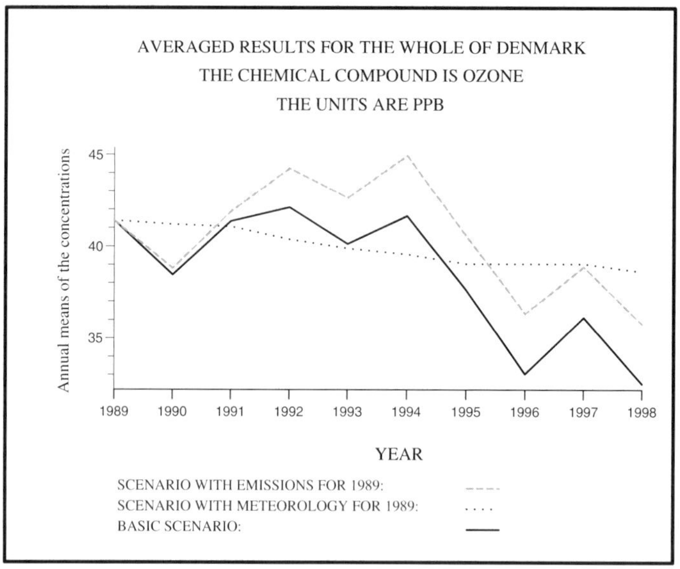

Figure 9.4.f: Averaged (over the Danish cells) annual means of O_3 concentrations obtained by three scenarios.

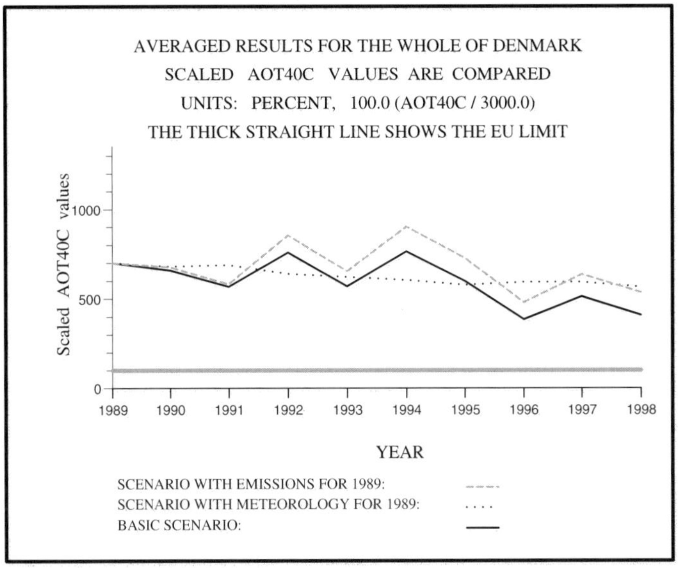

Figure 9.4.g: AOT40C (AOT40 for crops) values obtained by three scenarios.

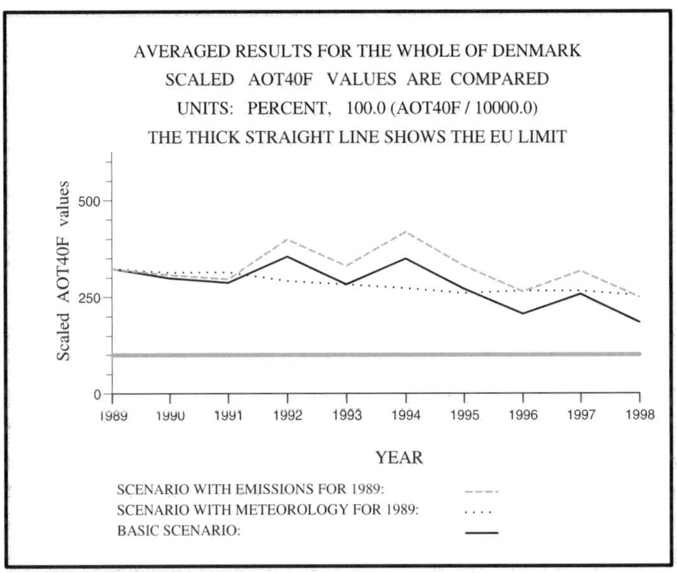

Figure 9.4.h: AOT40F (AOT40 for forest trees) values obtained by three scenarios.

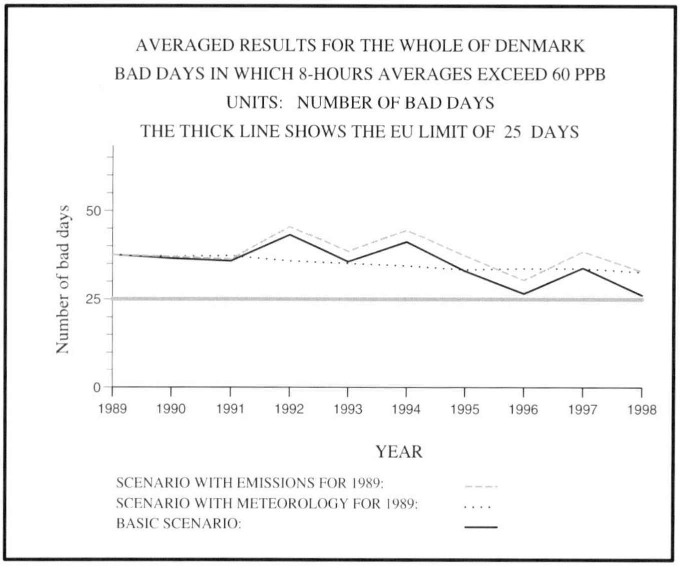

Figure 9.4.i: Comparison of numbers of days in which the 8-hour averaged ozone concentrations exceed at least once the limit of 60 ppb obtained by three scenarios. The units are days.

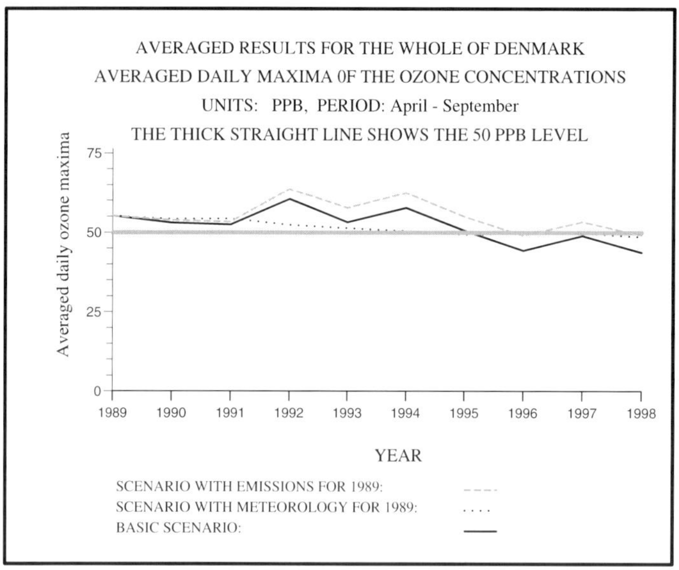

Figure 9.4.j: Averaged daily maxima of ozone concentrations obtained by three scenarios. The units are ppb.

Some results are given in Fig. 9.4 for the same compounds as those used in the previous section an in Fig. 9.3. Three major conclusions can be derived by studying these plots:

- The comparison of the results obtained by the Basic Scenario with the results obtained by using the second scenario (in which the emissions are kept constant) shows that

  (a) the variations of the concentrations (from one year to another) are similar (in the sense that when the concentrations from one year to the next year are increasing for the Basic Scenario they are also increasing for the second scenario, while when they are decreasing for the Basic Scenario, they are also decreasing for the second scenario), and

  (b) while a trend of decreasing can be seen for the Basic Scenario, no trend can be seen for the second scenario.

  Of course, this should be expected when the emissions are not varied and the fact that the model is giving results which are in agreement with the expected results indicates that the model produces qualitatively correct results.

- The comparison of the results obtained by the Basic Scenario with the results obtained by using the third scenario (in which the meteorology is kept constant) shows that

  (a) there is a clear trend of reductions of the concentrations for both scenarios, and

  (b) the results obtained by using the Basic Scenario show a clear variability from one year to another, while the curves representing the variations obtained by the third scenario are very smooth.

  Again, this should be expected and the fact that the results are as they should be expected to be indicates that the model is able to compute qualitatively correct results.

- **The immediate consequence from the previous two conclusions is that using meteorology from one year only is not sufficient for reliable conclusions.** It is absolutely necessary to run the model over a long period of time in order to investigate the model response to the variability of the meteorology. This is why a time period of ten years is simulated in the three climatic scenarios. It is perhaps more appropriate to use even a longer period of time (say, 25 to 30 years).

9.5 Results from the climatic scenarios

The results obtained by using the particular tool chosen by us, the Unified Danish Eulerian Model, were validated (by using measurement data) in Section 9.3. It was shown, by using a scenario with constant emissions and a scenario with constant meteorology, that it is necessary to run the model on a long time-interval. In this way, the necessary preliminary work is completed and we are ready to start the investigation of the influence of the climate changes on pollution levels. The three climatic scenarios from Section 9.2 are used in this part of the climatic study.

We shall first compare the temporal variations of the mean Danish concentrations of the major pollutants (averaged over the Danish cells of the grid) for the three climatic scenarios with the variation obtained when the Basic Scenario is used. After that we shall consider the changes of some quantities related to ozone levels for the whole of Europe. In the first case we shall consider the period of ten years, as in the previous sections, while results for a representative year (year 1997) will be used in the comparisons related to the second case.

9.5.1 Temporal variations of the concentrations

The results obtained when the three climatic scenarios were run for ten hypothetical years, each of them corresponding to one year in the period 1989 to 1998 (see Section 9.2), are compared with the corresponding results obtained by using the Basic Scenario in Fig. 9.5 for the same compounds as those used in the previous two sections. The following major conclusions can be drawn from this comparison.

- The results obtained by the three climatic scenarios are very similar to these obtained by the Basic Scenario. For some pollutants, as for example SO_2 it is very difficult to distinguish the different curves. The differences are slightly more considerable for ozone, but here the changes are also within a few percent.
- The annual Danish ozone concentrations are reduced when the three climatic scenarios are used (see Fig. 9.5.f). The reductions, are rather small.
- The quantities related to high ozone concentrations are slightly increased when the climatic scenarios are used (see Fig. 9.5.g to Fig. 9.5.j). In the next subsection we shall show that the increase of these quantities when the climatic scenarios are used is rather considerable in some parts of Europe.

9.5.2 Changes in other parts of Europe

Changes in Denmark were studied until now. It is also interesting to investigate the changes in other parts of Europe. High ozone levels can cause damages to plants, animals and human health when these exceed certain critical levels. Therefore, we shall concentrate our attention to the quantities related to high ozone levels. We choose again the quantities studied in the previous sections:

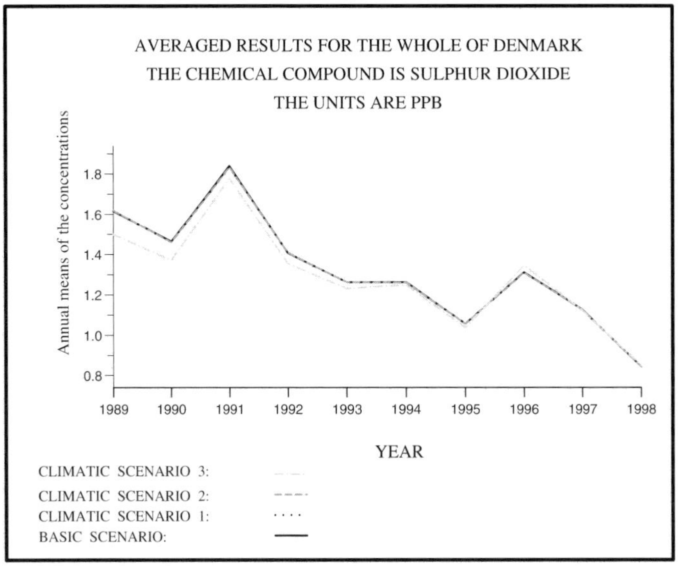

Figure 9.5.a: Comparison of SO_2 concentrations (climatic scenarios vs the Basic Scenario).

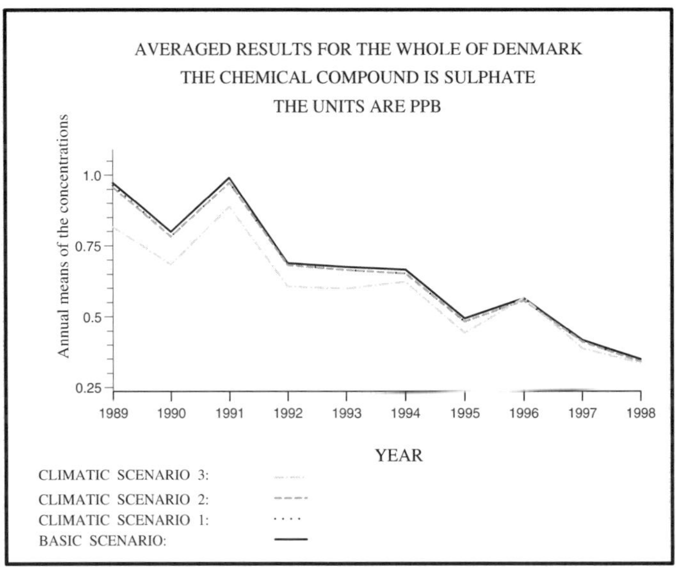

Figure 9.5.b: Comparison of $SO_4^{=}$ concentrations (climatic scenarios vs the Basic Scenario).

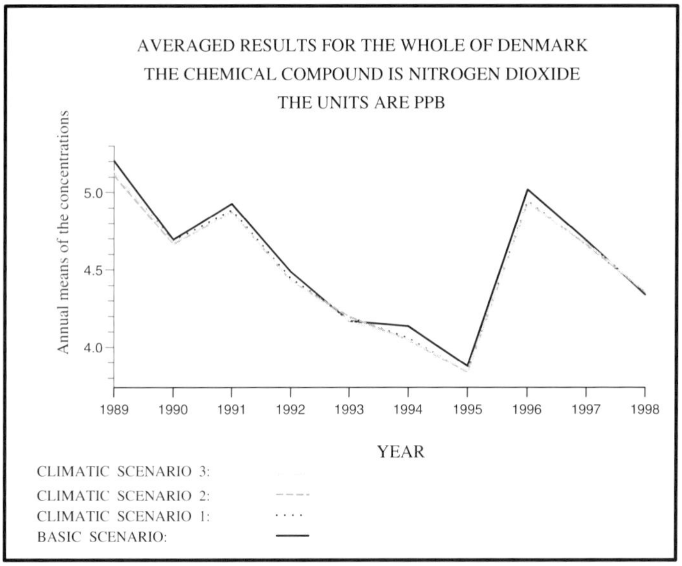

Figure 9.5.c: Comparison of NO_2 concentrations (climatic scenarios vs the Basic Scenario).

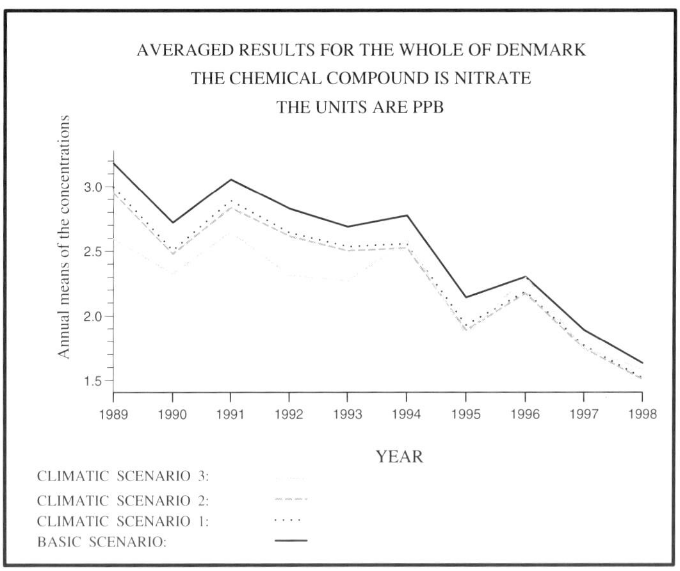

Figure 9.5.d: Comparison of $HNO_3 + NO_3^-$ concentrations (climatic scenarios vs the Basic Scenario).

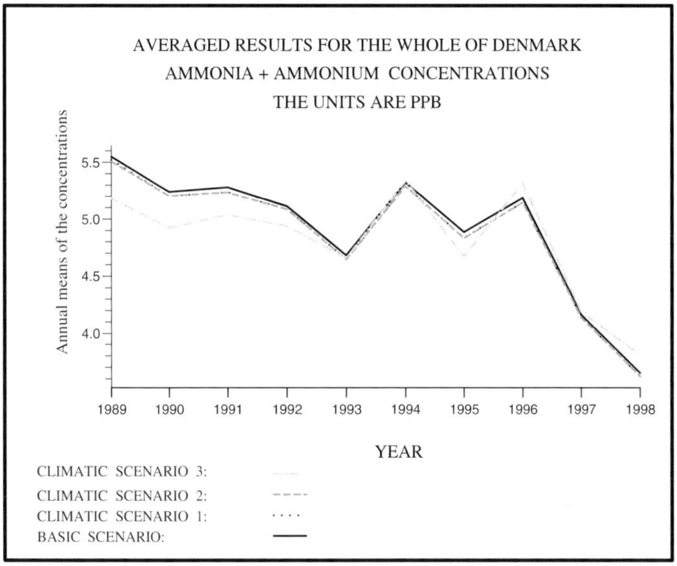

Figure 9.5.e: Comparison of $NH_3 + NH_4^+$ concentrations (climatic scenarios vs the Basic Scenario).

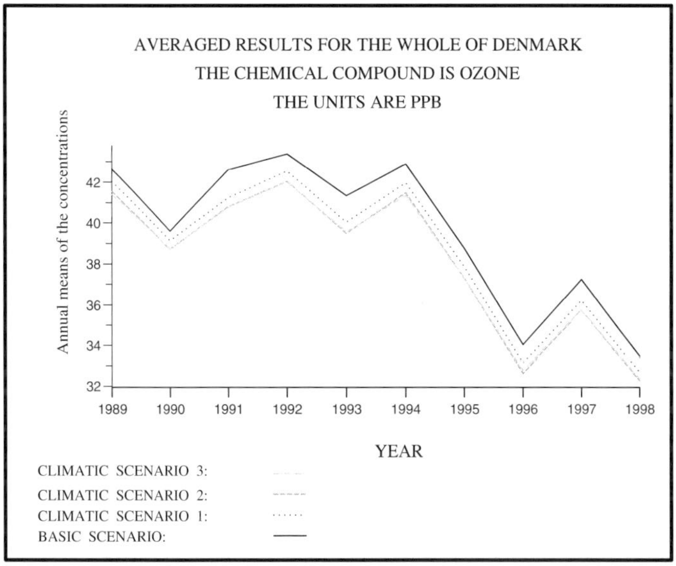

Figure 9.5.f: Comparison of O_3 concentrations (climatic scenarios vs the Basic Scenario).

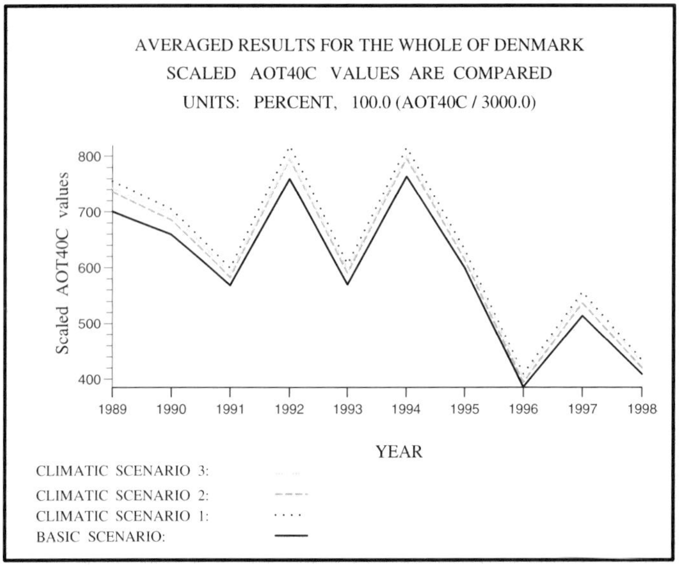

Figure 9.5.g: AOT40C (AOT40 for crops) values (climatic scenarios vs the Basic Scenario). The units are ppb.hours.

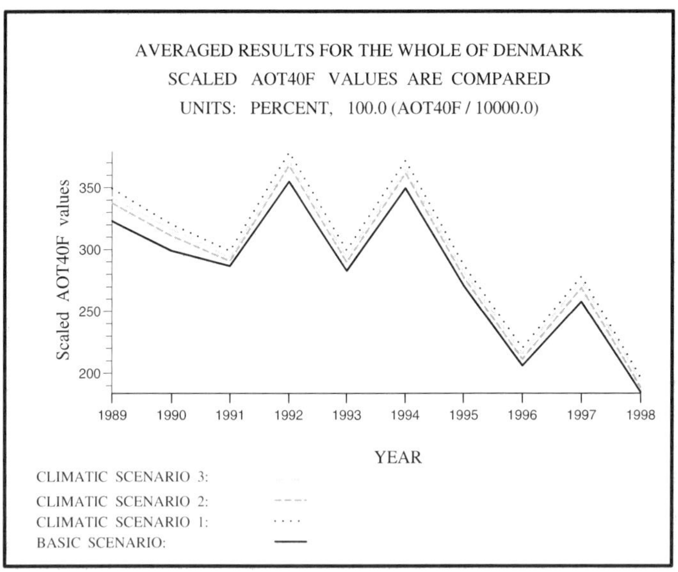

Figure 9.5.h: AOT40F (AOT40 for forest trees) values (climatic scenarios vs the Basic Scenario). The units are ppb.hours.

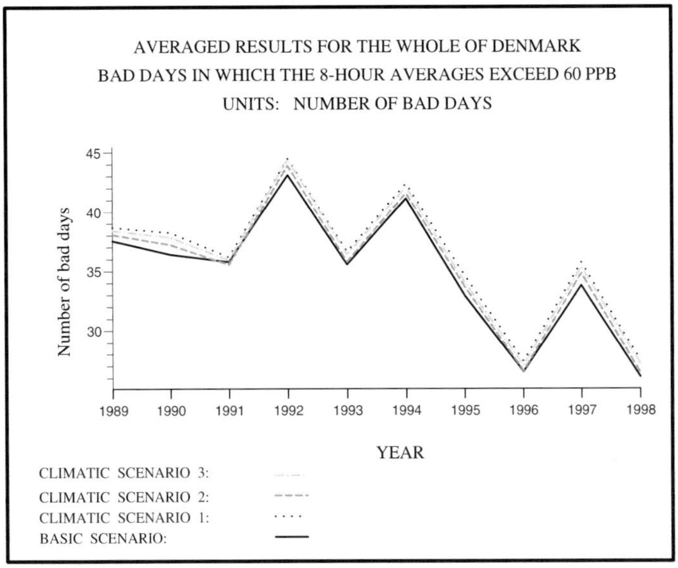

Figure 9.5.i: Comparison of numbers of days in which the 8-hour averaged ozone concentrations exceed at least once the limit of 60 ppb (climatic scenarios vs the Basic Scenario) The units are days.

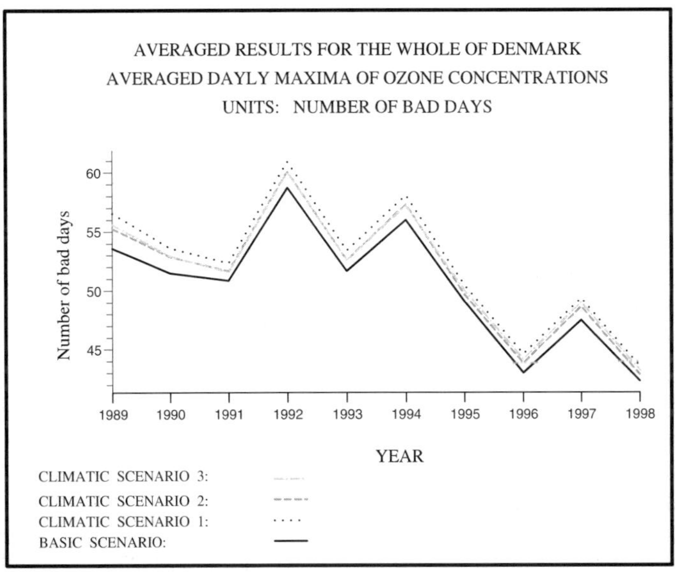

Figure 9.5.j: Averaged daily maxima of ozone concentrations (climatic scenarios vs the Basic Scenario). The units are ppb.

- AOT40C values (causing damages to crops when the critical level of 3000 ppb.hours is exceeded),
- AOT60F values (causing damages to forest trees when the critical level of 10000 ppb.hours is exceeded),
- numbers of "bad" days in which the 8-hour average of the ozone concentrations exceeds at least once the critical value of 60 ppb (causing health damages for certain groups of humans; the number of "bad" days should not exceed 25 after year of 2010, the long-term aim is to reduce this number to zero in the European Union [96]), and
- averaged daily maxima of ozone concentrations.

Some results for 1997, obtained when the Basic Scenario and Climatic Scenario 3 are used, are compared in Fig. 9.6. (a colour version of this figure is given in Appendix A (see Fig. 9.6.col there). In all plots of this figure the relative changes, in percent, of the considered ozone levels are presented (i.e. at each cell of the grid the result obtained by using Climatic Scenario 3 is divided by the results obtained by the Basic Scenario and the obtained ratio is multiplied by 100).

It is seen from Fig. 9.6, and from its colour version 9.6.col in Appendix A, that the quantities related to high ozone concentrations are in general increased when Climatic Scenario 3 is used. For some parts of Europe the increases are rather considerable.

9.5.3 Why are the high ozone levels increased?

It has already been mentioned, in Subsection 9.5.1, that while the annual means of the ozone concentrations in Denmark are reduced when the climatic scenarios are used, the quantities related to high ozone concentrations tend to be increased (compare the results given in Fig. 9.5.f with the corresponding results in Fig. 9.5.g to Fig. 9.5.j). For 1997, the increases of the quantities related to high ozone concentrations are rather considerable in some parts of Europe (see Fig. 9.6 and Fig. 9.6.col). This particular behaviour of quantities related to ozone deserves some additional explanation. It is useful to consider two typical situations: the diurnal variation of the ozone concentrations in summer time and in winter time.

(A) Diurnal ozone variation in summer time.

In a typical day during summer, the ozone concentration have a maximum during the afternoon and a minimum during the night. The photochemical reactions play an essential role in the production of ozone, while the destruction of ozone is the dominating process when these reactions are deactivated (which happens during the night). If the temperature is higher and the cloud cover is lower, then one should expect more ozone to be produced. The climate scenarios (and especially Climate

Scenario 3) simulate meteorological conditions in summer days, according to the predictions for climate changes, with both higher temperatures and lower cloud cover in comparison with these quantities for the Basic Scenario. Therefore, the conclusion is that we should expect the climatic scenarios to produce more ozone during daytime in summer, because the photochemical reactions are producing ozone in a higher rate. Some experiments were carried out in order to confirm this conclusion. In Fig. 9.7.a, we show the averaged diurnal variation of the ozone concentrations at the Danish site Frederiksborg for July 1997. It is seen that during the day Climatic Scenario 3 produces more ozone than the Basic Scenario. It is also seen that the patterns of the diurnal variation for the two scenarios are very similar to the pattern of the diurnal variations of the measured concentrations.

(B) Diurnal ozone variation in winter time.

The ozone concentrations in summer are higher than the ozone concentrations in winter. This means that the photochemical reactions (which cause production of ozone) are more active in clear summer days, while they do not play an essential role in cloudy winter days. In some cloudy winter days the photochemical reactions can be fully deactivated. When this happens one cannot anymore observe considerable increase of the ozone concentrations during the day. This is illustrated in Fig. 9.7.b, where the averaged diurnal variation of the ozone concentrations at the Danish site Frederiksborg for July 1997 is given for the Basic Scenario, Climatic Scenario 3 and the observations. It is again seen that the patterns of the diurnal variation for the two scenarios are very similar to the pattern of the diurnal variations of the measured concentrations. In this case it is even more important that the ozone concentrations produced by the Basic Scenario are higher than those produced by Climatic Scenario 3.

It is important to emphasize here that the relevant period for obtaining the quantities shown in Fig. 9.6 and Fig. 9.6.col is an extended summer period (either the period from April to September or the period from May to July). Moreover, all quantities shown in Fig. 9.6 and in Fig. 9.6.col are related to high ozone concentrations (occurring mainly during the afternoon). Therefore, all these quantities are very sensitive to small variations of the high ozone concentrations. For example, an increase of the high ozone concentrations with 1-2 ppb might lead to an exceedance of the 8-hour limit of 60 ppb (which will cause a switch from a "good" day to a "bad" one).

The results shown in Fig. 9.7.a to Fig. 9.7.b together with the analysis presented above explain the difference in the behaviour of the annual means of the ozone concentrations and in the behaviour of the quantities related to high ozone concentrations. It should be mentioned here that some more details about these phenomena can be found in Havasi and Zlatev [131].

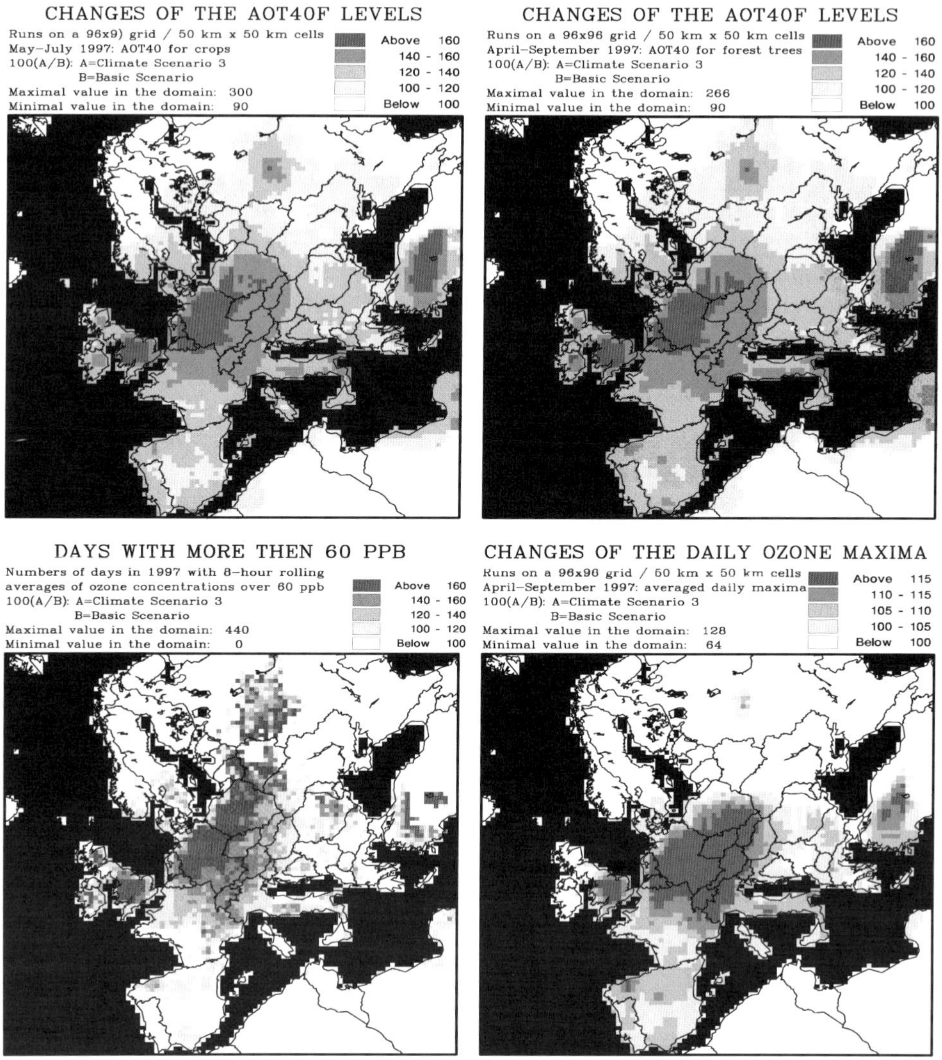

Figure 9.6: Relative changes (in percent) of quantities related to high ozone levels when Climatic Scenario 3 is used instead of the Basic Scenario. The areas in black are representing the water areas in the model domain (i.e. the black colour is not refering to "changes of AOT40C levels", "changes of AOT40F levels", "days with more than 60 ppb" or "changes of daily ozone maxima"). A colour version of this figure is given in Appendix A (see Fig. 9.6.col there).

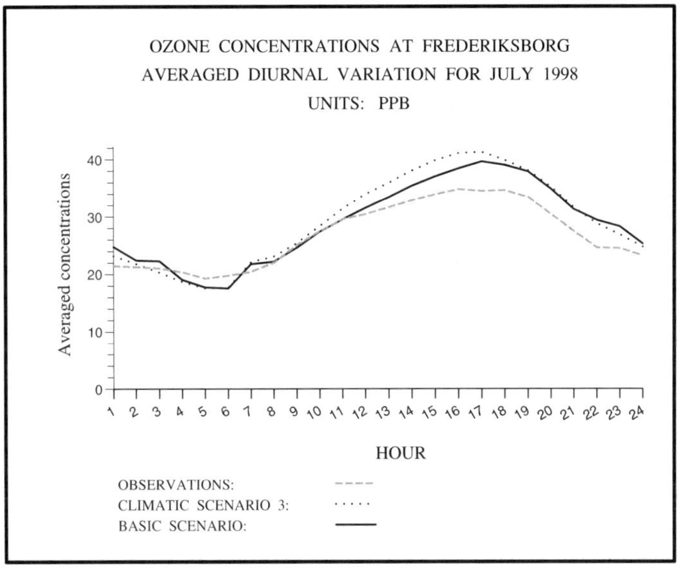

Figure 9.7.a: Averaged diurnal patterns of the ozone concentrations at the Danish measurement station Frederiksborg for July 1998.

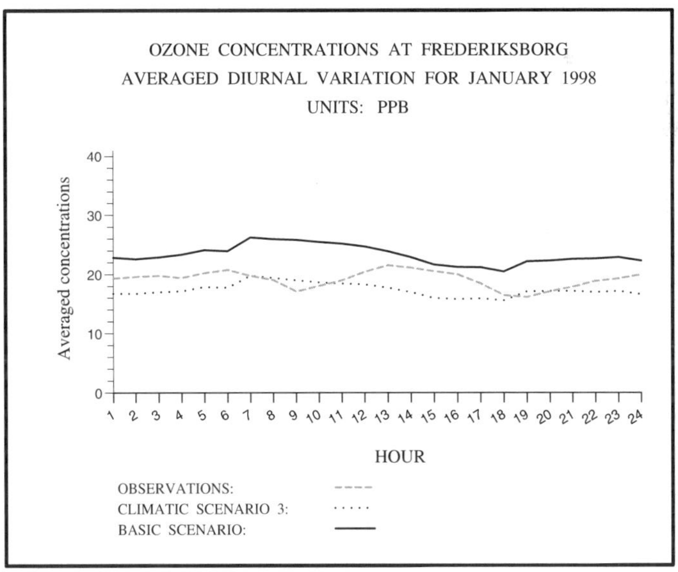

Figure 9.7.b: Averaged diurnal patterns of the ozone concentrations at the Danish measurement station Frederiksborg for February 1998.

9.6 Conclusions and plans for future work

The results presented in this chapter indicate that the annual values of the concentrations of the pollutants will not be influenced strongly by the climate changes. However, the influence of the climate changes on some quantities (such as AOT40C, AOT40F, days in which the rolling 8-hour averages of the ozone concentrations exceed at least once the critical limit of 60 ppb and daily maxima of the ozone concentrations) related to high ozone levels) is rather considerable. This is important, because recent investigations show that these quantities have damaging effects when certain critical levels are exceeded, see European Commission ([96], [95]) or Zlatev et al. [296].

Studying effects of climatic changes on air pollution levels is connected with a great number of uncertainties. While it is clear that it is not possible to eliminate all uncertainties, it is highly desirable to reduce some of them. One can reduce the uncertainties caused by numerical errors (which arise when a coarse grid is used) by refining the grid. This implies that runs by using the high resolution options of UNI-DEM are desirable. Also the time interval of ten years (which was used in this study) is, as already mentioned, not sufficiently long, i.e. run on time-intervals of 20 to 30 years are more desirable. Finally, the use of more scenarios can reveal more details about the studied relationships. The application of

- fine resolution options of UNI-DEM,
- longer time-intervals, and
- more scenarios,

leads to huge computational tasks. This shows that **great computational challenges are also present when the effect of future climatic changes on high pollution levels are to be studied in more details**.

It should be mentioned here that studying the future pollution levels is a hot topic and many results related to future pollution levels have been published in the last decade; see, for example, Jonson et al. [148].

Chapter 10

Implementation of variational data assimilation

The variational data assimilation approach could be viewed as an attempt to adjust globally the results obtained by a given model to a set of available observations. This approach has the theoretical advantage of providing consistency between the dynamics of the model and the final results of the assimilation. The idea was probably applied for the first time (within the field of atmospheric sciences) in 1971 in a paper written by Morel et al. [184]. They defined and implemented a procedure in which the assimilating model is repeatedly integrated forward and backward in time, the observations being constantly introduced in the model. The heuristic idea behind that procedure, supported by a number of numerical results, was that repeated calculations and corrections of model results would converge to a solution which would be compatible, at least to a certain degree of accuracy, with the observations.

Results from some preliminary investigations (based mainly on running simple tests) will be presented in this chapter. Several conclusions will be drawn. Data assimilation techniques will normally lead to huge computational tasks, which implies that one should try to find out, a priori, whether it us profitable or not to apply variational data assimilation in a particular study. The important problem of formulating a set of conditions under which the data assimilation approach will probably not give good results (these conditions have to be checked carefully before applying the data assimilation procedure in a comprehensive study similar to the studies, which were presented in Chapter 8 and Chapter 9), will be discussed in Section 6 and Section 7.

Some results, which were obtained by using an atmospheric chemical scheme containing 56 chemical compounds, will be presented in the end of Section 7.

A general discussion of the possibilities of improving the models by using data assimilation techniques as well as some plans for future work will be outlined in the end of the chapter.

10.1 Basic ideas

Assume that observations are available at time-points t_p , where $p \in \{0, 1, 2, \dots, P\}$. These observations can be taken into account in an attempt to improve the results obtained by a given model. This can be done by minimizing the value of the following functional (see, for example, Lewis and Derber [164]):

$$J\{\bar{c}_0\} = \frac{1}{2} \sum_{p=0}^{P} < W(t_p)\,(\bar{c}_p - \bar{c}_p^{obs})\,,\ \bar{c}_p - \bar{c}_p^{obs} >, \tag{10.1}$$

where

- the functional $J\{\bar{c}_0\}$ is depending on the initial value $\bar{c}_0$ of the vector of the concentrations at time t_0 (because the model results $\bar{c}_p$ depend on $\bar{c}_0$),
- $W(t_p)$ is a matrix containing some weights, and
- $<,>$ is an inner product in an appropriately defined Hilbert space (it will be assumed in this chapter that the usual vector space is used, i.e. it is assumed that $\bar{c} \in \mathbf{R^s}$ where s is the number of chemical species which are involved in the model).

It is seen that the functional $J\{\bar{c}_0\}$ depends on both the weights and the differences between calculated by the model concentrations $\bar{c}_p$ and observations $\bar{c}_p^{obs}$ at the time-levels t_p at which observations are available. $W(t_p)$ will be assumed to be the identity matrix I in this study, but in general weights are to be defined in some way and used in the computations.

The task is to find an improved initial field $\bar{c}_0$, which minimizes the functional $J\{\bar{c}_0\}$. This can be achieved by using some optimization algorithm. Most of the optimization algorithms are based on the application of the gradient of $J\{\bar{c}_0\}$. The adjoint equation has to be defined and used in the calculation of the gradient of the functional $J\{\bar{c}_0\}$.

It should be stressed that other formulations of the functional $J\{\bar{c}_0\}$ can also be used (see, for example, Daescu and Navon [54], Daescu et al. [55], Le Dimet et al. [159], Elbern and Schmidt, [88], [89], Elbern et al., [90], [91], [92], Talagrand and Courtier, [245]).

It is assumed here that data assimilation techniques are applied to improve an initial field of concentrations, but data assimilation can be applied for other purposes too. Other applications are

- improving emission fields,
- checking boundary conditions

and, in a more general context,

- checking the sensitivity of the concentrations to variation of different parameters (the major purpose being the improvement of some chemical and physical mechanisms used in the description of the model under consideration).

More details about data assimilation techniques and their implementation in different models can be found in Akella and Navon [3], Alekseev and Navon [4], Daescu et al. [55], Elbern et al., [92], Sandu et al., [209], [210], Wang et al., [264]. The relationship between the use of data assimilation and the use of a Kalman filter is studied in Li and Navon, [166].

10.2 Calculating the gradient of the functional

It is convenient to explain the basic ideas that are used when the gradient of the functional $J\{\bar{c}_0\}$ is calculated by the following very simple example. Assume that observations are available only at five time-points: t_0, t_1, t_2, t_3 and t_4. The gradient of the functional can be calculated by using the scheme which is shown in Fig. 10.1 (a colour version of Fig. 10.1, Fig. 10.1.col, is given in Appendix A). Assume that some tool, *model*, by which the values of the unknown vectors $\bar{c}(t_0)$, $\bar{c}(t_1)$, $\bar{c}(t_2)$, $\bar{c}(t_3)$ and $\bar{c}(t_4)$ can be calculated, is available. The tool *model* can, for example, be some air pollution model, but in some simpler cases *model* can be some ordinary solver for systems of PDEs or ODEs. Under this assumption, the calculations have to be performed, **for this particular example with** $P = 4$, in five steps.

- **Step 1.** Use the *model* to calculate $\bar{c}_1$ (performing integration, in a forward mode, from time-point t_0 to time-point t_1). Calculate the adjoint variable $\bar{q}_1 = \bar{c}_1 - \bar{c}_1^{obs}$. Form the adjoint equation (corresponding to the *model* used in the forward mode; adjoint equations will be discussed in the next section). Perform backward integration (by applying the adjoint equation) from time-point t_1 to time-point t_0 to calculate the vector $\bar{q}_0^1$, where the lower index shows that $\bar{q}_0^1$ is calculated at time-point t_0 , while the upper index shows that $\bar{q}_0^1$ is obtained by using $\bar{q}_1 = \bar{c}_1 - \bar{c}_1^{obs}$ as an initial vector in the backward integration.

- **Step 2 to Step 4.** Perform the same type of calculations, as those in Step 1 to obtain $\bar{q}_0^2$, $\bar{q}_0^3$ and $\bar{q}_0^4$. More precisely, the following operations are to be carried out for $p = 2, 3, 4$:

  (a) use the forward mode to proceed from time-point t_{p-1} to time-point t_p,
  (b) form the adjoint variable $\bar{q}_p = \bar{c}_p - \bar{c}_p^{obs}$,
  (c) use the adjoint equation in a backward mode from time-point t_p to time-point t_0 to calculate $\bar{c}_0^p$.

- **Step 5.** The sum of the vectors $\bar{q}_0^1$, $\bar{q}_0^2$, $\bar{q}_0^3$, $\bar{q}_0^4$ obtained in Step 1 to Step 4 and vector $\bar{q}_0^0 = \bar{q}_0 = \bar{c}_0 - \bar{c}_0^{obs}$ gives an approximation to the required gradient of the functional $J\{\bar{c}_0\}$.

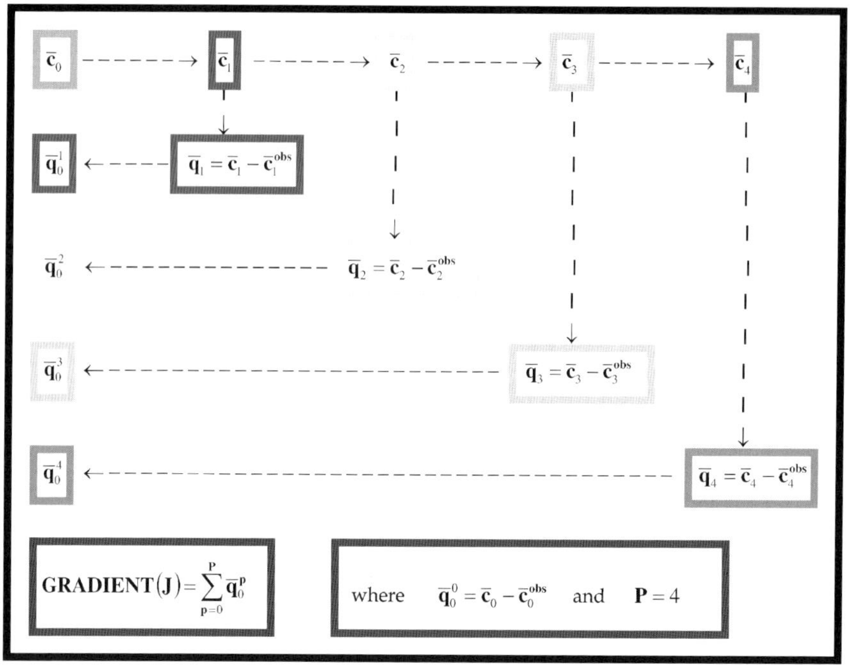

Figure 10.1: Computing the gradient of $J\{\bar{c}_0\}$ when observations in five time-points are available. A Colour version of this figure is given in Appendix A (see Fig. 10.1.col there).

It is clear that the above procedure can easily be extended for any number P of time-points at which observations are available.

It is seen from the scheme shown in Fig. 10.1 and Fig. 10.1.col that the gradient of the functional $J\{\bar{c}_0\}$ can be calculated by performing one forward step from time-point t_0 to time-point t_P and P backward steps from time-points t_p, $p = 1, 2, \ldots, P$, to time-point t_0. This explains the main idea, on which the data assimilation algorithms are based, in a very clear way, but it is expensive when P is large. In fact, the computational work can be reduced, performing only once the backward calculations. This possibility will be discussed in one of the following sections. It should be mentioned here that the reduction of the computational work by performing only once the backward calculations is leading to an increase of the storage requirements when non-linear problems are to be handled. Also this issue will be further discussed in the following sections.

10.3 Forming the adjoint equations

The adjoint equation has to be formed and used during the backward mode (see Fig. 10.1 and Fig. 10.1.col). The formation of the adjoint equation in two important cases, the case where the operator is linear and the case where the operator is non-linear will be discussed in this section.

Assume that the model is linear and, furthermore, that the model is written in the following general form:

$$\frac{\partial \bar{c}}{\partial t} = A\bar{c}. \tag{10.2}$$

Let $\bar{q} = \bar{c} - \bar{c}^{obs}$ be the adjoint variable (it is assumed here that some interpolation rules are used in order to get the continuous variable $\bar{c}^{obs}$ from the available discrete values of the observations). Then the adjoint equation, to equation (10.2) is defined by

$$\frac{\partial \bar{q}}{\partial t} = -A^T \bar{q}. \tag{10.3}$$

where the superscript T means that if we form the matrix by which the adjoint operator is represented after some kind of discretization, then this matrix is transposed to the matrix that represents operator A when the continuous equation (10.2) is discretized by applying some numerical algorithm.

It must be noted here that the notation A^* is normally used instead of A^T. The operator A^* is the conjugate operator of operator A. The reasons for using the notation A^T can be explain as follows:

- we should like to use the same notation both for the operator and for the matrix induced by some kind of discretization of the right-hand-side of equation (10.2), and
- if we agree to use the same notation for the operator A and for the matrix induced when when the right-hand-side of (10.2) is discretized, then it is natural to apply the notation for the transposed matrix, which is normally used in the numerical linear algebra studies.

Let us consider now the non-linear case. If the model under consideration is non-linear, then it is first necessary to produce a linearized version of the model. In other words, the non-linear model

$$\frac{\partial \bar{c}}{\partial t} = B(\bar{c}). \tag{10.4}$$

is rewritten as

$$\frac{\partial (\delta \bar{c})}{\partial t} = B'(\bar{c})\, \delta \bar{c}. \tag{10.5}$$

where $\delta \bar{c}$ is some small variation of $\bar{c}$ and B' is a linear operator obtained by differentiation of B. The adjoint equation of the linear model (10.5) can be formed in the following way:

$$\frac{\partial \bar{q}}{\partial t} = -\left[B'(\bar{c})\right]^T \bar{q}. \tag{10.6}$$

Note that the adjoint equation is linear both in the linear case and in the non-linear case, see (10.3) and (10.6). Much more important is the fact that in the non-linear case the adjoint operator depends on the values of $\bar{c}$. This causes difficulties (all values of $\bar{c}$ calculated during the forward mode, see Fig. 10.1 and Fig. 10.1.col, have to be saved because they are needed in the backward mode). The linear case, (10.2)-(10.3) is not causing difficulties related to the storage which has to be used with data assimilation procedures.

More details about adjoint equations can be found in Marchuk [178], Marchuk et al., [179], Wang et al., [264].

We shall finish this section with two remarks.

Remark 10.1. Adjoint equations of the continuous models have been discussed above. Sometimes it is more appropriate to form the adjoint equations of the discretised model (see, for example, Lewis and Derber [164]). Such formulations will be given in one of the next sections.

Remark 10.2. Very often splitting techniques are used in connection with large-scale models (Hundsdorfer and Verwer [142] and Zlatev [276], [282]; see also Chapter 2). If splitting is used, then one normally forms the adjoint equations corresponding to the obtained (by the splitting procedure chosen) sub-models; see, for example, Elbern and Schmidt, [88] or Sandu et al., [209] .

10.4 Algorithmic representation of a data assimilation algorithm

A data assimilation algorithm can be represented by applying the procedure described in Fig. 10.2. An optimization procedure is needed for the calculations that are to be carried out in the loop "DO ITERATIONS". In many optimization subroutines, the direction of the steepest descent is to be found and then the value of parameter ρ that gives the largest decrease in the direction found is to be used to improve the current solution. In practice, however, it is only necessary here to find a good standard minimization subroutine. In our experiments we used the subroutine E04DGF from the NAG Numerical Library [186]. This subroutine uses a preconditioned conjugate gradient methods and is based on the algorithm PLMA, described by Gill and Murray [120] and in Section 4.8.3 of Gill et al. [121]. It should be relatively easy to call another appropriate subroutine.

```
C
      Initialize scalar variables, vectors and arrays; set gradient to zero
C
      DO ITERATIONS = 1, MAX_ITERATIONS
C
         DO LARGE_STEPS = 1, P_STEP
C
            F_START=(LARGE_STEPS-1)*P_LENGTH+1
            F_END=LARGE_STEPS*P_LENGTH
            DO FORWARD_STEPS = F_START,F_END
               Perform forward steps by using the model
            END DO FORWARD_STEPS
C
            Form the adjoint equation and update the value of the functional
            DO BACKWARD_STEPS = F_END,1,-1
               Perform backward steps by using the adjoint equation
            END DO BACKWARD_STEPS
            Update the value of the gradient of the functional
C
         END DO LARGE_STEPS
C
         Compute an approximation of the optimization parameter ρ
         Update the initial values field:
         NEW_FIELD=OLD_FIELD + ρ*GRADIENT
         Check the stopping criteria
         Exit from loop DO ITERATIONS
         when the stopping criteria are satisfied
C
      END DO ITERATIONS
C
      Perform output operations and stop the computations
C
```

Figure 10.2: An algorithm for performing variational data assimilation by performing multiple backward calculations. P_STEP is equal to P. P_LENGTH is equal to the number of time-steps that are to be carried out between two time-points t_p and t_{p+1} at which observations are available.

```
C
      Initialize scalar variables, vectors and arrays; set gradient to zero
C
      DO ITERATIONS = 1, MAX_ITERATIONS
C
         DO FORWARD_STEPS = 1, P_STEP*P_LENGTH
            Perform forward steps by using the model
         END DO FORWARD_STEPS
C
         DO BACKWARD_STEPS = P_STEP*P_LENGTH, 1, -1
            Perform backward steps by using the adjoint equation
         END DO BACKWARD_STEPS
C
         Compute an approximation of the optimization parameter ρ
         Update the initial values field:
         NEW_FIELD=OLD_FIELD + ρ*GRADIENT
         Check the stopping criteria
         Exit from loop DO ITERATIONS
         when the stopping criteria are satisfied
C
      END DO ITERATIONS
C
      Perform output operations and stop the computations
C
```

Figure 10.3: An algorithm for performing variational data assimilation by performing the backward mode only once. P_STEP is equal to P. P_LENGTH is equal to the number of time-steps that are to be carried out between two time-points t_p and t_{p+1} at which observations are available.

Nearly all optimization subroutines (as mentioned in the previous sections of this chapter) need the value of the functional $J\{\bar{c}_0\}$ and its gradient. The calculation of these values is performed in the loop "DO LARGE_STEPS" in Fig. 10.2. This is a major part of the computational work and is based on scheme described in Fig. 10.1 and in Fig. 10.1.col. Let us reiterate here that the repeated backward steps can be avoided. An algorithm, in which only one backward step is carried out is given in Fig. 10.3. It is seen from Fig. 10.3 that one can perform the forward mode on the whole time-interval and then carry out the backward mode from the end of the interval to the starting point. The theory on which this algorithm is based is discussed in the remaining part of this section.

We shall show now that a single backward mode, from the end-point to the start-point of the integration interval, can be carried out instead of multiple backward modes that are used in the algorithm shown in Fig. 10.2. Let $W(t_p), p = 0, 1, \ldots, P$, be the identity matrix. Assume that the ordinary inner product is used. Then (10.1) can be rewritten as

$$J\{\bar{c}_0\} = \frac{1}{2} \sum_{p=0}^{P} \left(\bar{c}_p - \bar{c}_p^{obs}\right)^2. \tag{10.7}$$

It is also possible to consider $J\{\bar{c}_0\}$ as a function of the concentrations $\bar{c}_p$, $p = 0, 1, \ldots, P$, i.e.

$$J\{\bar{c}_0\} = J\{\bar{c}_0, \bar{c}_1, \ldots, \bar{c}_P\}. \tag{10.8}$$

It is clear that $J\{\bar{c}_0, \bar{c}_1, \ldots, \bar{c}_P\} : \quad \mathbf{R^{s(P+1)}} \quad \rightarrow \quad \mathbf{R}$. Denote by $\mathbf{M_p}$: $\mathbf{R^s} \quad \rightarrow \quad \mathbf{R^s}$ the operator which is used at the pth step of the forward mode; i.e. $\bar{c}_p = \mathbf{M_p}\, \bar{c}_{p-1}$. The following relationship is obtained by using this definition of operator $\mathbf{M_p}$ in (10.8) and, furthermore, assuming that there exists an operator $\mathbf{M_0}$, which satisfies the equality $\mathbf{M_0}\, \bar{c}_0 = \bar{c}_0$:

$$J\{\bar{c}_0, \bar{c}_1, \ldots, \bar{c}_P\} = J\{\bar{c}_0, \mathbf{M_1}\, \bar{c}_0, \mathbf{M_2}\, \mathbf{M_1}\, \bar{c}_0, \ldots, \mathbf{M_P}\, \ldots,\ \mathbf{M_2}\, \mathbf{M_1}\, \bar{c}_0\} \tag{10.9}$$

Let $\mathbf{M} : \quad \bar{c}_0 \quad \rightarrow \quad (\bar{c}_0,\ \bar{c}_1,\ \ldots,\ \bar{c}_P)$. It is obvious that (i) $\mathbf{M} : \mathbf{R^s} \ \rightarrow \ \mathbf{R^{s(P+1)}}$ and (ii) the following relationship holds

$$\mathbf{M} = (\mathbf{M_0},\ \mathbf{M_1 M_0},\ \mathbf{M_2 M_1 M_0}\ \ldots,\ \mathbf{M_P} \cdots \mathbf{M_2 M_1 M_0}). \tag{10.10}$$

The minimization of $\mathbf{J}$ requires the evaluation of the gradient of the composition $\mathbf{J} \circ \mathbf{M}$, because

$$\mathbf{Grad}\{\mathbf{J}(\bar{c}_0)\} = (\mathbf{J}\ \circ\ \mathbf{M})'\ (\bar{c}_0). \tag{10.11}$$

By using the chain rule in the differentiation of the right-hand-side, one can obtain:

$$\mathbf{Grad}\{\mathbf{J}(\bar{\mathbf{c}}_0)\} = (\mathbf{M}')^{\mathbf{T}}\ (\bar{c}_0)\ (\mathbf{J}'(\mathbf{M}\ (\bar{c}_0))). \tag{10.12}$$

Here $\mathbf{M}'\ (\bar{\mathbf{c}}_0)$ stands for the derivative of $\mathbf{M}$ and the superscript T refers to the adjoint operator (i.e. to the transpose of the corresponding matrix). This mean $(\mathbf{M}'\ (\bar{\mathbf{c}}_0))^T$ should be applied to the vector

$$\mathbf{J}'(\mathbf{M}\ (\bar{\mathbf{c}}_0)) = \mathbf{J}'(\bar{\mathbf{c}}_0, \bar{\mathbf{c}}_1, \ldots, \bar{\mathbf{c}}_P) \in \mathbf{R^{s(P+1)}} \tag{10.13}$$

in order to compute the gradient in the left-hand-side of equality (10.12).

As stated above, $\mathbf{M}'\ (\bar{\mathbf{c}}_0) : \mathbf{R}\ \rightarrow\ \mathbf{R^{s(P+1)}}$. Therefore, $(\mathbf{M}'\ (\bar{\mathbf{c}}_0))^T : \mathbf{R^{s(P+1)}}\ \rightarrow\ \mathbf{R}$. Consider $\bar{\mathbf{q}} = (\bar{\mathbf{q}}_0,\ \bar{\mathbf{q}}_1, \ldots,\ \bar{\mathbf{q}}_P)$, where $\bar{\mathbf{q}}_p = \bar{\mathbf{c}}_p - \bar{\mathbf{c}}_p^{obs}$, $\mathbf{p} = \mathbf{1},\ \mathbf{2},\ \ldots,\ \mathbf{P}$. It is readily seen that $\bar{\mathbf{q}} \in \mathbf{R^{s(P+1)}}$ and that the following equality holds:

$$(\mathbf{M}')^{\mathbf{T}}\ (\bar{q}_0) = (\bar{q}_0,\ \mathbf{M_1^T}(\bar{q}_1),\ \mathbf{M_1^T}\ \mathbf{M_2^T}(\bar{q}_2),\ \ldots,\ \mathbf{M_1^T}\ \mathbf{M_2^T} \cdots \mathbf{M_P^T}(\bar{q}_P)) \tag{10.14}$$

with $\mathbf{M}_\mathbf{p}^\mathbf{T}$, $\mathbf{p} = \mathbf{1}, \mathbf{2}, \ldots, \mathbf{P}$. The rule for forming the transposed matrix of a product of matrices is used in (10.14).

It follows from (10.7) and (10.8) that

$$\begin{aligned}
\mathbf{J}'(\bar{\mathbf{c}}_\mathbf{0}, \bar{\mathbf{c}}_\mathbf{1}, \ldots, \bar{\mathbf{c}}_\mathbf{P}) &= (\bar{\mathbf{q}}_\mathbf{0}, \bar{\mathbf{q}}_\mathbf{1}, \ldots, \bar{\mathbf{q}}_\mathbf{P}) \\
&= (\bar{\mathbf{c}}_\mathbf{0} - \bar{\mathbf{c}}_\mathbf{0}^\mathbf{obs}, \bar{\mathbf{c}}_\mathbf{1} - \bar{\mathbf{c}}_\mathbf{1}^\mathbf{obs}, \ldots \bar{\mathbf{c}}_\mathbf{P} - \bar{\mathbf{c}}_\mathbf{P}^\mathbf{obs}) \\
&\in \mathbf{R}^{\mathbf{s}(\mathbf{P}+\mathbf{1})}.
\end{aligned} \tag{10.15}$$

Now the desired gradient of the functional $J\{\bar{c}_0\}$ can be obtained by using (10.12), (10.14) and (10.15):

$$\begin{aligned}
\mathbf{Grad}\{\mathbf{J}(\bar{\mathbf{c}}_\mathbf{0})\} &= \mathbf{M}'(\bar{\mathbf{c}}_\mathbf{0})^\mathbf{T}(\mathbf{J}'(\bar{\mathbf{c}}_\mathbf{0})) \\
&= (\bar{\mathbf{c}}_\mathbf{0} - \bar{\mathbf{c}}_\mathbf{0}^\mathbf{obs}) \\
&+ \mathbf{M}_\mathbf{1}^\mathbf{T}(\bar{\mathbf{c}}_\mathbf{1} - \bar{\mathbf{c}}_\mathbf{1}^\mathbf{obs}) \\
&+ \mathbf{M}_\mathbf{1}^\mathbf{T}\mathbf{M}_\mathbf{2}^\mathbf{T}(\bar{\mathbf{c}}_\mathbf{2} - \bar{\mathbf{c}}_\mathbf{2}^\mathbf{obs}) \\
&+ \ldots \\
&+ \mathbf{M}_\mathbf{1}^\mathbf{T}\mathbf{M}_\mathbf{2}^\mathbf{T}\cdots\mathbf{M}_\mathbf{P}^\mathbf{T}(\bar{\mathbf{c}}_\mathbf{P} - \bar{\mathbf{c}}_\mathbf{P}^\mathbf{obs}).
\end{aligned} \tag{10.16}$$

The right-hand-side of (10.16) can be rewritten as

$$\begin{aligned}
(\bar{\mathbf{c}}_\mathbf{0} - \bar{\mathbf{c}}_\mathbf{0}^\mathbf{obs}) &+ \mathbf{M}_\mathbf{1}^\mathbf{T}(\bar{\mathbf{c}}_\mathbf{1} - \bar{\mathbf{c}}_\mathbf{1}^\mathbf{obs} \\
&+ \mathbf{M}_\mathbf{2}^\mathbf{T}(\bar{\mathbf{c}}_\mathbf{2} - \bar{\mathbf{c}}_\mathbf{2}^\mathbf{obs} \\
&+ \mathbf{M}_\mathbf{3}^\mathbf{T}(\bar{\mathbf{c}}_\mathbf{3} - \bar{\mathbf{c}}_\mathbf{3}^\mathbf{obs} + \ldots))\cdots),
\end{aligned} \tag{10.17}$$

which allows us to reduce considerably the computational work. It is necessary, according to (10.17) to use the algorithm shown in Table 10.1 instead of computing one by one the term in the right-hand-side of (10.16).

The algorithm shown in Table 10.1 has to be implemented when the second loop in Fig. 10.3 is to be executed. In this way, the number of backward loops is

Table 10.1: Algorithm for calculating the backward mode only once.

Step	Action
1	Set $i = P$ and $GRAD = 0$
2	Evaluate the difference $\bar{\mathbf{c}}_{\mathbf{i}} - \bar{\mathbf{c}}_{\mathbf{i}}^{\mathbf{obs}}$ and add it to $GRAD$
3	Apply $\mathbf{M}_{\mathbf{i}}^{\mathbf{T}}$ to the updated $GRAD$
4	Decrease the value of i by one
5	If $i \geq 0$ go to Step 2
6	The needed gradient $GRAD$ is calculated; exit

reduced from P (see Fig. 10.2) to only one. For large values of P this will result in a substantial reduction of the computational time. It is necessary to save all values $\bar{\mathbf{c}}_{\mathbf{p}}$, $p = 1, 2, \ldots, P$, which are calculated during the forward mode when the algorithm from Fig. 10.3 is used. If the problem solved is non-linear, then this has to be done anyway. For linear problems, however, this is not needed when the algorithm in Fig. 10.2 is used. This means that **for linear problems** there are two alternatives:

- the algorithm from Fig. 10. 2 is more profitable when the reduction of the storage used is more important than the reduction of the computational time, and
- the algorithm from Fig. 10.3, which is based on the algorithm shown in Table 10.1 is more profitable when the reduction of the computational time is more important than the reduction of the storage needed.

10.5 Variational data assimilation for some one-dimensional examples

It is relatively easy to understand the main ideas on which the variational data assimilation is based by studying the process for simple one-dimensional transport equations. Assume that the unknown variable $c \in \mathbf{R}$ is the concentration of a given chemical compound and V is some given function ("wind velocity field"), which may depend on t, x and c . Numerical experiments with

- V is a constant,
- V depending on t and/or x, and
- $V = c$

have been carried out. In all three cases the analytic solution was known, which allowed us to check easily the correctness of both the calculated results and the implementation of the algorithms. Only results obtained by using V depending on x (i.e. only results obtained in the solution of one-dimensional transport equations) will be presented in this section.

Any of the three cases can be described mathematically by the following partial differential equation (PDE):

$$\frac{\partial c}{\partial t} = -V\frac{\partial c}{\partial x}, \quad x \in [a,b], \quad t \in [0,T], \quad c(x,0) = f(x). \tag{10.18}$$

If the spatial interval, the initial value conditions and function V are chosen to be

$$[a,b] = [0.2\pi], \quad f(x) = \sin(x), \quad V(x) = \frac{6x(2\pi - x)}{(2\pi)^2}, \tag{10.19}$$

then (10.18) becomes identical to one of the examples, which are studied in Lewis and Derber [164]. Its analytical solution is given by

$$c(x,t) = \sin\left(\frac{2\pi x}{x + (2\pi - x)\exp\left(\frac{3t}{\pi}\right)}\right). \tag{10.20}$$

Assume that

$$x_i - x_{i-1} = \triangle x \quad for \quad i = 1, 2, \ldots, N_x \tag{10.21}$$

and

$$t_n - t_{n-1} = \triangle t \quad for \quad n = 1, 2, \ldots, N_t \tag{10.22}$$

Consider the grids:

$$\mathbf{G_x} = \{x_i | i = 0, 1, \ldots, N_x, \ x_0 = 0, \ x_{N_x} = 2\pi\} \tag{10.23}$$

and

$$\mathbf{G_t} = \{t_n | n = 0, 1, \ldots, N_t, \ t_0 = 0, \ t_{N_t} = 2\pi\} \tag{10.24}$$

Let $x_i \in \mathbf{G_x}$ and $t_n \in \mathbf{G_t}$. Different numerical methods can be used to calculate approximations

$$c_{i,n} \approx c(x_i, t_n) \tag{10.25}$$

of the exact solution at point (x_i, tn).

The formula (10.26), which is given below, can be obtained by using the notation $w_i = (Vx_i\triangle t)/4\triangle x)$ and simple finite differences:

$$-w_i c_{i-1,n+1} + c_{i,n+1} + w_i c_{i+1,n+1} = w_i c_{i-1,n} + c_{i,n} - w_i c_{i+1,n}. \tag{10.26}$$

Dirichlet boundary conditions can be introduced by the following formulae (note that these boundary conditions are the values of the analytical solution in the end-points of the x-interval):

$$c_{0,n} = 0, \quad c_{N_x,n} = 0, \quad n = 0, 1, \ldots, N_t. \tag{10.27}$$

By applying (10.27), formula (10.26) can be rewritten for $i = 1$ as

$$c_{1,n+1} + w_1 c_{2,n+1} = c_{1,n} - w_1 c_{2,n} \tag{10.28}$$

and for $i = N_x - 1$ as

$$-w_{N_x-1} c_{N_x-2,n+1} + c_{N_x-1,n+1} = w_{N_x-1} c_{N_x-2,n} + c_{N_x-1,n}. \tag{10.29}$$

Denote

$$\bar{c}_n = (c_{1,n}, c_{2,n}, \ldots, c_{N_x-1,n})^T. \tag{10.30}$$

By using the notation introduced by (10.30), it is possible to rewrite (10.26), (10.28) and (10.29) in a matrix form:

$$(I - A)\bar{c}_{n+1} = (I + A)\bar{c}_n \quad \Rightarrow \quad \bar{c}_{n+1} = D\bar{c}_n, \tag{10.31}$$

$$D \stackrel{\text{def}}{=} (I - A)^{-1}(I + A), \tag{10.32}$$

where I is the identity matrix and A is a matrix which has non-zero elements only on the two diagonals that are adjacent to the main diagonal (and, more precisely, $-w_i$ on the diagonal below the main diagonal and w_i on the diagonal above the main diagonal).

If $\bar{c}_n$ has already been calculated, then (5.12) can be used to proceed with the calculation of $\bar{c}_{n+1}$. Thus, if an initial field, $\bar{c}_0$, is given, then (10.31) can be used to calculate successively, step-by-step, approximations of the exact solution.

As mentioned above, it is necessary to calculate the gradient $Grad\{J\}$ of the functional $J\{\bar{c}_0\}$ from (10.1) in order to find an improved initial field $\bar{c}_0$. Assume as in the first section of this chapter that observations are available at times $\{t_p \mid p = 0, 1, \ldots, P\}$. Assume also that the calculations by formula (10.31) for some $n+1 = p$ have been completed. It is then necessary to form the adjoint variable

$$\bar{q}_p = \bar{W}_p\,(\bar{c}_p - \bar{c}_p^{obs}) \tag{10.33}$$

and to use it as a starting value in the integration of the adjoint equation backward from $t = t_p$ to $t = t_0$ (let us reiterate that we have assumed that $\bar{W}_p = I$ for all values of p). The backward calculations can be carried out by using the following formula, which is the discrete adjoint equation corresponding to (10.31):

$$\bar{q}_n = -D^T \bar{q}_{n+1}, \tag{10.34}$$

where D is defined in (10.32).

Table 10.3: Results obtained in the checks of the performance of the variational data assimilation algorithm when the initial values are perturbed by random errors generated by using $\alpha = 0.1$. $\mathbf{ERR_0}$ is the error induced by the random perturbations, $\mathbf{ERR_{global}}$ is the global error when the forward integration is performed with the perturbed initial field, $\mathbf{ERR_0^{impr}}$ is the error in the improved by the variational data assimilation algorithm initial field, $\mathbf{ERR_{global}^{impr}}$ is the global error when the forward integration is performed with the improved by the variational data assimilation algorithm initial field.

$\mathbf{N_x}$	$\mathbf{N_t}$	$\mathbf{ERR_0}$	$\mathbf{ERR_{global}}$	$\mathbf{ERR_0^{impr}}$	$\mathbf{ERR_{global}^{impr}}$
11	100	$4.55 * 10^{-2}$	$2.93 * 10^{-1}$	$1.94 * 10^{-1}$	$1.94 * 10^{-1}$
101	1000	$4.47 * 10^{-2}$	$7.53 * 10^{-2}$	$2.62 * 10^{-3}$	$2.62 * 10^{-3}$
1001	10000	$4.99 * 10^{-2}$	$1.10 * 10^{-1}$	$2.64 * 10^{-5}$	$2.64 * 10^{-5}$
10001	100000	$4.99 * 10^{-2}$	$1.44 * 10^{-1}$	$2.93 * 10^{-7}$	$2.93 * 10^{-7}$

10.6.3 Results obtained with different values of parameter α

The results given in Table 10.3 are obtained, as stated in the previous subsection, by using $\alpha = 0.1$, which means that small errors, of about 5%, are introduced in the initial solution. Therefore, it is necessary to investigate the case where the perturbations are larger. Larger perturbations can be created by increasing the value of α . Results that are obtained with $\mathbf{N_x} = 1001$, $\mathbf{N_t} = 10001$ and different values of parameter α are given in Table 10.4. It is clearly seen that the magnitude of the errors in the forward mode obtained with the improved initial values are practically not depending on the size of parameter α .

10.6.4 Results obtained when the perturbations create some bias

The perturbations used in previous three subsections are not creating any bias. Bias could be simulated by using the following formulae:

$$c_{i,0} = \{1 + \mathbf{DRAND(0)}\ \alpha\} \sin(x_i), \tag{10.36}$$

$$c_{i,0} = \{1 - \mathbf{DRAND(0)}\ \alpha\} \sin(x_i), \tag{10.37}$$

The first formula, (10.36), is producing over-estimated values of the initial solution. The second formula, (10.37), is giving under-estimated values of the initial solution.

Table 10.4: Results obtained in the checks of the performance of the variational data assimilation algorithm with $N_x = 1001$, $N_t = 10001$ and α varying from 0.1 to 1.0. ERR_0, ERR_{global}, ERR_0^{impr} and ERR_{global}^{impr} are defined in Table 10.3.

α	ERR_0	ERR_{global}	ERR_0^{impr}	ERR_{global}^{impr}
0.1	$4.99 * 10^{-2}$	$1.10 * 10^{-1}$	$2.64 * 10^{-5}$	$2.64 * 10^{-5}$
0.2	$9.94 * 10^{-2}$	$2.31 * 10^{-1}$	$2.64 * 10^{-5}$	$2.64 * 10^{-5}$
0.3	$1.47 * 10^{-1}$	$3.28 * 10^{-1}$	$2.64 * 10^{-5}$	$2.64 * 10^{-5}$
0.4	$2.00 * 10^{-1}$	$4.14 * 10^{-1}$	$2.64 * 10^{-5}$	$2.64 * 10^{-5}$
0.5	$2.46 * 10^{-1}$	$5.83 * 10^{-1}$	$2.64 * 10^{-5}$	$2.64 * 10^{-5}$
0.6	$2.99 * 10^{-1}$	$6.79 * 10^{-1}$	$2.64 * 10^{-5}$	$2.64 * 10^{-5}$
0.7	$3.43 * 10^{-1}$	$9.47 * 10^{-1}$	$2.64 * 10^{-5}$	$2.64 * 10^{-5}$
0.8	$3.96 * 10^{-1}$	$8.55 * 10^{-1}$	$2.64 * 10^{-5}$	$2.64 * 10^{-5}$
0.9	$4.49 * 10^{-1}$	$1.07 * 10^{0}$	$2.64 * 10^{-5}$	$2.64 * 10^{-5}$
1.0	$4.90 * 10^{-1}$	$1.02 * 10^{0}$	$2.64 * 10^{-5}$	$2.64 * 10^{-5}$

Table 10.5: Results obtained in the checks of the performance of the variational data assimilation algorithm with $N_x = 1001$, $N_t = 10001$ and α varying from 0.1 to 1.0. Bias is produced by using formula (10.36). ERR_0, ERR_{global}, ERR_0^{impr} and ERR_{global}^{impr} are defined in Table 10.3.

α	ERR_0	ERR_{global}	ERR_0^{impr}	ERR_{global}^{impr}
0.1	$1.00 * 10^{-1}$	$1.77 * 10^{-1}$	$2.64 * 10^{-5}$	$2.64 * 10^{-5}$
0.2	$1.99 * 10^{-1}$	$3.63 * 10^{-1}$	$2.64 * 10^{-5}$	$2.64 * 10^{-5}$
0.3	$2.96 * 10^{-1}$	$5.37 * 10^{-1}$	$2.64 * 10^{-5}$	$2.64 * 10^{-5}$
0.4	$4.00 * 10^{-1}$	$7.08 * 10^{-1}$	$2.64 * 10^{-5}$	$2.64 * 10^{-5}$
0.5	$4.96 * 10^{-1}$	$8.97 * 10^{-1}$	$2.64 * 10^{-5}$	$2.64 * 10^{-5}$
0.6	$5.98 * 10^{-1}$	$1.00 * 10^{0}$	$2.64 * 10^{-5}$	$2.64 * 10^{-5}$
0.7	$6.98 * 10^{-1}$	$1.33 * 10^{0}$	$2.64 * 10^{-5}$	$2.64 * 10^{-5}$
0.8	$7.89 * 10^{-1}$	$1.40 * 10^{0}$	$2.64 * 10^{-5}$	$2.64 * 10^{-5}$
0.9	$8.95 * 10^{-1}$	$1.65 * 10^{0}$	$2.64 * 10^{-5}$	$2.64 * 10^{-5}$
1.0	$9.87 * 10^{-1}$	$1.88 * 10^{0}$	$2.64 * 10^{-5}$	$2.64 * 10^{-5}$

Results obtained by producing perturbations in the initial value by using (10.36) are given in Table 10.5. The error in the improved initial field as well as the global error in the forward mode do not depend on the value of α (and, as a matter of fact, the rounded to the third significant decimal numbers are the same for all values of

α). Results obtained by using (10.37) instead of (10.36) are quite similar to the results shown in Table 10.5.

10.6.5 Results obtained by using reduced number of observations

The results given in the previous sections were obtained by using "observations" (exact values of the solution) at all spatial grid-points x_i, $i = 0, 1, \ldots, N_x$, at the time-points $p = 1, 2, \ldots, P$ (i.e. at every point where backward calculations are initiated). In the practical situations the number of observations will be limited. Therefore, it is necessary to investigate the accuracy of the results when the number of observations is reduced. Two different cases should be investigated:

- the adjoint variable is set to zero if there is no observation at spatial grid-points under consideration, and
- an attempt to get observations at all spatial grid-points (by interpolation) is made.

Case A - The adjoint variable is set to zero when no observation is available

It is easy to apply this possibility. Some results are given in Table 10.6. It is assumed that observations are available at every second spatial grid-point. This is a very restrictive assumption when refined grids are used. Nevertheless, it is seen that the results are much less accurate than those presented in Table 10.3 (where observations in all grid-points were used in the calculations). The results indicate that if the spatial grid is fine (100 or more grid-points), then the results can be improved by using data assimilation by a factor approximately equal to 2. The results shown in Table 10.7 indicate that this statement seems to be true also when the magnitude of the perturbations is increased by varying parameter α. Comparing the results in Table 10.7 with the corresponding results in Table 10.4, it is seen that

- the results in the third and the fourth columns are identical in the two tables, and
- while the errors given in the fifth and the sixth columns in Table 10.4 do not depend on parameter α, the corresponding errors in Table 10.7 grow when the value of α is increased.

Table 10.6: Results obtained in the checks of the performance of the variational data assimilation algorithm when the initial values are perturbed by random errors produced by using $\alpha = 0.1$. Observations are available in every second spatial grid-point. The adjoint variable is set to zero at the points where observations are not available. ERR_0, ERR_{global}, ERR_0^{impr} and ERR_{global}^{impr} are defined in Table 10.3.

N_x	N_x	ERR_0	ERR_{global}	ERR_0^{impr}	ERR_{global}^{impr}
11	100	$4.55 * 10^{-2}$	$2.93 * 10^{-1}$	$1.01 * 10^{-1}$	$2.28 * 10^{-1}$
101	1000	$4.47 * 10^{-2}$	$7.53 * 10^{-2}$	$2.47 * 10^{-2}$	$3.97 * 10^{-2}$
1001	10000	$4.99 * 10^{-2}$	$1.10 * 10^{-1}$	$2.54 * 10^{-2}$	$5.56 * 10^{-2}$

Table 10.7: Results obtained in the checks of the performance of the variational data assimilation algorithm with $N_x = 1001$, $N_t = 10001$ and α varying from 0.1 to 1.0. Observations are available in every second spatial grid-point. The adjoint variable is set to zero at the points where observations are not available. ERR_0, ERR_{global}, ERR_0^{impr} and ERR_{global}^{impr} are defined in Table 10.3.

α	ERR_0	ERR_{global}	ERR_0^{impr}	ERR_{global}^{impr}
0.1	$4.99 * 10^{-2}$	$1.10 * 10^{-1}$	$2.54 * 10^{-2}$	$5.56 * 10^{-2}$
0.2	$9.94 * 10^{-2}$	$2.31 * 10^{-1}$	$5.00 * 10^{-2}$	$1.16 * 10^{-1}$
0.3	$1.47 * 10^{-1}$	$3.28 * 10^{-1}$	$7.38 * 10^{-1}$	$1.62 * 10^{-1}$
0.4	$2.00 * 10^{-1}$	$4.14 * 10^{-1}$	$1.00 * 10^{-1}$	$2.07 * 10^{-1}$
0.5	$2.46 * 10^{-1}$	$5.83 * 10^{-1}$	$1.28 * 10^{-1}$	$2.94 * 10^{-1}$
0.6	$2.99 * 10^{-1}$	$6.79 * 10^{-1}$	$1.52 * 10^{-1}$	$3.43 * 10^{-1}$
0.7	$3.43 * 10^{-1}$	$9.47 * 10^{-1}$	$1,72 * 10^{-1}$	$4.72 * 10^{-1}$
0.8	$3.96 * 10^{-1}$	$8.55 * 10^{-1}$	$2.00 * 10^{-1}$	$4.26 * 10^{-1}$
0.9	$4.49 * 10^{-1}$	$1.07 * 10^{0}$	$2.31 * 10^{-1}$	$5.37 * 10^{-1}$
1.0	$4.90 * 10^{-1}$	$1.02 * 10^{0}$	$2.46 * 10^{-1}$	$5.15 * 10^{-1}$

Case B - Using interpolation to obtain observations at all spatial points

The results in given in the previous paragraph show that it is not advisable to set the adjoint variable to zero when no observation is available at the spatial grid-point under consideration. It will be shown now that it might be more profitable, in some situations at least, to try to calculate approximations to the missing observations by using some interpolation rules. For our purposes, the application of simple rules for linear interpolation is quite sufficient and such rules will be used in this section. However, some more advanced interpolation rules (as, for example, cubic splines) may be more successful in some other situations.

The results in Table 10.8 are obtained using the same parameters as those used to calculate the results in Table 10.6 with the only exception that linear interpolation is used to obtain approximations of the observations in the spatial grid-points at which "observations" are not available.

Some results for $\mathbf{ERR_{global}^{impr}}$ obtained when both the number of "available observations" and parameter α are varied are given in Table 10.9. It is seen that the results become poorer when the number of observations is decreased. However, the accuracy of the results is very insensitive to variations of α (i.e. the variations of the magnitude of the perturbations inserted in the initial values).

Table 10.8: Results obtained in the checks of the performance of the variational data assimilation algorithm when the initial values are perturbed by random errors produced by using $\alpha = 0.1$. Observations are available in every second spatial grid-point. Interpolation is used at the points where observations are not available. $\mathbf{ERR_0}$, $\mathbf{ERR_{global}}$, $\mathbf{ERR_0^{impr}}$ and $\mathbf{ERR_{global}^{impr}}$ are defined in Table 10.3.

$\mathbf{N_x}$	$\mathbf{N_x}$	$\mathbf{ERR_0}$	$\mathbf{ERR_{global}}$	$\mathbf{ERR_0^{impr}}$	$\mathbf{ERR_{global}^{impr}}$
11	100	$4.55 * 10^{-2}$	$2.93 * 10^{-1}$	$1.01 * 10^{-1}$	$2.28 * 10^{-1}$
101	1000	$4.47 * 10^{-2}$	$7.53 * 10^{-2}$	$2.61 * 10^{-3}$	$3.17 * 10^{-3}$
1001	10000	$4.99 * 10^{-2}$	$1.10 * 10^{-1}$	$2.64 * 10^{-5}$	$3.21 * 10^{-5}$
10001	100000	$4.99 * 10^{-2}$	$1.44 * 10^{-1}$	$2.64 * 10^{-7}$	$2.64 * 10^{-7}$

10.6.6 Varying the number of time-points at which observations are available)

In all experiments in the previous subsections the number of time-steps per a backward step was fixed (i.e. after performing 10 time-steps the adjoint variable was formed and a new backward step was calculated by performing backward integration to the initial time-point t_0). This might be very expensive (in terms of computational time). On the other hand (and this may be much more important in practice), observations are often available at a few time-points. Therefore, it is worthwhile to investigate the behaviour of the variational data assimilation when the number of backward steps is varied. It is desirable to be able to use as many as possible time-steps per backward step (or, in other words, to reduce the number of backward steps as much as possible). Some results, which are obtained by using

- $\mathbf{N_x} = 1001$,

Table 10.9: Values of $\mathbf{ERR^{impr}_{global}}$ (see the caption in Table 10.3) obtained by using the variational data assimilation algorithm with $\mathbf{N_x} = 1001$, $\mathbf{N_t} = 10001$ and α varying from 0.1 to 1.0. The number of observations N_OBS is varied in this experiment.

	Number of observations **N_OBS**				
α	**1000**	**500**	**125**	**20**	**5**
0.1	$2.64 * 10^{-5}$	$3.21 * 10^{-5}$	$3.39 * 10^{-4}$	$1.34 * 10^{-2}$	$2.05 * 10^{-1}$
0.2	$2.64 * 10^{-5}$	$3.21 * 10^{-5}$	$3.39 * 10^{-4}$	$1.34 * 10^{-2}$	$2.05 * 10^{-1}$
0.3	$2.64 * 10^{-5}$	$3.21 * 10^{-5}$	$3.39 * 10^{-4}$	$1.34 * 10^{-2}$	$2.05 * 10^{-1}$
0.4	$2.64 * 10^{-5}$	$3.21 * 10^{-5}$	$3.39 * 10^{-4}$	$1.34 * 10^{-2}$	$2.05 * 10^{-1}$
0.5	$2.64 * 10^{-5}$	$3.21 * 10^{-5}$	$3.39 * 10^{-4}$	$1.34 * 10^{-2}$	$2.05 * 10^{-1}$
0.6	$2.64 * 10^{-5}$	$3.21 * 10^{-5}$	$3.39 * 10^{-4}$	$1.34 * 10^{-2}$	$2.05 * 10^{-1}$
0.7	$2.64 * 10^{-5}$	$3.21 * 10^{-5}$	$3.39 * 10^{-4}$	$1.34 * 10^{-2}$	$2.05 * 10^{-1}$
0.8	$2.64 * 10^{-5}$	$3.21 * 10^{-5}$	$3.39 * 10^{-4}$	$1.34 * 10^{-2}$	$2.05 * 10^{-1}$
0.9	$2.64 * 10^{-5}$	$3.21 * 10^{-5}$	$3.39 * 10^{-4}$	$1.34 * 10^{-2}$	$2.05 * 10^{-1}$
1.0	$2.64 * 10^{-5}$	$3.21 * 10^{-5}$	$3.39 * 10^{-4}$	$1.34 * 10^{-2}$	$2.05 * 10^{-1}$

- $\mathbf{N_t} = 10001$,
- $\alpha = 0.1$,
- five different values of **N_OBS**, and
- different numbers of backward steps (**P_STEP**),

are given in Table 10.10.

10.6.7 Conclusions from the experiments with one-dimensional problems

The test-example defined by (10.18)-(10.20) is very simple. Therefore, it is only possible to make conclusions about the cases in which

- it is quite certain that the data assimilation algorithm is not performing well (it is quite clear that if the algorithm does not work for such a simple example, then it will not work when more complicated situations are to be handled), and
- there is some hope that the data assimilation algorithm will perform well (but it should be emphasized that more experiments with test examples that are closer to real cases are necessary in order to confirm such expectations).

Table 10.10: Values of ERR_{global}^{impr} (see the caption in Table 10.3) obtained by using the variational data assimilation algorithm with $N_x = 1001$, $N_t = 10001$, $\alpha = 0.1$, five values of N_OBS and different numbers of backward steps P_STEP = P.

	Number of observations N_OBS				
P_STEP	1000	500	125	20	5
1000	$2.64 * 10^{-5}$	$3.21 * 10^{-5}$	$3.39 * 10^{-4}$	$1.34 * 10^{-2}$	$2.05 * 10^{-1}$
100	$2.66 * 10^{-5}$	$3.20 * 10^{-5}$	$3.48 * 10^{-4}$	$1.34 * 10^{-2}$	$2.06 * 10^{-1}$
10	$2.90 * 10^{-5}$	$3.04 * 10^{-5}$	$3.91 * 10^{-4}$	$1.50 * 10^{-2}$	$2.15 * 10^{-1}$
1	$5.27 * 10^{-5}$	$5.27 * 10^{-5}$	$5.87 * 10^{-4}$	$2.75 * 10^{-2}$	$3.55 * 10^{-1}$

(A) When will the data assimilation algorithm not work?

The experiments in this section indicate that there are at least two cases in which the data assimilation algorithm will not lead to a substantial improvement of the results. These cases are:

- the discretization is crude, and
- the number of observations is very small.

This implies that two requirements should be imposed when data assimilation algorithms are to be used:

- the discretizations should be sufficiently accurate, and
- the number of available observations should be sufficiently large.

We are not discussing here the reliability of the observations, i.e. it was assumed that the observations are reliable. In the real case the reliability of the observations is an important topic, which should be taken into account when data assimilation algorithms are to be used.

(B) When will the data assimilation algorithm likely work well?

It is clear from the experiments in this section that one should expect the data assimilation algorithm to work satisfactorily well in the cases where the following conditions are satisfied:

- If there are sufficiently many observations, then the results will be substantially improved even if the initial values are quite crude (see Table 10.4). This is also true when there is some bias of the errors in the initial values (see Table 10.5).

- Assume that the number of observations is less than the number of spatial grid-points (which will be always the case in real situations). In this case, it might be useful to calculate (by applying some interpolation rules) approximate observations in the grid-points where there are not observations (compare the results in Table 10.6 and Table 10.7 with the results in Table 10.8 and Table 10.9).

- The results seem to be rather insensitive to the number of backward steps that are actually used (see Table 10.10). It is important to check further whether this nice property holds in more complicated cases, because the data assimilation procedure will become extremely expensive if many backward steps are needed.

- It is important to improve the performance of the data assimilation algorithm with regard to computational time (it is seen that there are problems with the computational time even for the simple one-dimensional example treated in this paper). Parallel algorithms, similar to those used in Alexandrov et al. [5] and Dimov et al. [71] (see also Chapter 7) might lead to a considerable improvement of the performance.

(C) How to continue the verification of the algorithm for transport equations?

It is clear that, before applying data assimilation algorithms in particular studies involving large-scale air pollution models, it is necessary to eliminate the cases in which these algorithms will not work well. In this section it was shown, by using a simple example, that there are at least two such cases.

The problem defined by (10.18)-(10.20) is a simple one-dimensional transport equation. In the further investigations two and three-dimensional test-examples should also be studied.

Some numerical examples, which are similar in some sense to the chemical reaction schemes that are used in large-scale air pollution models (as well as in some large-scale models that appear in other fields of science and engineering) will be treated in the next section of this chapter. We shall conclude the chapter by studying the impact of operational splitting, which is practically unavoidable in real applications (see Chapter 2), on the data assimilation algorithms.

10.7 Treatment of some simple non-linear tests-problems

The program EXAMINATOR, which has been discussed and tested in Chapter 2 to examine the performance of different numerical methods combined with different splitting procedures in the case when the problems are solved without using data assimilation techniques, will be applied in this section to study the performance of the same 30 combinations of numerical methods and splitting procedures in the case where some data assimilation technique is applied.

The test-example studied at the end of Chapter 2, will also be used in this chapter. In Chapter 2 the interaction of the splitting procedures and the numerical methods was studied. Now the performance of the data assimilation modules for for non-linear problems will be investigated by using this test-example.

The test-example is defined by (2.51) and (2.52). Let us reiterate here that this test-example is non-linear, the mass conservation law holds and the components of the solution are non-negative. This means that the problem formulated by (2.51) and (2.52) has the same properties as the chemical schemes. Moreover, the exact solution is known, given by (2.53), and, therefore, it is possible to use the same techniques as in the previous sections of this chapter, i.e. to perturb the initial values and to study the ability of the data assimilation modules to calculate a good approximation of the initial values.

The same numerical methods, as those used in Chapter 2, will be used in the solution of the problem (2.51) and (2.52). In other words, the numerical methods, which are used to produce the results given in the tables presented in the remaining part of this chapter, are:

- the Backward Euler Method defined by (10.38),
- the Implicit Mid-point Rule defined by (2.59) , and
- the Fully Implicit Three-stage Runge-Kutta Method defined (2.60) - (2.67).

Three other well-known numerical methods for integration of systems of ordinary differential equations

- the Modified Diagonally Implicit Runge Kutta Method (Zlatev [273]),
- the Two-stage Rosenbrock Method (Hundsdorfer and Verwer [142]), and
- the Trapezoidal Rule (Hairer and Wanner [127])

were also used in the experiments.

Most of the results presented in the remaining part of this chapter were obtained by using the first three numerical methods. However, a few results obtained when all these six numerical methods were used will also be shown and discussed.

The same splitting procedures, as those used in Chapter 2, will be used in this chapter, i.e. the sequential splitting, the symmetric splitting, the weighted sequential splitting and the weighted symmetric splitting. Tests where no splitting procedure is applied will also be carried out. The simple operators defined by (2.55) and (2.56) are used in all splitting procedures.

Two major issues will be discussed in this section: (i) the ability of the data assimilation modules to improve the accuracy of the solution and (ii) the interaction between numerical methods and splitting procedures.

10.7.1 Application of the Backward Euler Method

The forward mode of the data assimilation procedure is carried out by using (2.58) from Chapter 2. For convenience this formula is given below:

$$c_{n+1} = c_n + \Delta t f(c_{n+1}). \tag{10.38}$$

The fact that the test-example (2.51) is an autonomous system of ODEs is exploited in (10.38). If the problem is non-autonomous, then $f(t_{n+1}, c_{n+1})$ should be used in (10.38) instead of $f(c_{n+1})$ (the alternative is to transform the non-autonomous system of ODEs to an autonomous one by adding the equation $t' = 1$).

The adjoint equation, which has to be used in the backward mode, is defined as follows:

$$q_n = q_{n+1} - \Delta t \left[-\frac{\partial f(c_n)}{\partial c} \right]^T q_n. \tag{10.39}$$

Also the adjoint equation (10.39) is written for autonomous systems of ODEs. As above, $f(c_n)$ should be replaced by $f(t_n, c_n)$ when the system of ODEs is non-autonomous (the alternative is to transform, as above, the non-autonomous system of ODEs to an autonomous one).

The accuracy of the results obtained when the data assimilation modules are used with different splitting procedures is shown in Table 10.11. In fact, the greatest errors obtained with the improved solution are given in Table 10.11. This means that the following actions are performed:

- the initial solution is perturbed by using $\alpha = 1.0$ in (10.35), which means that the size of perturbations being about 50% of the values of the initial solution,
- the accuracy of the initial solution is recovered by the data assimilation modules, and, finally,
- a forward run is performed with the improved initial solution.

The greatest errors obtained during this final run are given Table 10.11.

Increasing the number of time-steps by a factor of two leads, as in Chapter 2, to a reduction of the time-step size by a factor of two. Therefore, one should expect that the accuracy will be improved when the number of time-steps is increased. The results in Table 10.11 shows that the accuracy is indeed improved when more time-steps are used.

It is also important to investigate the rate of the improvement of the accuracy. The rates of improvement (i.e. the ratios between the greatest errors found in two successive runs) are given in Table 10.12. The Backward Euler Method is a first-order numerical method. Therefore, we should expect, see also (2.70), that the combined method (the Backward Euler Method with the selected splitting procedure) will always perform as a first-order method. It is seen from Table 10.12

Table 10.11: Performance of the data assimilation modules for non-linear problems. Accuracy results achieved when the Backward Euler Method is used with five splitting procedures. The abbreviations *Steps*, *"Splitting 0"*, *"Splitting 1"*, *"Splitting 2"*, *"Splitting 3"* and *"Splitting 4"* are used for (i) the number of time-steps and (ii) the cases where no splitting, sequential splitting, symmetric splitting, weighted sequential splitting and weighted symmetric splitting are used in the computations.

Steps	Splitting 0	Splitting 1	Splitting 2	Splitting 3	Splitting 4
40	2.09E-2	2.08E-2	1.40E-1	2.60E-1	1.81E-1
80	1.09E-2	1.09E-2	7.40E-2	1.26E-2	9.22E-2
160	5.60E-3	5.60E-3	3.80E-2	6.26E-2	4.64E-2
320	2.84E-3	2.84E-3	1.93E-2	3.12E-2	2.33E-2
640	1.43E-3	1.48E-3	9.70E-3	1.56E-2	1.17E-2
1280	7.16E-4	7.16E-4	4.86E-3	7.81E-3	5.85E-3
2560	3.59E-4	3.59E-4	2.44E-3	3.91E-3	2.93E-3
5120	1.79E-4	1.79E-4	1.22E-3	1.95E-3	1.46E-3
10240	8.98E-5	8.98E-5	6.10E-4	9.76E-4	7.31E-4
20480	4.49E-5	4.49E-5	3.05E-4	4.88E-4	3.66E-4
40960	2.24E-5	2.24E-5	1.52E-4	2.44E-4	1.83E-4
81920	1.12E-5	1.12E-5	7.63E-5	1.22E-4	9.15E-5
163840	5.61E-6	5.61E-6	3.81E-5	6.10E-5	4.58E-5
327680	2.81E-6	2.81E-6	1.91E-5	3.05E-5	2.28E-5

that the combined method behaves as a first-order numerical method for all five splitting procedures (including here the case when no splitting is used).

The results given in Table 10.11 and Table 10.12 are nearly identical to the results obtained when the data assimilation modules are not used (i.e. when the exact values of the initial solution are used and only the forward mode is carried out); compare the results in Table 10.11 and Table 10.12 with the corresponding results in Table 2.7 and Table 2.8. This means that when the perturbed values of the initial solution are improved, the errors of made by the Backward Euler Method and the selected splitting procedure are becoming the majors sources for errors during the forward mode with the improved solution.

10.7.2 Application of the Implicit Mid-point Rule

The forward mode of the data assimilation procedure is carried out by using (2.59) from Chapter 2. For convenience this formula, representing the Implicit Mid-point Rule, is given below:

$$c_{n+1} = c_n + \triangle t f(0.5(c_n + c_{n+1})). \tag{10.40}$$

Table 10.12: Performance of the data assimilation modules for non-linear problems. The ratios of the errors between successive runs that are obtained when the Backward Euler Method is used with five splitting procedures. The abbreviations *Steps*, *"Splitting 0"*, *"Splitting 1"*, *"Splitting 2"*, *"Splitting 3"* and *"Splitting 4"* are used for (i) the number of time-steps and (ii) the cases where no splitting, sequential splitting, symmetric splitting, weighted sequential splitting and weighted symmetric splitting are used in the computations.

Steps	Splitting 0	Splitting 1	Splitting 2	Splitting 3	Splitting 4
40	-	-	-	-	-
80	1.908	1.908	1.894	2.061	1.966
160	1.951	1.951	1.947	2.015	1.983
320	1.975	1.975	1.974	2.004	1.992
640	1.987	1.987	1.987	2.001	1.996
1280	1.994	1.994	1.993	2.000	1.998
2560	1.997	1.997	1.997	2.000	1.999
5120	1.999	1.999	1.998	2.000	2.000
10240	1.999	1.999	1.999	2.000	2.000
20480	2.000	2.000	2.000	2.000	2.000
40960	2.000	2.000	2.000	2.000	2.000
81920	2.000	2.000	2.000	2.000	2.000
163840	2.000	2.000	2.000	2.000	2.000
327680	2.000	2.000	2.000	2.000	2.000

The fact that the test-example (2.51) is an autonomous system of ODEs is exploited in (10.40). If the system of ODEs is non-autonomous, then $f(t_n + 0.5\Delta t, 0.5(c_n + c_{n+1}))$ should be used in (10.40) instead of $f(0.5(c_n + c_{n+1}))$ (the transformation of the non-autonomous system of ODEs to an autonomous one can be used).

The adjoint equation, which has to be used in the backward mode, is defined as follows:

$$q_n = q_{n+1} - \Delta t \left[-\frac{\partial f(0.5(c_n + c_{n+1}))}{\partial c} \right]^T q_n. \tag{10.41}$$

Also the adjoint equation (10.41) is written for autonomous systems of ODEs. As above, $f(0.5(c_n + c_{n+1}))$ should be replaced by $f(t_n + 0.5\Delta t, 0.5(c_n + c_{n+1}))$ when the system of ODEs is non-autonomous (again the non-autonomous system of ODEs can alternatively be transformed to an autonomous one).

The experiments were carried out as in the case where the Backward Euler Method is used. Accuracy results are given in Table 10.13. Ratios of errors obtained in successive runs (indicating the rate of convergence) are given in Table 10.14.

Table 10.13: Performance of the data assimilation modules for non-linear problems. Accuracy results achieved when the Implicit Mid-point Rule is used with five splitting procedures. The abbreviations *Steps*, *"Splitting 0"*, *"Splitting 1"*, *"Splitting 2"*, *"Splitting 3"* and *"Splitting 4"* are used for (i) the number of time-steps and (ii) the cases where no splitting, sequential splitting, symmetric splitting, weighted sequential splitting and weighted symmetric splitting are used in the computations.

Steps	Splitting 0	Splitting 1	Splitting 2	Splitting 3	Splitting 4
40	9.65E-04	2.00E-01	2.12E-02	6.67E-03	3.70E-03
80	2.40E-04	1.11E-01	5.47E-03	8.16E-04	9.64E-04
160	5.99E-05	5.88E-02	1.38E-03	1.17E-04	2.43E-04
320	1.50E-05	3.03E-02	3.45E-04	2.13E-05	6.10E-05
640	3.74E-06	1.54E-02	8.64E-05	4.49E-06	1.53E-05
1280	9.36E-07	7.75E-03	2.16E-05	1.03E-06	3.81E-06
2560	2.34E-07	3.89E-03	5.40E-06	2.45E-07	9.52E-07
5120	5.85E-08	1.95E-03	2.90E-06	3.33E-07	2.38E-07
10240	1.46E-08	9.75E-04	3.38E-07	2.15E-07	5.96E-08
20480	1.47E-07	4.88E-04	6.95E-07	3.29E-07	6.93E-07
40960	1.49E-06	2.44E-04	3.07E-08	8.85E-07	1.51E-08
81920	4.25E-08	1.22E-04	4.42E-07	4.48E-08	4.36E-08
163840	1.27E-08	6.10E-05	1.25E-08	1.29E-08	1.27E-08
327680	5.47E-07	3.05E-05	5.48E-07	5.23E-07	5.51E-08

It is seen that the combined method behaves as a second order method for all splitting procedures (including the case where no splitting procedure is used) excepting the sequential splitting procedure. The combined method formed by *the Implicit Mid-point Rule + the sequential splitting procedure* behaves as a first-order method (which, of course, should be expected).

The following conclusions can be drawn by comparing the results in Table 10.13 and Table 10.14 with the corresponding results obtained in Chapter 2 (the results given in Table 2.9 and Table 2.9):

- the results for the sequential splitting are the same (when rounded to four significant digits), and
- the results for the other splitting procedures (including the case where no splitting procedure is applied) are the same for large time-step sizes, but the accuracy is not improved furthermore when the time-step size becomes very small.

It is not very clear why the accuracy cannot be further improved when the accuracy achieved become $O(10^{-7})$. Perhaps the optimization subroutine is limiting the accuracy that can be achieved when the data assimilation algorithm is used.

Table 10.14: Performance of the data assimilation modules for non-linear problems. The ratios of the errors between successive runs that are obtained when the Implicit Mid-point Rule is used with five splitting procedures. The abbreviations *Steps*, *"Splitting 0"*, *"Splitting 1"*, *"Splitting 2"*, *"Splitting 3"* and *"Splitting 4"* are used for (i) the number of time-steps and (ii) the cases where no splitting, sequential splitting, symmetric splitting, weighted sequential splitting and weighted symmetric splitting are used in the computations.

Steps	Splitting 0	Splitting 1	Splitting 2	Splitting 3	Splitting 4
40	-	-	-	-	-
80	4.021	1.800	3.879	8.179	3.835
160	4.005	1.889	3.968	6.944	3.962
320	4.001	1.941	3.992	5.502	3.991
640	4.000	1.970	3.998	4.752	3.998
1280	4.000	1.985	3.999	4.377	3.999
2560	4.000	1.992	4.000	4.189	4.000
5120	4.000	1.996	1.863	0.735	3.997
10240	4.000	1.998	8.587	1.548	4.003
20480	0.099	1.999	0.485	0.654	0.086
40960	0.099	2.000	22.646	0.372	45.840
81920	34.934	2.000	0.694	19.744	0.347
163840	3.346	2.000	3.527	3.478	3.433
327680	0.023	2.000	0.023	0.053	0.023

10.7.3 Application of the Fully Implicit Three-stage Runge-Kutta Method

The forward mode of the variational data assimilation procedure is carried out by using (2.60) - (2.63). For convenience these formulae are given below (without the values of the coefficients used in the formulae; these are listed in Chapter 2):

$$c_{n+1} = c_n + \Delta t[\alpha_1 f(Y_1) + \alpha_2 f(Y_2) + \alpha_3 f(Y_3)], \tag{10.42}$$

the quantities Y_1, Y_2 and Y_3 being calculated by

$$Y_1 = c_n + \Delta t[\gamma_{11} f(Y_1) + \gamma_{12} f(Y_2) + \gamma_{13} f(Y_3)], \tag{10.43}$$

$$Y_2 = c_n + \Delta t[\gamma_{21} f(Y_1) + \gamma_{22} f(Y_2) + \gamma_{23} f(Y_3)], \tag{10.44}$$

$$Y_3 = c_n + \Delta t[\gamma_{31} f(Y_1) + \gamma_{32} f(Y_2) + \gamma_{33} f(Y_3)]. \tag{10.45}$$

The fact that the test-example (2.51) is an autonomous system of ODEs is exploited in (2.60) - (2.63). If the system of ODEs is non-autonomous, then $f(t_n + \beta_i \triangle t, Y_i)$ should be used in (2.60) - (2.63) instead of $f(Y_i)$, where $i = 1, 2, 3$ and the coefficients β_i are defined in Chapter 2: see (2.68). As in the previous two cases (the application of the Backward Euler Method or the Implicit Mid-point Rule), the non-autonomous system of ODEs can be transformed to an autonomous one.

The adjoint equation, which has to be used in the backward mode, is defined as follows:

$$
\begin{aligned}
q_n = q_{n+1} \quad & - \quad \alpha_1 \triangle t \left[- \frac{\partial f(Y_1)}{\partial Y_1} \right]^T Q_1 \\
& - \quad \alpha_2 \triangle t \left[- \frac{\partial f(Y_2)}{\partial Y_2} \right]^T Q_2 \\
& - \quad \alpha_3 \triangle t \left[- \frac{\partial f(Y_3)}{\partial Y_3} \right]^T Q_3,
\end{aligned}
\tag{10.46}
$$

where

$$
\begin{aligned}
Q_1 = q_{n+1} \quad & - \quad \gamma_{11} \triangle t \left[- \frac{\partial f(Y_1)}{\partial Y_1} \right]^T Q_1 \\
& - \quad \gamma_{12} \triangle t \left[- \frac{\partial f(Y_2)}{\partial Y_2} \right]^T Q_2 \\
& - \quad \gamma_{13} \triangle t \left[- \frac{\partial f(Y_3)}{\partial Y_3} \right]^T Q_3,
\end{aligned}
\tag{10.47}
$$

$$
\begin{aligned}
Q_2 = q_{n+1} \quad & - \quad \gamma_{21} \triangle t \left[- \frac{\partial f(Y_1)}{\partial Y_1} \right]^T Q_1 \\
& - \quad \gamma_{22} \triangle t \left[- \frac{\partial f(Y_2)}{\partial Y_2} \right]^T Q_2 \\
& - \quad \gamma_{23} \triangle t \left[- \frac{\partial f(Y_3)}{\partial Y_3} \right]^T Q_3,
\end{aligned}
\tag{10.48}
$$

$$
Q_3 = q_{n+1} \quad - \quad \gamma_{31} \triangle t \left[- \frac{\partial f(Y_1)}{\partial Y_1} \right]^T Q_1
\tag{10.49}
$$

$$- \quad \gamma_{32}\,\Delta t \left[-\frac{\partial f(Y_2)}{\partial Y_2} \right]^T Q_2$$

$$- \quad \gamma_{33}\,\Delta t \left[-\frac{\partial f(Y_3)}{\partial Y_3} \right]^T Q_3,$$

Also in this case, the adjoint equations (10.46) - (10.49) are written for autonomous systems of ODEs. As above, $f(Y_i)$ should be replaced by $f(t_n - \beta_i \Delta t, Y_i)$ for $i = 1, 2, 3$ when the system of ODEs is non-autonomous. The non-autonomous system of ODEs can alternatively be transformed to an autonomous one.

The experiments were carried out as the corresponding experiments performed by using the Backward Euler Method and the Implicit Mod-point Rule. Accuracy results are given in Table 10.15. Ratios of errors obtained in successive runs (indicating the rate of convergence) are given in Table 10.16.

Table 10.15: Performance of the data assimilation modules for non-linear problems. Accuracy results achieved when the Fully Implicit Three-stage Runge Kutta Method is used with five splitting procedures. The abbreviations *Steps*, *"Splitting 0"*, *"Splitting 1"*, *"Splitting 2"*, *"Splitting 3"* and *"Splitting 4"* are used for (i) the number of time-steps and (ii) the cases where no splitting, sequential splitting, symmetric splitting, weighted sequential splitting and weighted symmetric splitting are used in the computations.

Steps	Splitting 0	Splitting 1	Splitting 2	Splitting 3	Splitting 4
40	2.60E-07	7.01E-01	1.95E-02	1.66E-02	2.62E-03
80	2.13E-06	1.08E-01	5.12E-03	3.39E-03	6.49E-04
160	4.27E-08	5.82E-02	1.30E-03	7.49E-04	1.62E-04
320	1.25E-06	3.01E-02	3.25E-04	1.75E-04	4.07E-05
640	1.89E-06	1.53E-02	8.13E-05	4.22E-05	1.02E-05
1280	4.54E-07	7.74E-03	2.03E-05	1.03E-05	2.54E-06
2560	8.81E-07	3.89E-03	5.08E-06	2.56E-06	6.36E-07
5120	7.74E-07	1.95E-03	1.27E-06	6.39E-07	1.69E-07
10240	1.29E-06	9.75E-04	7.27E-07	1.59E-07	4.80E-06
20480	1.28E-07	4.88E-04	1.88E-07	2.79E-07	1.48E-07
40960	1.02E-07	2.44E-04	1.60E-06	1.12E-06	5.43E-08
81920	8.52E-07	1.22E-04	1.61E-07	2.48E-09	1.18E-06
163840	1.08E-10	6.10E-05	1.62E-07	3.39E-07	3.64E-06
327680	1.73E-09	3.05E-05	3.10E-10	1.10E-06	2.18E-06

The results for the Fully Implicit Three-stage Runge-Kutta Method are in general similar to the results obtained with the Implicit Mid-point Rule. The combined method behaves again as a second order method for all splitting procedures (except-

Table 10.16: Performance of the data assimilation modules for non-linear problems. The ratios of the errors between successive runs that are obtained when the Fully Implicit Three-stage Runge Kutta Method is used with five splitting procedures. The abbreviations *Steps*, *"Splitting 0"*, *"Splitting 1"*, *"Splitting 2"*, *"Splitting 3"* and *"Splitting 4"* are used for (i) the number of time-steps, (ii) the cases no splitting, sequential splitting, symmetric splitting, weighted sequential splitting and symmetric splitting are used in the computations.

Steps	Splitting 0	Splitting 1	Splitting 2	Splitting 3	Splitting 4
40	-	-	-	-	-
80	0.122	6.417	3.815	4.892	4.037
160	49.892	1.866	3.950	4.531	3.995
320	0.034	1.930	3.987	4.283	3.997
640	6.604	1.964	3.997	4.145	3.999
1280	0.416	1.982	3.999	4.074	4.000
2560	0.515	1.991	4.000	4.037	4.000
5120	1.138	1.995	4.000	4.019	0.376
10240	0.602	1.998	1.762	4.009	0.352
20480	10.027	1.999	3.840	0.570	32.515
40960	1.259	1.999	0.118	452.413	2.714
81920	0.120	2.000	1.000	0.007	0.046
163840	7867.257	2.000	9.951	0.308	0.325
327680	0.063	2.000	522.113	0.249	1.666

ing the cases where no splitting procedure is used and where the sequential splitting procedure is applied). The combined method consisting of *the Fully Implicit Three-stage Runge-Kutta Method and the sequential splitting procedure* behaves as a first-order method. There is an exception: if no splitting procedure is applied, then the method should behave as a six-order method. It is, however, difficult to see such a behaviour when this particular example is treated, because accuracy of order $O(10^{-6})$ (which is, as mentioned above, the accuracy of the reference solution and, thus, the best accuracy that can be obtained) is achieved with the largest stepsize and, therefore, the rate of improving the accuracy oscillates around one.

10.7.4 Conclusions from the experiments with simple non-linear problems

The use of the data assimilation modules in the solution of non-linear problems can be combined

- with different numerical methods for integration of systems of ODEs, and

- with different splitting procedures (including here the case where no splitting procedure is used).

Results obtained by using three numerical methods and five splitting procedures were shown in Table 10.11 to Table 10.16. The results shown in these tables as well as the results of many other runs indicate that the following conclusion can be drawn:

- the use of a high-order numerical method (as the sixth order Fully Implicit Three-stage Runge-Kutta Method) is only justified when no splitting is used (if some splitting procedure is used, then the order of the combined method will in general be lower than the order of the numerical method),
- if the order of the splitting procedure is only one (this is the case for the sequential splitting procedure), then the use of the simple Backward Euler Method is probably the best choice, and
- if the splitting procedure is of order two, then it is probably best to use a second-order numerical method (as, for example the Implicit Mid-point Rule or the Trapezoidal Rule).

Note that these conclusions are similar to the conclusion which were made in Chapter 2. These conclusions can also be derived by using the relationship given in (2.70) or by applying the results presented in Table 2.13. In fact, the results given in Table 10.11 to Table 10.16 can be viewed as an experimental confirmation of the of the relationship (2.70) and of the results in Table 2.13.

Only the relationship between numerical methods and splitting procedures is studied here. Such a study was quite sufficient in Chapter 2. In the case where data assimilation modules are studied, there is another important factor, which should be taken into account: the performance of the optimization subroutine. Only one optimization subroutine is currently used in the data assimilation modules, the subroutine E04DGF from the NAG Numerical Library [186], which is based on the algorithm PLMA, described by Gill and Murray [120], as well as in Section 4.8.3 of Gill et al. [121]. It is also necessary to carry out experiments with other optimization subroutines.

10.7.5 Checking all combinations of splitting procedures and numerical methods.

More complete tests were performed by using the program EXAMINATOR. All 30 combinations of splitting procedures and numerical methods which were run in Chapter 2 without applying data assimilation algorithms (see the results given in Table 2.16, Table 2.17, and Table 2.18), were also run by using the data assimilation algorithm discussed in this chapter.

The experiments can shortly be described as follows. As in Chapter 2 and in the previous part of this section, the problem defined by (2.51) and (2.52) is considered.

Sixteen runs were performed for each splitting procedure and for each numerical method and the results are summarized in the three tables given below (i.e. the total number of runs was 480, but it should be emphasized here that the conclusions were drawn by using much more experiments). For each combination of a numerical method and a splitting procedure, the first run is performed with a time-step size $\triangle t = 0.25$. After that the stepsize is successively reduced (15 times) by a factor of two. Reducing the time-step size by a factor of two leads to an increase of the number of time-steps by a factor of two (the number of time steps for the first run is 40, while it becomes 5242880 for the last run). The initial values of the solution are perturbed by using $\alpha = 0.2$ in all runs. The initial solution is first improved and then the problems is solved with the improved initial solution. The results given in the three tables below (Table 10.17, Table 10.18 and Table 10.19) are related to the accuracy obtained by using the improved with the data assimilation solution initial values. In this way, it is possible to compare directly the results obtained without using data assimilation with the results obtained by using the data assimilation procedure.

The smallest and the largest errors obtained in the runs with 16 different time-step sizes (or, in other words, the left-hand-end-points and the right-hand-end-points of the intervals in which the errors vary) are given, for each combination of a splitting procedure and a numerical method, in Table 10.18 and Table 10.17 respectively. Let uis emphasize the fact that one can obtain the intervals in which the errors are varied for a given combination of a numerical method and a splitting procedure by taking the corresponding figures in these two tables.

The largest errors (i.e. the right-hand-end-points of the intervals in which the errors vary) are normally obtained when the largest time-step size, $\triangle t = 0.25$, is used. It was expected that the smallest errors will be obtained when the smallest stepsizes are used. However, this is not always true for the smallest errors (i.e. the left-hand-end-points of the intervals in which the errors vary), because if the errors become less than $O(10^{-8})$ (note that this limit is considerably lower than the corresponding limit in Chapter 2) and millions of time-steps are to be carried out, then the rounding error start to interfere with the other errors. Accuracy of order $O(10^{-8})$ is achieved very quickly, when the RK6 method is used without splitting).

The rates of decreasing the errors when the time-step sizes decrease are given in Table 10.19. These values are approximate values. For some combinations "numerical method with splitting procedure", the rates are lower in the beginning. If the errors become very small, then rates start to oscillate around 1. However, the figures in Table 10.19 give quite adequate information about the rates which can be achieved by different combinations of a numerical method and a splitting procedure.

Similar conclusions, as those made in Chapter 2 after Table 2.18 can be drawn by using the results given in Table 10.17 to Table 10.19. Instead of repeating these conclusions, we shall concentrate our attention on some similarities and some differences between runs carried out without using the data assimilation procedure and runs carried out with the data assimilation procedure.

Table 10.17: The largest errors obtained in the runs with 16 different time-step sizes (i.e. the right-hand-end-points of the intervals in which the errors vary) are given in this table for different splittings and different numerical methods. The initial solution is perturbed by using $\alpha = 0.2$ and the perturbed solution is improved by applying data assimilation. Results obtained by running the problem with the improved initial solution are given in this table. The abbreviations *"Splitting 0"*, *"Splitting 1"*, *"Splitting 2"*, *"Splitting 3"* and *"Splitting 4"* are used for the different splitting, while the abbreviations *"BE"*, *"IMR"*, *"RK2"*, *"RK6"*, *"ROS2"* and *"TR"* are denoting the Backward Euler Method, the second-order RK method, the six-order RK method, the second-order Rosenbrock method and the Trapezoidal Rule, respectively.

Method	Splitting 0	Splitting 1	Splitting 2	Splitting 3	Splitting 4
BE	2.1E-02	2.1E-02	1.4E-01	2.6E-1	1.8E-01
IMR	9.6E-04	2.0E-01	2.1E-02	6.7E-3	3.7E-03
RK2	8.6E-04	1.4E-01	3.7E-02	1.2E-1	2.7E-02
RK6	2.6E-07	7.1E-01	2.0E-02	1.7E-2	2.6E-02
ROS2	8.6E-03	1.4E-01	3.7E-02	4.9E-2	3.9E-02
TR	9.6E-04	4.6E-01	2.1E-02	6.7E-3	3.7E-03

- The largest errors (i.e. the right-hand-end-points of the intervals in which the error vary), which are given in Table 10.17, are very similar to those given in Table 2.16 (in fact, many of the results becomes equal when these are rounded to the second significant digit). However, the accuracy obtained when the time-step size becomes very small is much better when the exact initial solution is used (i.e. when no data assimilation is applied); compare the results in Table 2.17 with those in Table 10.18. It is not very clear what is the reason for this behaviour. This is probably, as mentioned above, due to the fact that the optimization subroutine used with the data assimilation procedure has difficulties to find a very accurate optimal solution when the required accuracy is very high.

- The rates of convergence of the combined methods (splitting procedure with numerical method) are also very similar in the two cases (runs without applying data assimilation and run with applying data assimilation). This is clearly seen by comparing the results in Table 2.18 with the results in Table 10.19. The only difference is the case where the RK6 method is run without using splitting. If the data assimilation algorithm is used, then the highest accuracy which can be achieved is around $O(10^{-9})$, see Table 10.18. Accuracy that is close to this limit is achieved in the very beginning of the runs, when the largest value of the time-stepsize Δt is used. Therefore, the accuracy cannot be further improved too much. In fact, it is improved by a factor of 48 after

Table 10.18: The smallest errors obtained in the runs with 16 different time-step sizes (i.e. the left-hand-end-points of the intervals in which the errors vary) are given in this table for different splittings and different numerical methods. The initial solution is perturbed by using $\alpha = 0.2$ and the perturbed solution is improved by applying data assimilation. Results obtained by running the problem with the improved initial solution are given in this table. The abbreviations *"Splitting 0"*, *"Splitting 1"*, *"Splitting 2"*, *"Splitting 3"* and *"Splitting 4"* are used for the different splitting, while the abbreviations *"BE"*, *"IMR"*, *"RK2"*, *"RK6"*, *"ROS2"* and *"TR"* are denoting the Backward Euler Method, the second-order RK method, the six-order RK method, the second-order Rosenbrock method and the Trapezoidal Rule, respectively.

Method	Splitting 0	Splitting 1	Splitting 2	Splitting 3	Splitting 4
BE	3.5E-07	1.8E-07	1.2E-06	1.9E-06	1.4E-06
IMR	8.0E-09	1.9E-06	5.9E-09	3.7E-09	2.5E-08
RK2	1.6E-08	1.9E-06	5.3E-08	1.1E-08	1.2E-08
RK6	1.1E-10	1.9E-06	3.1E-10	2.5E-09	5.2E-10
ROS2	6.5E-08	1.9E-06	4.9E-10	9.9E-09	4.5E-08
TR	1.2E-08	2.8E-06	2.0E-08	6.3E-09	2.2E-08

the second run. After that the error starts to oscillate (but there is a trend of improving slowly the accuracy). If no data assimilation is used, then it is improved by a factor of about 64 three times and only after that the error starts to oscillate (again there is a trend of improving slowly the accuracy).

10.7.6 Experiments with an atmospheric chemistry scheme

A series of experiments with an atmospheric chemical scheme containing 56 chemical compounds have been carried out in order to check the ability of the data assimilation algorithm discussed in the previous sub-sections of this chapter to improve a perturbed initial solution. Some results from these experiments will be presented here. The program EXAMINATOR is used in these experiments. The results obtained by using the Backward Euler Method without splitting will be shown.

The atmospheric chemical scheme, which is used in this subsection, is similar to the chemical scheme used in the EMEP models (see Andersson-Sköld and Simpson, [12], Bartnicki et al. [17] and Simpson et al. [224]) and in the Swedish model MATCH (see Robertson [205] and Zlatev et al. [285]). Some experiments in which this scheme was used in UNI-DEM have been carried out (see Chapter 1 and Chapter 7). Now the scheme will be used in some runs with the data assimilation algorithm.

Table 10.19: The approximate rates of convergence (measured by the ratio of the error found at run i and the error found at run $i-1$, where $i = 2, 3, \ldots, 16$). The initial solution is perturbed by using $\alpha = 0.2$ and the perturbed solution is improved by applying data assimilation. Results obtained by running the problem with the improved initial solution are given in this table. The abbreviations *"Splitting 0"*, *"Splitting 1"*, *"Splitting 2"*, *"Splitting 3"* and *"Splitting 4"* are used for the different splitting, while the abbreviations *"BE"*, *"IMR"*, *"RK2"*, *"RK6"*, *"ROS2"* and *"TR"* are denoting the Backward Euler Method, the second-order RK method, the six-order RK method, the second-order Rosenbrock method and the Trapezoidal Rule, respectively.

Method	Splitting 0	Splitting 1	Splitting 2	Splitting 3	Splitting 4
BE	2	2	2	2	2
IMR	4	2	4	4	4
RK2	4	2	4	4	4
RK6	48	2	4	4	4
ROS2	4	2	4	4	4
TR	4	2	4	4	4

The atmospheric chemistry scheme was used as a box model as in Chapter 4 and Chapter 5. However, a scheme with only 35 compounds was used in Chapter 4 and Chapter 5, while a more advanced chemical scheme with 56 compounds is used in this chapter.

The runs were organized as in Section 4.4. The most important issues related to the runs can be summarized as follows:

- The time-interval was sufficiently large, containing changes from day-time to nigh-time and from night-time to day-time.
- Diurnal variations of the temperature were simulated using a cosine function.
- A reference solution was calculated by applying a time-step size $\triangle t = 10^{-5}$ and used in the attempts to evaluate the results.
- Formula (4.40) was used in a similar way as in Section 4.4 to control the accuracy achieved during the computations.
- The first run is performed with a time-step size $\triangle t = 150$ seconds. After that nine additional runs are performed by using smaller time-step sizes (reducing successively the time-step size by a factor of two).
- Only the Backward Euler Method without any splitting was used in experiments until now.

- as well as to the determination of the reliable weights in formula (10.1).

This means that two independent sets of representative and reliable measurements (or, more preferably, two independent sets of measurements stations) are needed. The requirement for representative and reliable measurements does not refer only to the accuracy of the measurements (in spite of the fact that inaccurate measurements can cause great problems). There might be also problems with the unfortunate locations of the measurement stations (strong emission sources close to the station, some problems with the terrain around the station, etc.).

It should also be noted that the variational data assimilation technique is not an universal tool that can be used in all situations. If different scenarios are used to study the response of the model to the variation of different parameter (emissions, meteorological conditions, climatic changes, etc), then no measurements are available. This means that we have to rely on the model in such studies. This fact indicates that it is absolutely necessary to use data assimilation techniques not only to improve directly the model results by incorporating measurements, but also in order to improve some of the physical and chemical mechanisms implemented in the model. The hope is that the improved in this way model will give more reliable results also in situations where no measurements are available.

The general conclusion is that there are a lot of unresolved problems connected with the practical implementation of variational data assimilation modules. This is why, variational data assimilation issues will also be discussed in the next chapter together with other open problems.

Chapter 11

Discussion of some open questions

Many challenging numerical and computational problems, which appear when comprehensive air pollution studies are to be carried out, were studied in the previous chapters. It was also emphasized in all chapters of this book (excepting Chapter 8 and Chapter 9) that the ideas are applicable not only in connection with different air pollution models, but also when many mathematical models arising in other fields of science and engineering are to be handled on high-speed modern computers. It is relevant now to ask the question:

> *Are the mathematical models used at present sufficient for studying the most important problems, the solution of which is highly desirable by the modern society?*

The general answer to this question is, of course, no. While many problems can successfully be studied and **are** successfully studied by the models that are available at present, it is also true that many other problems cannot be handled (or, at least, it is very difficult to treat them on the available at present computer architectures) with the models which are existing at present.

Several open problems will be discussed in this chapter. The successful solution of each of these problems will increase the ability of the scientists and engineers to handle situations which cannot be handled by the existing at present models.

11.1 Avoiding the use of splitting procedures

The use of splitting procedures was discussed in Chapter 2. It has been pointed out there that the use of any splitting procedure will in general lead to splitting errors. It is difficult to keep these errors into control. Therefore one must try to avoid splitting errors if this is possible. Avoiding the use of splitting procedures is **one of the greatest computational challenges** in the treatment of very large air pollution models. This is also true for many environmental and ecological models as well as for many models arising in other fields of science and engineering.

The discretization of an air pollution models (or, in other words, of the system of partial differential equations by which the large-scale air pollution model is described mathematically) will lead to the solution of a very large system of non-linear algebraic equations when no splitting is used. This system can contain many millions of equations. The largest systems solved by us contained about 387 million equations. Such big systems arise in the case where the considered three-dimensional space domain is discretized by using a $(480 \times 480 \times 10)$ grid and the number of chemical species is 168; see Table 1.1. It is true that the Jacobian matrix of the non-linear system of algebraic equations is a banded matrix. This means that the application of the version of the Newton iterative method, which has been selected, will lead to the solution of several systems of linear algebraic equations per time-step, whose coefficient matrices are banded. The treatment of these systems, however, will cause difficulties. The use of direct methods can be very inefficient (especially if three-dimensional models are used), because the bandwidth is very large. Iterative methods can be applied, but there is a danger (because of the stiffness) that they will converge only when the time-step size is very small which increases the computational work. Preconditioning can be applied in an attempt to keep the time-step size large, but most of the preconditioned iterative methods cannot be easily parallelized. Some of the methods described in the books of Dongarra et al. [80] and van der Vorst [256] might be very useful in the efforts to resolve the difficult computational problems.

Some methods for general sparse matrices, which exploit the cache memories of the existing now computers, were described in Chapter 6. Some examples, which demonstrate the efficiency of these algorithms in the attempt to exploit the cache memories of some SUN computers, were also presented in Chapter 6. These methods can in an obvious way be designed for parallel computations. This was illustrated in Chapter 7. However, much more efforts are needed in the attempts to implement efficiently the methods for handling general sparse matrices in the case where comprehensive air pollution studies involving

- the use of fine grid,
- the integration of the model over long time-intervals

and

- runs of many scenarios

are to be performed.

This short analysis indicates that it is probably worthwhile to try to exploit some special properties of the matrices that arise after the discretization of large air pollution models in the efforts to treat efficiently such models without using splitting procedures. The use of some kind of partitioning of the chemical species (like the procedure discussed in Chapter 4 and Chapter 5) followed by some reordering of the arising matrices may give substantial improvements. However, it is not clear how to extend the partitioning procedures, which were applied in connection only with the chemical schemes, for the case where the other physical and chemical processes are not separated, by the splitting procedure chosen, from the chemical scheme.

11.2 Need for reliable error control

It is worthwhile to control the size of the errors caused by the numerical algorithms and/or by the splitting procedure used. This is a very challenging and still an open problem.

In nearly all existing large-scale air pollution models, which are run operationally, the error control is **indirect**. This means that the errors from the applied in the model numerical methods have been checked by running some appropriate numerical tests. The well-known Crowley-Molenkampf test, which is also called the rotation test (see [48] and [183]), is normally used to check the reliability of the numerical methods used in the horizontal advection (Zlatev [282]). In Hov et al., [141], the Crowley-Molenkampf test was extended to cover additionally terms describing horizontal diffusion and chemistry.

The reliability of the splitting procedures has been tested by performing runs with different time-steps and comparing the results (this approach has been discussed in Chapter 1 and after that applied in Chapter 2 and Chapter 10).

It is clear that it is much more desirable to insert a device for performing error control **directly** in the code. In this way, the error test will be carried out at the end of every time-step in order to decide whether the results are acceptable or not and furthermore to decide what has to be done if they are not. This is a very difficult task, especially when splitting procedures are used. It is not clear at present how to solve this task. At the same time, it is also clear that one should continue to work on the solution of the problem with the automatic error control in large scale air pollution models.

The problem of finding ways to control automatically the errors caused by the numerical methods is relevant not only for large scale air pollution models, but also for many other large-scale applications which arise in different fields of science and engineering. Therefore, it is important to study carefully the results obtained in other fields that are related to the attempts to resolve this problem and to apply some of these results in the solution of problems arising in air pollution modelling.

11.3 Running of air pollution models on fine grids

Two of the greatest challenges in the 21st centuries, which are related to the requirement to keep our environment sufficiently clean, can be formulated (see also Chapter 1) in the following way:

- Requirement for more detailed information about the pollution levels and the possible damaging effects.

- Requirement for more reliable information (especially for the worst cases).

More advanced mechanisms are to be used in description of the physical and chemical processes. Moreover, the models are to be discretized on fine grids in order to satisfy these two requirements. The application of these two actions leads to very large computational problems the treatment of which is very difficult, also when modern high-speed computers are used.

11.4 Transition from regional to urban scale

It is sometimes necessary to study the air pollution in urban area. The grid-cells relevant for urban areas are normally $2\ km \times 2\ km$ or $1\ km \times 1\ km$. The procedure used at present is often called "nesting". One first calculates results by using a coarse grid. Then the model is run on a smaller sub-domain by using a finer grid. On the boundaries of the smaller domain the concentrations obtained in the run on the large domain are applied (by using interpolation for the intermediate grid-points). This procedure can be repeated several times (i.e. multiple nesting can be applied). It is intuitively clear that the information passed from the coarser grid to the finer grid (or from the coarser grids to the finer grids when multiple nesting is used) is not sufficient. Moreover, there is no feed-back from the finer grid to the coarser grid (or from the finer grids to the coarser grids when multiple nesting is used) when the nesting procedure is implemented in the simple manner sketched above (which is often called "one-way nesting").

One can apply "two-way nesting" in order to resolve the problem with feed-back. After each time-step on the coarser grid the model is also run on the finer grid. Also here multiple nesting can be used (in this case, at each time-step runs on all grids are carried out). Appropriate results from the coarser grid (coarser grids when multiple nesting is applied) are again used as boundary conditions on the finer grid (finer grids when multiple nesting is applied). However, one additional action is carried out: the information obtained on the finer grid (finer grids) is projected back to the coarser grid (coarser grids) at each time-step. It is clear that the results obtained by using two-way nesting (two-way multiple nesting) are more reliable than the results obtained by using a corresponding one-way nesting procedure. The two-way nesting can be viewed as an attempt to achieve feed-back from the fine grid (finer grids) to the coarse grid (coarser grids). However, the problems related to the fact that the information passed from the coarser grid to the finer grid (finer grids) is not sufficient remains also when two-way nesting is used.

The correct way to perform the transition from a regional scale to an urban scale is the use of **local refinement** of the grid. If this is done, then the transition from the coarse grid to the fine grid can be carried out by reducing gradually the size of the grid-cells. This means that the main problem of the nesting procedure (insufficient information in the transition from coarser grids to fine grids) will be treated in a more reliable way. Moreover, the use of local refinement ensures feed-back from the urban area to the remaining parts of the space domain. Another advantage of the local refinement of the grid is the fact that local refinement can be specified simultaneously around several urban areas. There are some difficulties. It will, for instance, be not very easy to get the needed input data for the area of transition from the coarse grid to the fine grid (to the finer grids).

11.5 Static and dynamic local refinement

In the examples given in the previous section the refinement of the grid depends on the sub-domain which is of interest and, therefore, can be made in advance, i.e. before the actual computations. This means that the refinement is fixed and not changed during the run. Such refinements of the grid are called static.

Sometimes it is desirable to refine dynamically, when this is judged by the code to be necessary, the grid in some sub-domains (not a priori defined) during the run. In other words, dynamic refinement of the grid might in some situations be necessary. The application of dynamic refinement of the grid is an useful approach in the cases where some sharp gradients of some concentrations appear during the computations. Tools for performing dynamic refinement of the grid are available (first and foremost, advanced packages based on the use of finite elements in the discretization of the space derivatives). However, two major problems arise in the attempts to implement such tools in large-scale air pollution models (and also in large-scale mathematical models arising in other areas of science and engineering):

- the codes become very time-consuming (even when modern high speed computers are used)

and

- it is not very easy to adjust the input data for the refined grid when the grid is refined dynamically in sub-domains which are not defined in advance before the run.

The second difficulty is very serious and causes difficult problems in the practical implementation of the dynamic refinement procedures in large air pollution codes. This is the reason for not using dynamic refinement in the existing operational large-scale air pollution model (however, there are some exceptions, see, for example, Tomlin et al. [250]). On the other hand, the idea of using dynamic refinement is very attractive and, therefore, it is highly desirable to resolve the very challenging task of implementing and using on a routine basis efficient tools for performing automatically dynamic refinement of the grid. This is especially true for the case when large-scale air pollution models are to be run over long time-intervals.

11.6 Need for advanced optimization techniques

Reductions of human-made (anthropogenic) emissions are needed in the attempts to reduce the high pollution levels to some prescribed acceptable levels or to keep them under the prescribed acceptable levels. However, reductions of human-made (anthropogenic) emissions are normally very expensive. Therefore, it is necessary to try to find out

- where to reduce the emissions, and
- by how much to reduce them.

Normally, one tries to find answers to these questions by running different emission scenarios and analyzing the results obtained in the selected scenarios. It is clear that the answers found in this way may not be optimal. Therefore, it may be more worthwhile to try to formulate an optimization problem and to get optimal answers by solving this problem. However, the optimization problems related to finding optimal reductions of emissions in a large area (space domain covering the whole of Europe) will lead to very large computational tasks. It will probably not be possible to handle such optimization problems on the existing at present computers. However, the computers are becoming faster and faster. This is why one should expect that such problems will become tractable in the near future. It is necessary to resolve this problem, because one of the big advantages of air pollution modelling is to support decision-makers in finding **where** and by **how much** to reduce the emissions in order to reach the requirements for achieving acceptable pollution levels.

The application of different variational data assimilation techniques in large-scale mathematical models (not necessarily large-scale air pollution models) is also leading to the solution of optimization problems (see the discussion in the beginning of Chapter 10). It should again be stressed here that the use of variational data assimilation techniques is becoming more and more popular in the area of air pollution modelling. The computational challenges, which arise when different variational data assimilation procedures are to be used, are discussed in some more detail in the next section.

Optimization techniques can be used to resolve many other problems related to environmental issues. Some topics in this context are discussed in Dimov and Zlatev, [75]).

11.7 Use of data assimilation techniques

Several ideas related to the application of variational data assimilation techniques in large-scale air pollution models were discussed in the previous chapter. Many test-results were also presented in Chapter 10. The major problems arising when data assimilation techniques are used will be summarized in this section.

Initial concentration fields are needed when air pollution models are to be treated numerically. The problem of finding good initial concentration fields is very difficult. Normally, one starts with some background concentrations and runs the model over a certain sufficiently large period (say, five days) to generate initial concentration fields. It is expected that the initial fields so found are good, but there is no guarantee that this is so. Moreover, this is an unnatural solution in the case where an air pollution forecast for the next two-three days is to be produced (because the time to start-up the model, five days, is much larger than the period of two-three days that is needed for the forecast).

An alternative approach is based on the use of time series of measurements to analyze the initial data that is to be used in the air pollution model under consideration. The experiments indicate the remarkable fact that time series of only a few key chemical species are sufficient to produce initial fields (also for species that are not measured; by exploiting their coupling with the measured species).

There are several different ways of implementing the data assimilation approach. Since the mid-eighties the approach based on the principle of optimal control theory is becoming more and more popular. In this approach the initial values are viewed as control parameters. Then a distance function is defined. This function provides weighted and accumulated distances between the available measurements and the values of the corresponding state variables calculated by the model during a predefined assimilation window. A minimization procedure is applied to minimize the distance function (i.e. optimization methods, see the previous section, are also needed here). This requires some knowledge of local gradients with respect to the initial state. To obtain these gradients one has to use adjoint equations. The procedure is normally called a four-dimensional variational data assimilation.

It must be emphasized here that data assimilation is a very powerful approach. Not only can it be used to analyze the initial data (as sketched above), but also to obtain optimal estimations of

- emission fields,
- deposition rates, and
- other model parameters.

However, one must also be careful. The measurement data which are used must be reliable. Even when the reliability of the measurements is validated, one should check carefully whether the measurement data are representative or not. An unfortunate location of a measurement station very close to a very strong emission source will make the results not very suitable for usage in a large-scale air pollution model, although the results might be very suitable for other purposes (as, for example, in the case where the exposure of the population around the source to high pollution levels is to be monitored).

More details about different implementations of data assimilation techniques in air pollution models can be found in Chapter 10 as well as in the works of D. N.

Daescu and his co-workers (see [53], [54], [55] and [209]) as well as in the works of H. Elbern and his co-workers (see Elbern and Schmidt [88] [89] and Elbern et al. [90], [91], [92]).

It is necessary to emphasize here the fact that also the introduction of a data assimilation module in an air pollution model leads to a very considerable increase of the computational work and, thus, is a great computational challenge.

11.8 Special modules for treatment of particles

It is highly desirable to incorporate special modules for treatment of particles in state-of-the-art air pollution models. If this is properly done, then the air pollution model will be able to provide sufficient information about both the chemical composition and the size distribution of the particles.

The Modal Aerosol Dynamics Model for Europe (MADE) is such an advanced module for the treatment of particles (see, for example, Ackermann et al., [1], [2]). It is based on the Regional Particulate Model (RPM) developed by Binkowski and Shankar, [21]. MADE was incorporated in the EURAD models system and used in simulations of tropospheric aerosols over Europe (see Ackermann et al. [1]). This first version was limited to sub-micron particles consisting of inorganic matters and water. The MADE module was after that further extended to include coarse particle size range and a more complete fine particle chemistry (see, for example, Ackermann et al., [2]). Finally, an extension for organic particles was developed and tested in Schell et al. [215].

Also the incorporation of an advanced module for treatment of particles is increasing the computational work very considerably and, again, this is another great computational challenge.

11.9 Use of computational grids

Grid computing (or the use of computational grids) is based on applying the resources of many computers in a network to a single task at the same time - usually to a scientific or technical problem that requires a great number of computer processing cycles or access to large amounts of data. *"A computational grid is a hardware and software infrastructure that provides dependable, consistent, pervasive and inexpensive access to high-end computational capabilities"* (see, Foster and Kesselman, [105]). Grid computing can be thought of as a powerful tool for performing comprehensive computations on a distributed and large-scale computer cluster. It requires the use of software that can divide and farm out pieces of a program to as many as several thousand computers.

Grid computing appears to be a very promising tool both for improving the performance of the existing large-scale air pollution models and for the development and successful treatment of sets of many models, which are different but have to be coupled in order to study some important interrelations between different processes. The sets of models can, for example, consist of economical models, models controlling the energy development in a given region of the world, models controlling the sustainability of the industrial and agricultural development, models controlling the air pollution levels, etc. It is quite clear that

- the air pollution models are only a part of such big sets of mathematical models (and not necessarily the most time-consuming part), and
- the computational tasks that are to be resolved, when such big sets of models are to be used (perhaps by running hundreds of carefully chosen scenarios), are enormous.

The great potential power of the computational grids can make it possible to resolve the enormous tasks related to the treatment of big sets of coupled models when this power is efficiently exploited. Several reasons for this expectation are listed below.

- Grid computing is a way to handle tasks that cannot be approached without an enormous amount of computing power.
- Grid computing suggests that the resources of many computers can be co-operatively and perhaps synergistically managed as a collaboration toward a common objective.
- Grid computing allows the users to apply in a more efficient way the distributed storage of the data, which are needed as input data in big applications (in the case of large-scale air pollution models the most relevant input data are the sets of meteorological data and emission data) when these data are produced and kept in different sites (which is practically always the case).
- A computational grid is potentially able to make more cost-effective use of a given amount of computer resources.

More information about the grid computing could be found in the papers of Foster and Kesselman [105] and Foster et al. [106]. Information about computational grids can also be found in the Internet; see, for example, Grid Computing Info Centre, [123]. In this website, one can find a definition of a computational grid (which differ from the definition of Foster and Kesselman given above) and some information about international activities related to computational grids (including conferences, projects, applications, grid portals, grid programming environments, etc.).

An example for a possible need of a computational grids in connection with a comprehensive set of air pollution tasks (including the use of advanced weather forecast models, which have to be run together, at every time-step and on the same spatial grid, with the air pollution models) was defined and shortly discussed in the end of Chapter 7. Such tasks cannot be treated at present without imposing some crude and as a rule non-physical simplifications.

11.10 Applicability of the methods to other large mathematical models

It should be emphasized, also in this chapter, that the application of splitting techniques and discretization methods in the computer treatment of large-scale models arising in different fields of science and engineering leads in a natural way to the numerical solution of systems of PDEs which are very similar to the systems of PDEs (1.1) arising in air pollution modelling. For several environmental models this was demonstrated in Section 7 of Chapter 1. We concentrated our attention in the discussion of a particular air pollution model, the Unified Danish Eulerian Model (UNI-DEM) with regard to the following important issues:

- the choice of numerical methods (different numerical methods were discussed in Chapter 2 to Chapter 5),
- efficient ways of performing matrix computations (this important issue was mainly discussed in Section 6),
- the development of standard parallelization devices (the parallelization techniques were discussed in detail in Chapter 7)
- the application of a variational data assimilation algorithm (implantation problems and test-results related to data assimilation were presented in Chapter 10), and
- the discussion of selected application studies for air pollution problems (comprehensive air pollution studies were discussed in Section 1, Section 8 and Section 9).

This has been done only in order to facilitate the presentation of the results (without presenting long discussions about the treatment of different special cases). However, the problems solved by using UNI-DEM are, in fact, large systems of partial differential equations that appear in many other fields of science and engineering. Therefore, the methods and the devices used when UNI-DEM is handled on modern computers can also be used in connection with many other large-scale models. There are short sections in this book, which provide explanations about the applicability of the numerical methods and the parallelization devices used in the computer treatment of UNI-DEM to other models (and, first and foremost, to environmental models). Such sections have been added in the end of the relevant chapters (Chapter 1 to Chapter 7).

Appendix A

Colour plots

Colour plots give usually better possibilities to see different details in the figures. In this appendix several colour plots are given in order to facilitate the understanding of the information which was presented in some of the plots in Chapter 8, Chapter 9 and Chapter 10.

It must be emphasized here that the following notation is used in the numbering of the figures in this appendix as well as in references in the previous chapters to figures from this appendix.

- The figures in this appendix have the same numbers as the corresponding figures in the previous chapters + an ending "col". For example, Figure 8.1.col is a colour version of Figure 8.1 in Chapter 8.

- When we refer (in the previous chapters) to a figure that has also a colour version, we give a clear message that the colour version exists and can be seen in Appendix A.

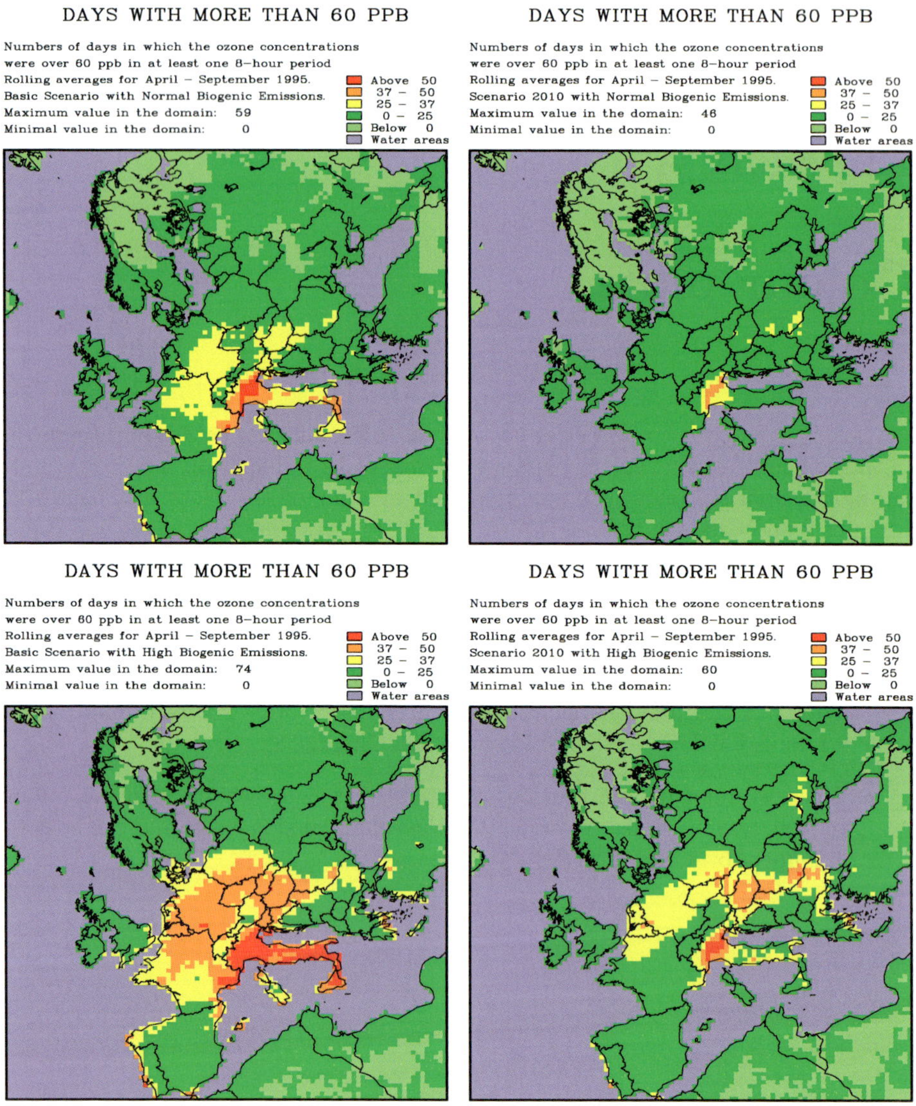

Figure 8.1.col: Distribution of "bad" days in Europe: (a) results for 1995 human-made emissions with normal biogenic emissions are given on the upper left-hand-side plot, (b) results for the 2010 human-made emissions with normal biogenic emissions are given on the upper right-hand-side plot, (c) results for 1995 human-made emissions with high biogenic emissions are given on the lower left-hand side plot and (d) results for 2010 human-made emissions with high biogenic emissions are given on the lower right-hand side plot.

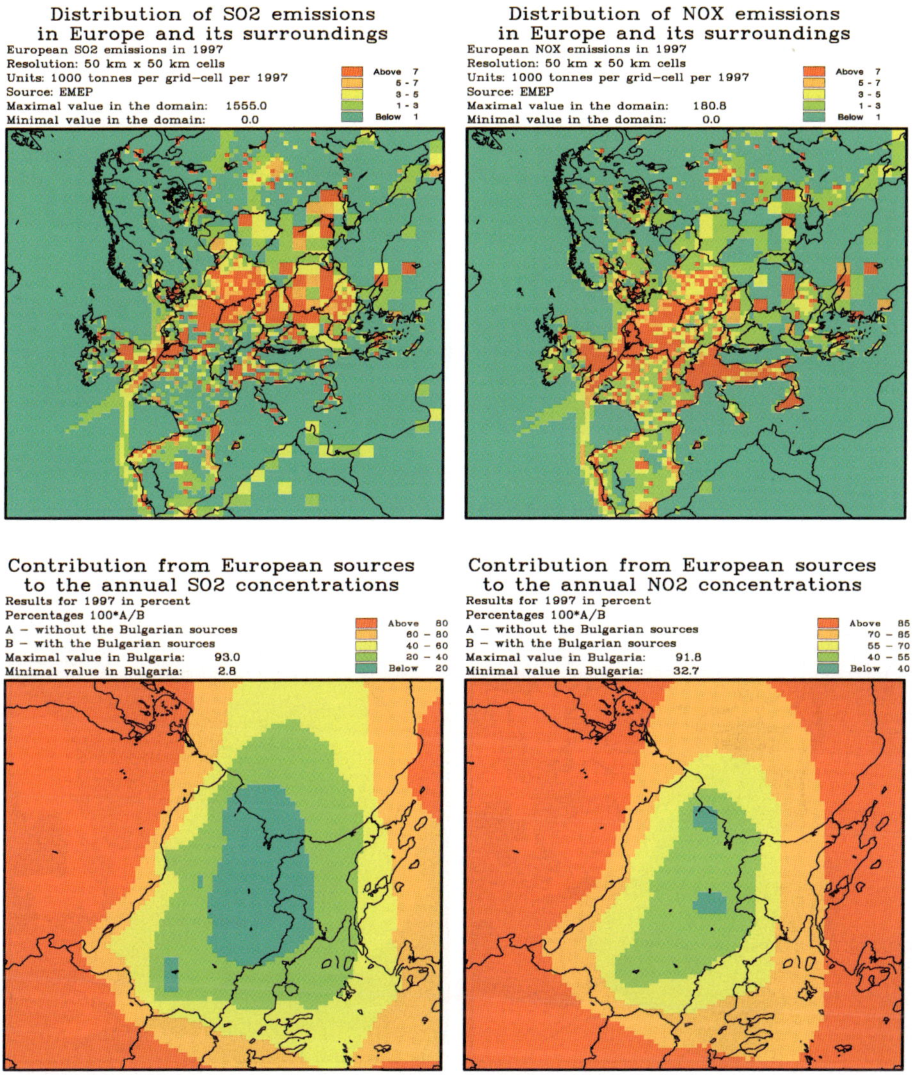

Figure 8.2.col: Influence of European emission sources to pollution levels in Bulgaria: (a) European SO_2 emission sources, (b) European NO_x emission sources, (c) contribution of the European SO_2 emission sources to the SO_2 concentrations in Bulgaria (in percent), (d) contribution of the European NO_x emission sources to the NO_2 concentrations in Bulgaria (in percent).

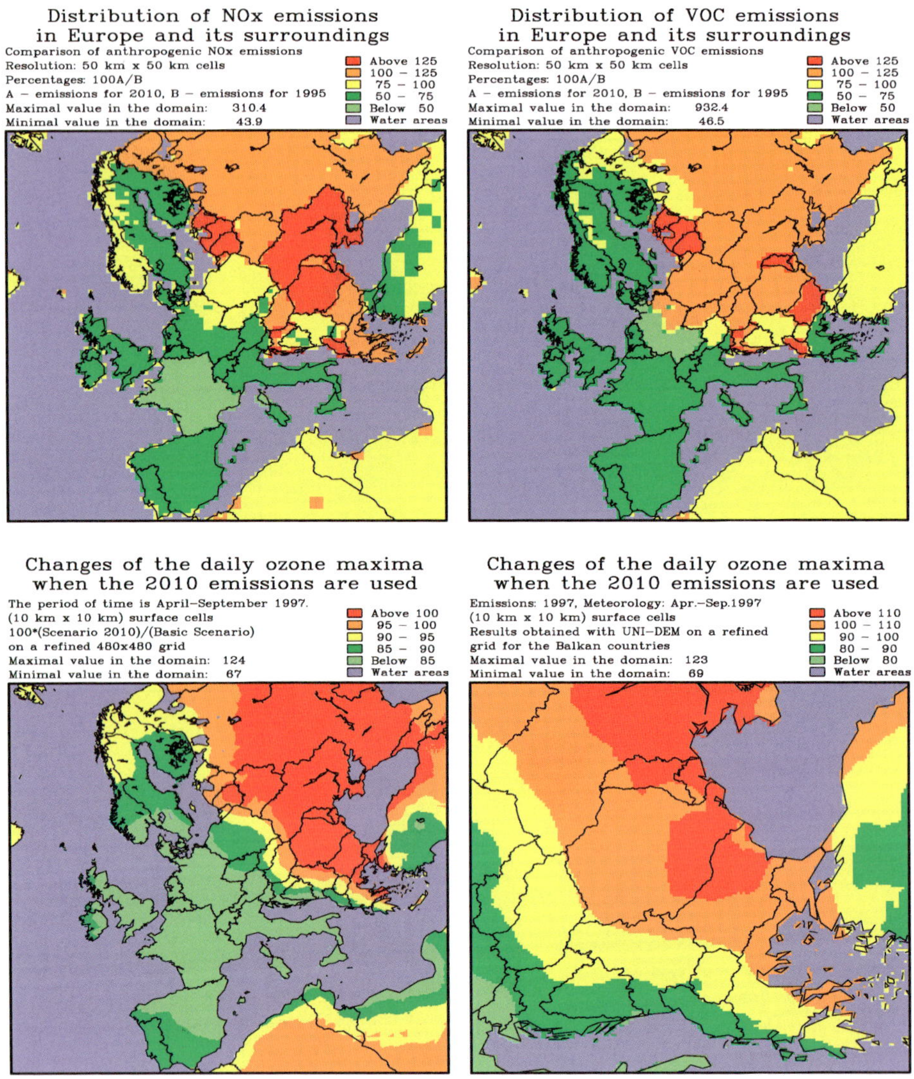

Figure 8.3.col: Changes of 1997 levels of the emissions and ozone concentrations according to Scenario 2010: (a) changes of the European NO_x emissions, (b) changes of the European VOC emissions, (c) changes of the ozone concentrations in different parts of Europe, (d) changes of the ozone concentrations in the Balkan countries.

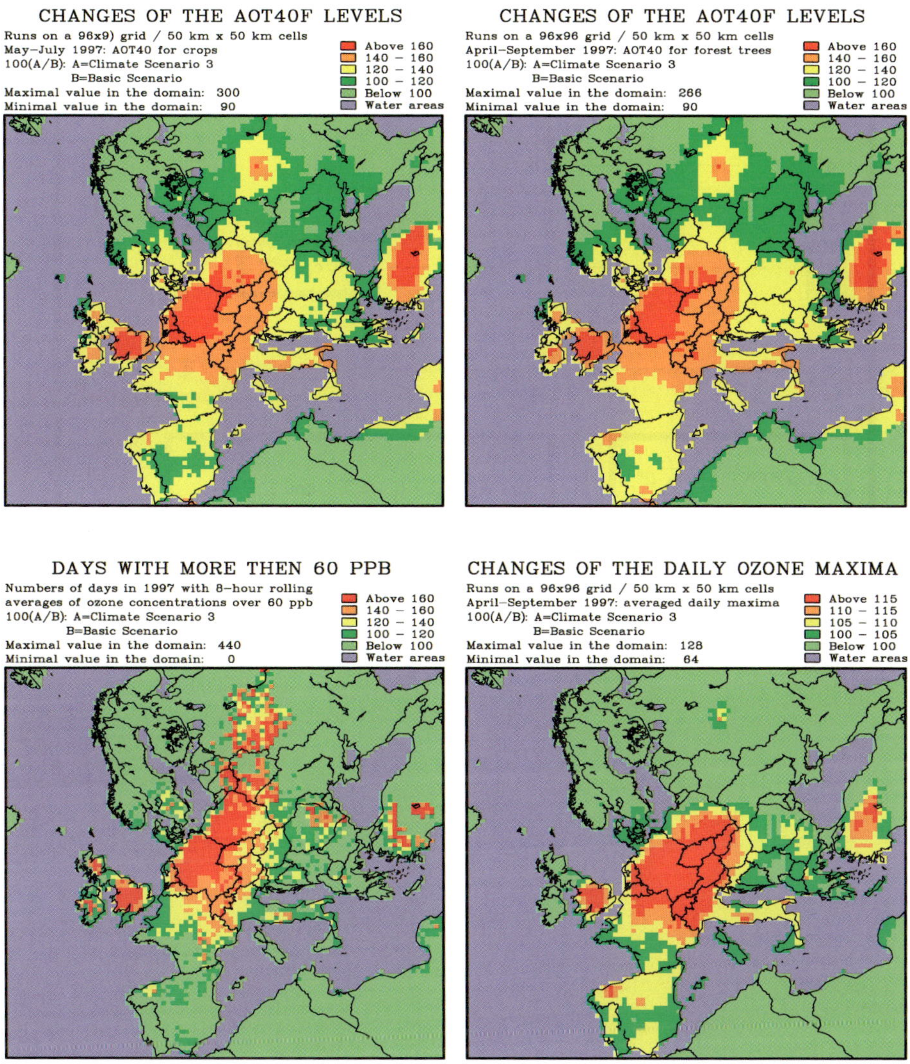

Figure 9.6.col: Relative changes (in percent) of quantities related to high ozone levels when Climatic Scenario 3 is used instead of the Basic Scenario.

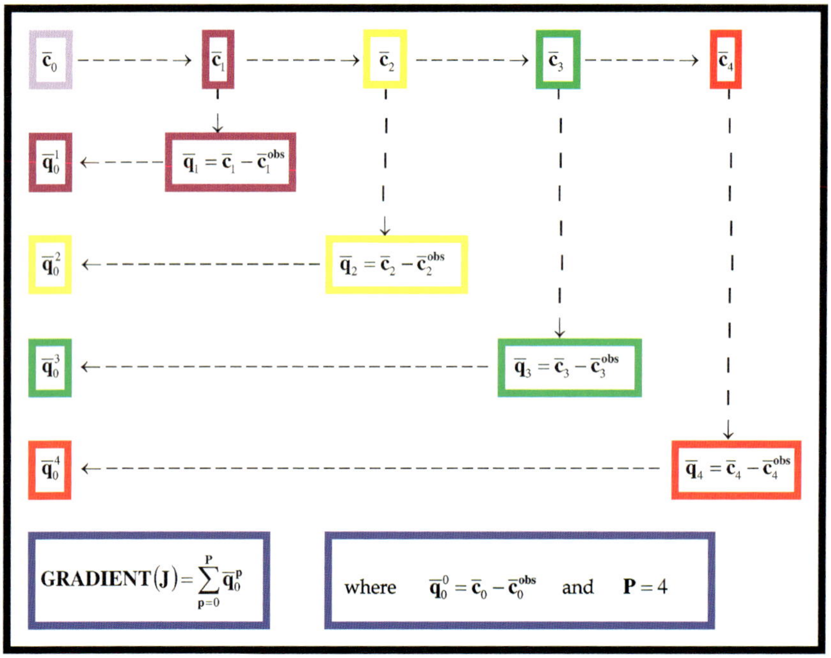

Figure 10.1.col: Computing the gradient of $J\{\bar{c}_0\}$ when observations in five time-points are available.

Bibliography

[1] Ackermann, I., Hass, H., Memmesheimer, M., Ebel, A., Binkowski, F. S. and Shankar, U. (1998). *Modal aerosol dynamics model for Europe: development and first applications.* Atmospheric Environment, 32: 2981-2999.

[2] Ackermann, I., Hass, H., Schell, B. and Binkowski, F. S. (1999). *Regional modelling of particulate matter with MADE.* Environmental Management and Health, 10: 201-208.

[3] Akella, S. and Navon, I. M. (2004). *Documentation of the TLM and adjoint models of the Lin-Rood spherical shallow water finite volume model.* Report. CSIT and Department of Mathematics, Florida State University, Tallahassee, Florida, USA.

[4] Alekseev, A. K. and Navon I. M. (2005). *A posteori pointwise error estimation for compressible fluid flows using adjoint parameters and Lagrange remainder.* International Journal for Numerical Methods in Fluids, 47: 45-74.

[5] Alexandrov, V., Owczarz, W., Thomsen, P. G. and Zlatev, Z. (2004). *Parallel runs of a large air pollution models on a grid of SUN computers.* Mathematics and Computers in Simulations, 65: 557-577.

[6] Alexandrov, V., Sameh, A., Siddique, Y. and Zlatev, Z. (1997). *Numerical integration of chemical ODE problems arising in air pollution models.* Environmental Modelling and Assessment, 2: 365-377.

[7] Aloyan, A. E. (1999). *Modelling of global and regional transport and transformation of air pollutants.* In: Large Scale Computations in Air Pollution Modelling (Zlatev, Z. Brandt, J. Builtjes, P. J. H., Carmichael, G., Dimov, I. Dongarra, J., van Dop, H., Georgiev, K. Hass, H. and San Jose, R., eds.), pp. 15-24. NATO Science Series, 2. Environmental Security - Vol. 57. Kluwer Academic Publishers, Dordrecht-Boston-London.

[8] Amann, M., Bertok, I., Cofala, J., Gyarfas, F., Heyes, C., Klimont, Z., Makowski, M., Schöpp, W. and Syri, S. (1999). *Cost-effective control of*

acidification and ground-level ozone. Seventh Interim Report, IIASA (International Institute for Applied System Analysis), Laxenburg, Austria.

[9] Ambelas Skjøth, C., Bastrup-Birk, A., Brandt, J. and Zlatev, Z. (2000). *Studying variations of pollution levels in a given region of Europe during a long time-period*. Systems Analysis Modelling Simulation, 37: 297-311.

[10] Anastasi, C., Hopkinson, L. and Simpson, V. J. (1991). *Natural hydrocarbon emissions in the United Kingdom*. Atmospheric Environment, 25A: 1403-1408.

[11] Anderson, E., Bai, Z, Bischof, C, Demmel, J, Dongarra, J., Du Croz, J., Greenbaum, A., Hammarling, S., McKenney, A., Ostrouchov, S. and Sorensen, D. (1992). *LAPACK: Users' Guide*. SIAM, Philadelphia.

[12] Andersson-Sköld, Y. and Simpson, D. (1999). *Comparison of the chemical schemes of the EMEP MSC-W and IVL photochemical trajectory models*. Atmospheric Environment, 33: 1111-1129.

[13] Axelsson, O. and Barker, V. A. (1984). *Finite element solution of boundary value problems*. Academic Press, Orlando, Florida, USA.

[14] Bagrinovskii, K. A. and Godunov, S. K. (1957). *Difference schemes for multidimensional problems*. Dokl. Akad. Nauk USSR, 115: 431-433.

[15] Barrett, R., Berry, M., Chan, T. F., Demmel, J., Donato, J., Dongarra, J., Eijkhout, V., Pozo, R., Romine, C. and van der Vorst, H. (1994). *Templates for the Solution of Linear Systems: Building Blocks for Iterative Methods*. SIAM, Philadelphia.

[16] Bartholy, J., Faragó, I. and Havasi, Á. (2001). *Splitting method and its application in air pollution modelling*. IDOJARAS, 105: 39-58.

[17] Bartnicki, J., Olendrzynski, K., Jonson, J. E. and Unger, S. (1998) *Description of the Eulerian Acid Deposition Model*. In: EMEP/MSC-W Report 1/98, Appendix A2. Meteorological Synthesizing Centre-West, Norwegian Meteorological Institute, P. O. Box 43-Blindern, N0313 Oslo 3, Norway.

[18] Bastrup-Birk, A., Brandt, J., Uria, I. and Zlatev, Z. (1997). *Studying cumulative ozone exposures in Europe during a seven-year period*. Journal of Geophysical Research, 102: 23917-23935.

[19] Bastrup-Birk, A., Brandt, J. and Zlatev, Z. (1998). *Using partitioned ODE solvers in large air pollution models*. Systems Analysis Modelling Simulation, 32: 3-17.

[20] Belitzkii, G. R. and Ljubich, Ju. I. (1984). *Matrix norms and their applications*. Naukova Dumka, Kiev (in Russian).

[21] Binkowski, F. and Shankar, U. (1995). *The regional particulate model 1: Model description and preliminary results.* Journal of Geophysical Research, 100: 26191-26209.

[22] Borrell, P, Borrell, P. M. and Seiler, W. (1990). *Transport and Transformation of Pollutants in the Troposphere.* SPB Academic Publishing, The Hague.

[23] Botchev, M., Faragó, I. and Havasi, Á. (2004). *Testing weighted splitting schemes on a one-column transport-chemistry model.* International Journal of Environmental Pollution, 22 (No. 1-2): 3-16.

[24] Bott, A. (1989) *A positive definite advection scheme obtained by nonlinear renormalization of the advective fluxes.* Monthly Weather Review, 117:1006-1015.

[25] Bouchet, V. S., Laprise, R., Torlaschi, E. and McConnel, J. C. (1999). *Studying ozone climatology with a regional climate model 1. Model description and evaluation.* Journal of Geophysical Research, 104: 30351-30371.

[26] Bouchet, V. S., Laprise, R., Torlaschi, E., McConnel, J. C. and Plummer, D. A. (1999). *Studying ozone climatology with a regional climate model 2. Climatology.* Journal of Geophysical Research, 104: 30373-30385.

[27] Brandt, J., Christensen, J., Frohn L. M., Berkowicz, R. and Zlatev, Z. (2000). *Optimization of operational air pollution forecast modelling from European to local scale.* In: Global and Regional Atmospheric Modelling (Barone, G., Builtjes, P. J. H. Giunta, G., eds.), pp. 269-279. Annalli, Facolta di Scienze Nautiche, Instituto Universario Navale, Napoli, Italy.

[28] Brandt, J., Christensen, J. and Zlatev, Z. (1999). *Real time predictions of transport, dispersion and deposition from nuclear accidents.* Environmental Management and Health, 10: 216-223.

[29] Brandt, J., Wasniewski, J. and Zlatev, Z. (1996). *Handling the chemical part in large air pollution models.* Appl. Math. and Comp. Sci., 6: 101-121.

[30] Brenan, K., Campbell, S. and Petzold, L. (1996). *Numerical solution of initial value problems in differential-algebraic equations.* SIAM, Philadelphia.

[31] Brost, R. A. (1988) *The sensitivity to input parameters of atmospheric concentrations simulated by a regional chemical model.* Journal of Geophysical Research, 93: 2371-2387.

[32] Brown, J., Wasniewski J. and Zlatev, Z. (1995). *Running air pollution models on massively parallel machines.* Parallel Computing, 21: 971-991.

[33] Bruegge, B., Riedel, E. Russell, A., Segall, E. and Steenkiste, P. (1995). *Heterogeneous distributed environmental modelling.* SIAM News, No. 28.

[34] Builtjes, P. J. H. (1992). *The LOTOS - LOng Term Ozone Simulation project.* TNO Report No. R92/240. TNO (Institute of Environmental Sciences, Energy Research and Process Information), P. O. Box 342, 7300 AH Apeldoorn, The Netherlands.

[35] Builtjes, P. J. H. (1999). *The EUROTRAC contribution to the development of tools for the study of environmentally relevant trace constituents.* In: Photo-oxidants, Acidification and Tools: Policy Applications of EUROTRAC Results (Borrell, P., Builtjes, P., Grennfelt, P. and Hov, Ø., eds.), pp. 143-194. Springer, Berlin.

[36] Burrage, K. (1995). *Parallel and sequential methods for ordinary differential equations.* Oxford University Press, Oxford-New York.

[37] Butcher, J. C. (1987). *The numerical analysis of ordinary differential equations. Runge-Kutta methods and general linear methods.* Wiley, New York.

[38] Carmichael, G. R. and Peters, L. P. (1984). *An Eulerian transport-transformation-removal model for SO_2 and sulphate: I. Model development.* Atmospheric Environment, 18: 937-952.

[39] Carmichael, G. R., Peters, L. P. and Kitada, T. (1986). *A second generation model for regional-scale transport-chemistry-deposition.* Atmospheric Environment, 20: 173-188.

[40] Carmichael, G. R., Sandu, A., Potra, F. A., Damian, V. and Damian, M. (1996). *The current state and the future directions in air quality modeling.* Systems Analysis Modelling Simulation, 25: 75-105.

[41] Carmichael, G. R., Sandu, A., Song, C. H., He, S. Phandis, M. J., Daescu, D., Damian-Iordache, V. and Potra, F. A. (1999). *Computational challenges of modelling interactions between aerosol and gas phase processes in large-scale air pollution models.* In: Large Scale Computations in Air Pollution Modelling (Zlatev, Z. Brandt, J. Builtjes, P. J. H., Carmichael, G., Dimov, I. Dongarra, J., van Dop, H., Georgiev, K. Hass, H. and San Jose, R., eds.), pp. 99-136. NATO Science Series, 2. Environmental Security - Vol. 57. Kluwer Academic Publishers, Dordrecht-Boston-London.

[42] Chang, J. S., Brost, R. A., Isaksen, I. S. A., Madronich, S., Middleton, P., Stockwell, W. R. and Walcek, C. J. (1987). *A three dimensional Eulerian acid deposition model: Physical concepts and formulation.* Journal of Geophysical Research, 92: 14691-14700.

[43] Chock, D. P. (1985). *A comparison of numerical methods for solving advection equations - II.* Atmospheric Environment, 19: 571-586.

[44] Chock, D. P. (1991). *A comparison of numerical methods for solving advection equations - III.* Atmospheric Environment, 25A: 853-871.

[45] Chock, D. P. and Dunker, A. M. (1983). *A comparison of numerical methods for solving advection equations - I.* J. Comp. Phys., 19: 571-586.

[46] Chock, D. P., Winkler, S. L. and Sun, P. (1994). *Comparison of stiff chemistry solvers for air quality models.* Environ. Sci. Technol., 28: 1882-1892.

[47] Cichota, R., Elias, E. A. and Lier, Q. J. van. (2004). *Testing a finite-difference model for soil hear transfer by comparing numerical and analytical solutions.* Environmental Modelling & Sortware, 19: 495-506.

[48] Crowley, W. P. (1968). *Numerical advection experiments.* Monthly Weather Review, 96: 1-11.

[49] Csomos, P., Faragó, I. and Havasi, Á. (2005). *Weighted sequential splittings and their analysis.* Comput. Math. Appl. (to appear).

[50] Dabdub, D. and Manohar, R. (1997). *Performance and portability of an air pollution model.* Parallel Computing, 23: 2187-2200.

[51] Dabdub, D. and Seinfeld, J. H. (1994). *Numerical advective schemes used in air quality models - sequential and parallel implementation.* Atmospheric Environment, 29: 403-410.

[52] Dabdub, D. and Seinfeld, J. H. (1996). *Parallel computation in atmospheric chemical modelling.* Parallel Computing, 22: 111-130.

[53] Daescu, D. N. and Carmichael, G. R. (2003). *An adjoint sensitivity method for the adaptive location of the observations in air quality modeling.* Journal of the Atmospheric Sciences, 60: 434-450.

[54] Daescu, D. N. and Navon, I. M. (2003). *An analysis of a hybrid optimization method for variational data assimilation.* International Journal of Computational Fluid Dynamics, 17: 299-306.

[55] Daescu, D. N., Sandu, A. and Carmichael, G. R. (2003). *Direct and adjoint sensitivity analysis of chemical kinetic systems with KKP: II. Numerical; validation and applications.* Atmospheric Environment, 37: 5097-5114.

[56] Dahlquist, G. (1963). *A special stability problem for linear multistep methods.* BIT, 3: 27-43.

[57] D'Ambra, P., Barone, G., di Serafino, D., Giunta, G., Murli, A. and Riccio, A. (1999). *PNAM: parallel software for air quality simulations in the Naples area.* Environmental Management and Health, 10: 209-215.

[58] Davis, T. A. and Davidson, E. S. (1988). *Pairwise reduction for the direct parallel solution of sparse unsymmetric sets of linear equations.* IEEE Trans. Comput., 37: 1648-1654.

[59] Davis, T. A. and Yew, P.-C. (1990). *A nondeterministic parallel algorithm for general unsymmetric sparse LU factorization.* SIAM J. Matrix Anal. Appl., 3: 383-402.

[60] Demmel, J. W. (1997). *Applied Numerical Linear Algebra.* SIAM, Philadelphia.

[61] Demmel, J. W., Eisenstat, S. C., Gilbert, J. R., Li, X. S. and Liu, J. W. H. (1999). *A supernodal approach to sparse partial pivoting.* SIAM Journal of Matrix Analysis and Applications, 20: 720-755.

[62] Demmel, J. W., Gilbert, J. R. and Li, X. S. (1999). *An asynchronous parallel supernodal algorithm for sparse Gaussian elimination.* SIAM Journal of Matrix Analysis and Applications, 20: 915-952.

[63] Deuflhard, P. (1983). *Order and stepsize control in extrapolation methods.* Numerische Mathematik, 41: 399-422.

[64] Deuflhard, P. (1985). *Recent progress in extrapolation methods for ordinary differential equations.* SIAM Review, 27: 505-535.

[65] Deuflhard, P., Hairer, E. and Zugch, M. (1987). *One step and extrapolation methods for differential-algebraic equations.* Numerische Mathematik, 51: 501-516.

[66] Deuflhard, P. and Nowak, U. (1986). *Efficient numerical simulation and identification of large reaction systems.* Ber. Bensendes Phys. Chem., 90: 940-946.

[67] Deuflhard, P., Nowak, U. and Wulkow, M. (1990). *Recent development in chemical computing.* Computers Chem. Engng., 14: 1249-1258.

[68] Dimov, I., Faragó, I., Havasi, Á. and Zlatev, Z. (2001). *Commutativity of the operators in splitting methods for air pollution models.* Annalles Univ. Sci. Budapest., 44: 129-150.

[69] Dimov, I., Faragó, I., Havasi, Á. and Zlatev, Z. (2004). *Operator splitting and commutativity analysis in the Danish Eulerian Model.* Mathematics and Computers in Simulation, 67: 217-233.

[70] Dimov, I., Faragó, I. and Zlatev, Z. (2003). *Parallel computations with large-scale air pollution models.* Problems in Programming, 5(3): 44-52.

[71] Dimov, I., Georgiev, K., Ostromsky, Tz. and Zlatev, Z. (2004). *Computational challenges in the numerical treatment of large air pollution models.* Ecological modelling, 179: 187-203.

[72] Dimov, I., Geernaert, G. and Zlatev, Z. (2003). *Influence of future climate changes in Europe on ozone critical levels.* In: Proceedings of Nordic Meteorological Meeting 23 (Joergensen, H. E., ed.). http://www.dams.dk/nmm2002/proceedings.htm

[73] Dimov, I., Ostromsky, Tz., Tzvetanov, I. and Zlatev, Z. (1999). *Economical estimations of the losses of crops due to high ozone levels.* Agricultural Economics and Management, 5: 48-52.

[74] Dimov, I. and Zlatev, Z. (1999). *Testing the sensitivity of air pollution levels to variations of some chemical rate constants.* In: Large-Scale Scientific Computations and Environmental Problems (Griebel, M., Iliev, O. P., Margenov, S. D. and Vassilevski, P. S., eds.), pp. 167-175. Vieweg, Braunschweig/Wiesbaden.

[75] Dimov, I. and Zlatev, Z. (2002). *Optimization problems in air pollution modelling.* In: Handbook on Applied Optimization (Pardalos, P. M. and Resende, M. G. C., eds.), pp. 943-956. Oxford University Press, Oxford-New York.

[76] Djouad, R. and Sportisse, B. (2003). *Solving reduced chemical models in air pollution modelling.* Applied Numerical Mathematics, 44: 49-61.

[77] Dongarra, J. J., Moler, C. B., Bunch, J. R. and Stewart, G. W. (1979). *LINPACK Users' Guide.* SIAM, Philadelphia, 1979.

[78] Dongarra, J. J., Du Croz, J. Duff, I. S. and Hammarling, S. (1988). *An extended set of FORTRAN Basic Linear Algebra Subprograms.* ACM Transaction on Mathematical Software, 14: 1-17.

[79] Dongarra, J. J., Du Croz, J. Duff, I. S. and Hammarling, S. (1990). *An extended set of FORTRAN Basic Linear Algebra Subprograms.* ACM Transaction on Mathematical Software, 16: 18-28.

[80] Dongarra, J. J., Duff, I. S., Sorensen, D. C. and van der Vorst, H. A. (1998). *Numerical Linear Algebra for High-Performance Computers.* SIAM, Philadelphia.

[81] Duff, I. S., Erisman, A. M. and Reid, J. K. (1986). *Direct methods for sparse matrices.* Oxford University Press, Oxford-London.

[82] Duff, I. S., Marone, M., Radicati, G. and Vittoli, C. (1997). *Level 3 Basic Linear Algebra Subprograms for sparse matrices: a user level interface.* ACM Transaction on Mathematical Software, 23: 379-401.

[83] Dyakonov, E. G. (1962). *Difference schemes with splitting operators for multidimensional stationary problems.* Journal of Computational Mathematics and Mathematical Physics, 2(4): 57-79 (in Russian).

[84] Ebel, A. (2000). *Chemical Transfer and Transport Modelling.* In: Transport and Chemical Transformation of Pollutants in the Troposphere (Borrell, P. and Borrell, P. M., eds.), pp. 85-128. Springer, Berlin.

[85] Ebel, A., Memmesheimer, M. and Jacobs, H. (1997). *Regional modelling of tropospheric ozone distributions and budgets.* In: Atmospheric Ozone Dynamics: I. Global Environmental Change, (Varotsos, C., ed.), pp. 37-57. NATO ASI Series, Vol. 53. Springer, Berlin.

[86] Eirola, T. and Nevanlinna, O. (1989). *Accelerating with rank-one updates.* Lin. Alg. Appl., 121: 511-520.

[87] Elbern, H. (1997). *Parallelization and load balancing of a comprehensive atmospheric chemistry model.* Atmospheric Environment, 31: 3561-3574.

[88] Elbern, H. and Schmidt, H. (1999). *A four-dimensional variational chemistry data assimilation scheme for Eulerian chemistry transport modelling.* Journal of Geophysical Research, 104: 18583-18598.

[89] Elbern, H. and Schmidt, H. (2002). *4D-var data assimilation and its numerical implications for case study analyses.* In: Atmospheric Modeling, (Chock, D. P. and Carmichael, G. R., eds.), pp. 165-183. Springer, New York.

[90] Elbern, H., Schmidt, H. and Ebel, A. (1997). *Variational data assimilation for tropospheric chemistry modelling.* Journal of Geophysical Research, 102: 15967-15985.

[91] Elbern, H., Schmidt, H. and Ebel, A. (1999). *Implementation of a parallel 4D-variational chemistry data-assimilation scheme.* Environmental Management and Health, 10: 236-244.

[92] Elbern, H., Schmidt, H., Talagrand, O. and Ebel, A. (2000). *4D-variational data assimilation with an adjoint air quality model for emission analysis.* Environmental Modelling & Software, 15: 539-548.

[93] EMEP (1998). *Transboundary acidifying air pollution in Europe: Part 1, Estimated dispersion of acidifying and eutrofying compounds and comparison with observations.* Status Report 1/98. EMEP MSC-W (Meteorlogical Synthesizing Centre - West), Norwegian Meteorological Institute, P. O. Box 43 - Blindern, N-0313 Oslo, Norway.

[94] EMEP Home Page. (2005). http//www.emep.int.

[95] European Commission (1999). *Ozone position paper.* European Commission, Directorate XI: "Environment, Nuclear Safety and Civil Protection", Brussels.

[96] European Parliament (2002). *Directive 2002/3/EC of the European Parlianent and the Council of 12 February 2002 relating to ozone in ambient air.* Official Journal of the European Communities, L67, 9.3.2002, pp. 14-30.

[97] Faragó, I. and Havasi, Á. (2002). *The mathematical background of operator splitting and the effect of non-commutativity.* In: Large Scale Scientific Computations III (Margenov, S., Yalamov P. and Wasniewski, J. eds.), pp. 264-271. Springer, Berlin.

[98] Faragó, I. and Havasi, Á. (2005) *On the convergence and local splitting error of different splitting schemes.* Progress in Computational Fluid Dynamics, to appear.

[99] Faragó, I., Havasi, Á. and Georgiev, K. (2005) *Advances in Air Pollution Modelling for Environmental Security.* Springer, Berlin.

[100] Faragó, I., Havasi Á. and Korotov, S. (2002) *Mathematical analysis of the weighted splitting.* Technical Report, University of Jyväskylä.

[101] Fingas, M. F. (1995). *A literature review of the physics and predictive modelling of oil spill evaporation.* Journal of Hazardous Materials, 42: 157-175.

[102] Fletcher, R. (1972) *FORTRAN subroutines for minimization by quasi-Newton methods.* Report No. R7125, A.E.R.E, Harwell, England.

[103] Fornberg, B. (1975). *On a Fourier method for the integration of hyperbolic equations.* SIAM J. Numer. Anal., 12: 509-528.

[104] Fornberg, B. (1996). *A practical guide to pseudospectral methods.* Cambridge Monographs on Applied and Computational Mathematics. Cambridge University Press, Cambridge.

[105] Foster, I. and Kesselman, C. (1998). *Computational Grids.* In: The Grid: Blueprint for a New Computing Infrastructure (Foster, I. and Kesselman, C., eds.), pp. 15-52. Morgan Kaufmann Publishers, San Francisco, California.

[106] Foster, I., Kesselman, C. and Tuecke, S. (2001). *The Anatomy of the Grid: Enabling Scalable Virtual Organizations.* Intl. J. Supercomputer Applications, 15(3): 1-25.

[107] Freund, R. W. (1993). *A transpose-free quasi-minimal residual algorithm for non-Hermitian linear systems.* SIAM J. Sci. Statist. Comput., 14: 470-482.

[108] Friedrich, R. (1997). *Tropospheric modelling and emission estimation.* Springer-Verlag, Berlin, pp. 179-357.

[109] Forster, C., Stohl, A., James, P. and Thouret, V. (2003) *The residence times for aircraft emissions in the stratosphere and emissions along actual flight tracks.* Journal of Geophysical Research, 108, No. D12, 8524.

[110] Gallivan, K., Hansen, P. C., Ostromsky, Tz. and Zlatev, Z. (1995). *A locally optimized reordering algorithm and its application to a parallel sparse linear system solver.* Computing, 54: 39-67.

[111] Gallivan, K. A., Sameh, A. H. and Zlatev, Z. (1990). *A parallel hybrid sparse linear system solver,* Computing Systems in Engineering, 1: 183-195.

[112] Geernaert, G. and Zlatev, Z. (2002). *Testing the influence of the biogenic emissions on high ozone levels in Denmark and in Europe.* In: Proceedings from the EUROTRAC-2 Symposium 2002 (Midgley, P. M. and Reuther, M., eds.). Margraf Verlag, Weikersheim.

[113] Geernaert, G. and Zlatev, Z. (2004). *Studying the influence of biogenic emissions on AOT40 values in Europe.* International Journal of Environment and Pollution, 22 (No. 1-2): 29-42.

[114] Geist, A., Beguelin, A. Dongarra, J., Jiang, W., Manchek, R. and Sunderam, V. (1994). *PVM: Parallel Virtual Machine, A User's Guide and Tutorial for Networking Parallel Computing.* MIT Press, Cambridge, Massachusetts.

[115] George, J. A. and Ng, E. (1985). *An implementation of Gaussian elimination with partial pivoting for sparse systems.* SIAM J. Sci. Statist. Comput., 6: 390-405.

[116] Georgiev, K. and Zlatev, Z. (1998). *Running an advection-chemistry code on message passing computers.* In: Recent Advances in Parallel Virtual Machine and Message Passing Interface (Alexandrov, V. and Dongarra, J., eds.), pp. 354-363. Springer, Berlin.

[117] Georgiev, K. and Zlatev, Z. (2000). *Parallel Sparse Matrix Algorithms for Air Pollution Models.* Parallel and Distributed Computing Practices, 2: 429-442.

[118] Gery, M. W., Whitten, G. Z., Killus, J. P. and Dodge, M. C. (1989). *A photochemical kinetics mechanism for urban and regional modeling.* Journal of Geophysical Research, 94: 12925-12956.

[119] Gilbert, J. R. and Peierls, T. (1988). *Sparse partial pivoting in time proportional to arithmetic operations.* SIAM J. Sci. Statist. Comput., 9, 862-874.

[120] Gill, P. E. and Murray. W. (1979). *Conjugate-gradient methods for large-scale nonlinear optimization.* Technical Report SOL 79-15. Department of Operations Research, Stanford University, Stanford, California, USA.

[121] Gill, P. E., Murray. W. and Wright, M. H. (1981). *Practical Optimization.* Academic Press, New York.

[122] Golub, G. H. and Van Loan, C. F. (1983). *Matrix computations* Johns Hopkins University Press, Baltimore, Maryland.

[123] Grid Computing Info Centre: GRID Infoware. (2005) *Grid Computing Info Centre Newsletter.* http://www.gridcomputing.com/.

[124] Gropp, W., Lusk, E. and Skjellum, A. (1994). *Using MPI: Portable programming with the message passing interface.* MIT Press, Cambridge, Massachusetts.

[125] Gupta, I., Dhage, S., Chandorkar, A. A. and Srivastav, A. (2004). *Numerical modeling for Thane creek.* Environmental Modelling & Software 19: 571-579.

[126] Hairer, E., Nørsett, S. P. and Wanner, G. (1987). *Solving Ordinary Differential Equations, I: Nonstiff Problems.* Springer, Berlin-Heidelberg-New York-London.

[127] Hairer, E. and Wanner, G. (1991). *Solving Ordinary Differential Equations, II: Stiff and Differential-algebraic Problems.* Springer, Berlin-Heidelberg-New York-London.

[128] Harrison, R. M., Zlatev, Z. and Ottley, C. J. (1994). *A comparison of the predictions of an Eulerian atmospheric transport chemistry model with measurements over the North Sea.* Atmospheric Environment, 28: 497-516.

[129] Hass, H., Ebel, A., Feldmann, H., Jacobs, H. J. and Memmesheimer, M. (1993). *Evaluation studies with a regional chemical transport model (EURAD) using air quality data from the EMEP monitoring network.* Atmospheric Environment, 27A: 867-887.

[130] Hass, H., Jacobs, H. J. and Memmesheimer, M. (1995). *Analysis of a regional model (EURAD) near surface gas concentration predictions using observations from networks.* Atmos. Phys., 57: 173-200.

[131] Havasi, Á. and Zlatev, Z. (2002). *Trends of Hungarian air pollution levels on a long time-scale.* Atmospheric Environment, 36: 4145-4156.

[132] Henrici, P. (1962). *Discrete variable methods in ordinary differential equations.* Wiley, New York.

[133] Hertel, O. Ambelas Skøth, C., Frohn, L. M., Vignati, E., Frydendall, J., de Leeuw, G., Schwarz, U. and Reis, S. (2002). *Assessment of the atmospheric nitrogen and sulphur inputs into the North Sea using a Lagrangian model.* Physics and Chemistry of the Earth, 27: 1507-1515.

[134] Hertel, O., Berkowicz, R., Christensen, J. and Hov Ø. (1993). *Test of two numerical schemes for use in atmospheric transport-chemistry models.* Atmospheric Environment, 27A: 2591-2611.

[135] Hesstvedt, E., Hov, Ø. and Isaksen, I. A. (1978). *Quasi-steady-state approximations in air pollution modelling: Comparison of two numerical schemes for oxidant prediction.* International Journal of Chemical Kinetics, 10: 971-994.

[136] Hjellbrekke, A. G. (1995). *Ozone measurements 1990-1992.* EMEP/CCC Report 4/95, Norwegian Institute for Air Research (NILU), POSTBOKS 64, N-2001 Lillestøm, Norway.

[137] Holm, E. and Erbes, G. (1991). *Solving the advection equation in flows with sharp gradients using Bott's flux scheme.* Report No. DM-59. Department of Meteorology, Stockholm University, Stockholm.

[138] Hongisto, M. (2003). *HILATAR, a limited area simulation model for acid contaminants. Part I. Model description and verification.* Atmospheric Environment, 37: 1535-1447.

[139] Hongisto, M., Sofiev, M. and Joffre, S. (2003). *HILATAR, a limited area simulation model for acid contaminants. Part II. Long-term simulation results.* Atmospheric Environment, 37: 1549-1560.

[140] Houghton, J. T., Ding, Y., Griggs, D. J., Noguer, M., van der Linden, P. J., Dai, X., Maskell, K. and Johnson, C. A., eds. (2001). *Climate Change 2001: The Scientific Basis.* Cambridge University Press, Cambridge-New York-Melbourne-Madrid-Cape Town.

[141] Hov, Ø., Zlatev, Z., Berkowicz, R., Eliassen, A. and Prahm, L. P. (1988). *Comparison of numerical techniques for use in air pollution models with non-linear chemical reaction.* Atmospheric Environment, 23: 967-983.

[142] Hundsdorfer, W. and Verwer, J. G. (2003) *Numerical solution of time-dependent advection-diffusion-reaction equations.* Springer, Berlin.

[143] Jacobson, M. Z. and Turco, R. P. (1994). *SMVGEAR: a sparse-matrix, vectorized Gear code for atmospheric models.* Atmospheric Environment, 28: 273-284.

[144] Jaffe, A. (1984). *Ordering the universe: The role of mathematics.* SIAM Rev., 26: 475-488.

[145] Jay, L. O., Sandu, A., Potra, F. A. and Carmichael, G. R. (1997). *Improved QSSA methods for atmospheric chemistry integration.* SIAM J. Sci. Comput., 18: 182-202.

[146] Jeltsch, R. and Nevanlinna, O. (1981). *Stability of explicit time discretizations for solving initial value problems.* Numerische Mathematik, 37: 61-91.

[147] Jennings, A. A. (2003). *Modelling sedimentation and scour in small urban lakes.* Environmental Modelling & Software 18: 281,291.

[148] Jonson, J. E., Sundet, J. and Tarrason, L. (2001). *Model calculations of present and future levels of ozone and ozone precursors with a global and a regional model.* Atmospheric Environment, 35: 525-537.

[149] Jonson, J. E., Tarrason, L. and Sundet, J. (1999). *Calculation of ozone and other pollution for the summer, 1996.* Environmental Management and Health, 10: 245-257.

[150] Kincaid, D., Oppe, T., Respess, J. and Young, D. (1984). *ITPACKV 2C: User's giude.* Report No. CNA-191. Center for Numerical Analysis, University of Texas at Austin, Austin, Texas.

[151] Kincaid, D., Respess, J., Young, D. and Grimes, R. (1982). *Algorithm 586, ITPACKV 2C: A FORTRAN package for solving large sparse systems by adaptive accelerated iterative methods.* ACM Trans. Math. Software, 8: 302-322.

[152] Kreiss, H. O. (1978). *Difference methods for stiff ordinary differential equations.* SIAM Journal on Numerical Analysis, 15: 21-58.

[153] Kreiss, H. O. and Oliger J. (1972). *Comparison of accurate methods for the integration of hyperbolic equations.* Tellus, 24: 199-215.

[154] Kuck, D. J., Davidson, E. S., Lawrie D. and Sameh, A., H. (1986). *Parallel supercomputing today and the CEDAR approach.* Science, 231: 967-974.

[155] Lambert, J. D. (1991). *Numerical Methods for Ordinary Differential Equations.* Wiley, Chichester-New York-Brisbane-Toronto-Singapore.

[156] Langner, J., Robertson, L., Persson, C. and Ullerstig, A. (1998). *Validation of the operational emergency response model at the Swedish Meteorological and Hydrological Institute using data from ETEX and the Chernobyl accident.* Atmospheric Environment, 32: 4325-4333.

[157] Lanser, D. and Verwer, J. G. (1999). *Analysis of operators splitting in advection-diffusion-reaction problems in air pollution modelling.* Journal of Computational and Applied Mathematics, 111: 201-216.

[158] Lawson, C., Hanson, R., Kincaid, D. and Krogh, F. (1979). *Basic Linear Algebra Subprograms for Fortran usage.* ACM Transactions on Mathematical Software, 5: 308-323.

[159] Le Dimet, F.-X., Navon, I. M. and Daescu, D. N. (2002). *Second order information in data assimilation.* Monthly Weather Review, 130: 629-648.

[160] Lee H, N. (1981). *An alternative pseudospectral model for pollutant transport, diffusion and deposition in the atmosphere.* Atmospheric Environment, 15: 1017-1024.

[161] Lee, H. S., Matthews, C. J., Braddock, R. D., Sander, G. C. and Gandola, F. (2004) *A MATLAB method of lines template for transport equations.* Environmental Modelling & Software, 19: 603-614.

[162] LeVeque, R. J. (1992). *Numerical methods for conservation laws.* Lecture Notes in Mathematics, ETH Zürich, Birkhäuser Verlag, Basel.

[163] LeVeque, R. J. (1992). *Finite volume methods for hyperbolic problems.* Cambridge Texts in Applied Mathematics, Cambridge University Press, Cambridge.

[164] Lewis, J. M. and Derber, J. C. (1985). *The use of adjoint equations to solve a variational adjustment problem with advective constraints.* Tellus, 37A: 309-322.

[165] Li, X. S. and Demmel, J. W. (1999). *Super_LU_DIST: A scalable distributed-memory sparse direct solver for unsymmetric linear system.* ACM Transactions on Mathematical Software, 29: 110-140.

[166] Li, Z. and Navon, I. M. (2001). *Optimality of 4D-Var and its relationship with the Kalman filter and the Kalman smoother.* Quarterly Journal of the Royal Meteorological Society, 127B: 661-684.

[167] Loon, M. van, (1993). *Testing interpolation and filtering techniques in connection with a semi-Lagrangian method.* Atmospheric Environment, 27A: 2351-2364.

[168] Losinskii, S. M. (1958). *Error estimates in the numerical integration of ordinary differential equations.* Communications of the High Schools - Mathematics, 5(6): 52-90 (in Russian)

[169] Lübkert, B. and Schöpp, W. (1989). *A model to calculate natural VOC emissions from forests in Europe.* Report WP-89-082, International Institute for Applied System Analysis (IIASA), Laxenburg, Austria.

[170] Lyness, J. N. (1974). *Computational techniques based on the Lanczos transformation.* Mathematics of Computation, 28: 81-123.

[171] Mao, X., Prommer, H., Barry, D. A., Langevin, C. D., Panteleit, B. and Li, L. (2005). *Three-dimensional model for multi-component reactive transport with variable density groundwater flow.* Environment Modelling & Software, to appear (available on line at www.sciencedirect.com).

[172] Marchuk, G. I. (1968). *Some application of splitting-up methods to the solution of mathematical physics problems.* Applik. Mat., 13(2).

[173] Marchuk, G. I. (1980). *Methods of computational mathematics.* Nauka, Moscow, 1980.

[174] Marchuk, G. I. (1982). *Mathematical Modeling for the Problem of the Environment.* Nauka, Moscow (in Russian).

[175] Marchuk, G. I. (1986). *Mathematical modelling for the problem of the environment.* Studies in Mathematics and Applications, No. 16, North-Holland, Amsterdam.

[176] Marchuk, G. I. (1988). *Methods of Splitting.* Nauka, Moscow, (in Russian).

[177] Marchuk, G. I. (1994). *Applications of Adjoint Equations to Problems of Global Change and Environmental Protection.* In: Proceedings of HERMIS'94, Vol. 1, (E.A. Lipitakis, ed.), pp. 1-20. Hellenic Mathematical Society, Athens.

[178] Marchuk, G. I. (1995): *Adjoint equations and analysis of complex systems.* Kluwer Academic Publishers, Dordrecht.

[179] Marchuk, G. I., Agoshkov, P. V. and Shutyaev, I. V. (1996). *Adjoint equations and perturbation algorithms in nonlinear problems.* CRC Press, New York.

[180] Markowitz, H. M. (1957). *The elimination form of the inverse and its applications to linear programming.* Management Science, 3: 255-267.

[181] McRae, G. J., Goodin, W. R. and Seinfeld, J. H. (1982). *Numerical solution of the atmospheric diffusion equations for chemically reacting flows.* Journal of Computational Physics, 45: 1-42.

[182] Merilees, P. E. (1974). *Numerical experiments with the pseudospectral method in spherical coordinates.* Atmosphere, 12:77-96.

[183] Molenkampf, C. R. (1968). *Accuracy of finite-difference methods applied to the advection equation.* Journal of Applied Meteorology, 7: 160–167.

[184] Morel, P., Lefevre, G. and Rabreau, G. (1971). *On initialization and non-synoptic data assimilation.* Tellus, 23: 197-206.

[185] NAG Library Fortran Manual (1979). *Mark 7, Vol. 3-4.* Numerical Algorithms Group (NAG), Banbury Road 7, Oxford, England.

[186] NAG Library Fortran Manual (2004). *E04 - minimizing and maximizing a function.* http://www.nag.co.uk. Numerical Algorithms Group (NAG), Banbury Road 7, Oxford, England.

[187] Odman, M. T., Kumar, N. and Russell, A. G. (1992). *A comparison of fast chemical kinetic solvers for air quality modeling.* Atmospheric Environment, 26A: 1783-1789.

[188] Owczarz, W. and Zlatev, Z. (2000). *Running a large-scale air pollution model on fast super computer.* In: Atmospheric Modeling (Chock, D. P. and Carmichael, G. R., eds.), pp. 185-204. Spinger, Berlin.

[189] Owczarz, W. and Zlatev, Z. (2001). *Running a large air pollution model on an IBM SMP computer.* International Journal of Computer Research, 10: 485-500.

[190] Owczarz, W. and Zlatev, Z. (2002). *Parallel matrix computations in air pollution modelling.* Parallel Computing, 28: 355-368.

[191] Penenko V. V. and Obraztsov, N. N. (1976). *A variational initialization model for the fields of meteorological elements.* Soviet Meteorol. Hydrol., 11: 1-11.

[192] Pepper, D. W. and Baker, A. J. (1979). *A simple one-dimensional finite element algorithm with multidimensional capabilities.* Numerical Heath Transfer, 3: 81-95.

[193] Pepper, D. W., Kern, C. D. and Long, Jr., P. E. (1979). *Modelling the dispersion of atmospheric pollution using cubic splines and chapeau functions.* Atmospheric Environment, 13: 223-237.

[194] Peters, L. K., Berkowitz, C. M., Carmichael, G. R., Easter, R. C., Fairweather, G., Ghan, S. J., Hales, J. M., Leung, L. R., Pennell, W. R., Potra, F. A., Saylor, R. D. and Tsang, T. T. (1995). *The current state and future direction of Eulerian models in simulating tropospheric chemistry and transport of trace species: A review.* Atmospheric Environment, 29: 189-222.

[195] Pissanetzki, S. (1984). *Sparse matrix technology.* Academic Press, New York.

[196] Priklonsky, V. I. Reible, D. D. and Tyler, J. M. (2005). *Consistent unconfined contaminated disposal facilities dike tidal flow and transport model.* Environmental Modelling and Software, to appear, (available online on www.sciencedirect.com).

[197] Prommer, H., Barry, D. A. and Zheng, C. (2003). *MPDFLOW/MT3DMS-based reactive multicomponent transport modelling.* Ground Water, 41: 247-257.

[198] Rancic, M. (1992). *Semi-Lagrangian piecewise by parabolic scheme for two dimensional horizontal advection of a passive scalar.* Monthly Weather Review, 120: 1394-1406.

[199] Rancic, M. and Williamson, D. L. (1991). *On shape preserving interpolation and semi-Lagrangian transport.* SIAM Journal of Scientific and Statistical Computing, 4: 656-687.

[200] Reed, M., Johansen, Ø, Brandvik, P. J., Dailing, P., Lewis, A., Fiocco, R., Mackay, D. and Prentki, R. (1999) *Oil spill modelling toward the close of the 20th century: overview of the state of the art.* Spill Science and Technology Bulletin, 5: 3-16.

[201] Richtmyer, R. D. and Morton, K. W. (1967). *Difference methods for initial value problems.* Interscience Publishers, New York, 1967.

[202] Roache, P. J. (1971). *Computational fluid dynamics.* Hermosa, Albuquerque, New Mexico, USA.

[203] Roache, P. J. (1978). *A pseudo-spectral FFT technique for non-linear problems.* Journal of Computational Physics, 27: 204-220.

[204] Robert, A. J., Yee, T. L. and Ritchie, H. (1985). *A semi-Lagrangian and semi-implicit numerical integration scheme for multi-level atmospheric models.* Monthly Weather Review, 113: 388-394.

[205] Robertson, L., Langner, J. and Engardt, M. (1999). *An Eulerian limited-area atmospheric transport model.* Journal of Applied Meteorology, 38: 190-210.

[206] Rosenbrock, H. H. (1963). *Some implicit processes for the numerical solution of ordinary differential equations.* Computer Journal, 5: 329-330.

[207] Saad, Y. and Schultz, M. H. (1986). *GMRES: a generalized minimal residual algorithm for solving nonsymmetric linear systems.* SIAM J. Sci. Statist. Comput., 7: 856-869.

[208] Sandnes Lenschow, H. and Tsyro, S. (2000). *Meteorological input data for EMEP/MSC-W air pollution models.* EMEP Note 2/00. Meteorological Synthesizing Centre - West, Norwegian Meteorological Institute, P. O. Box 43 - Blindern, N-0313 Oslo 3, Norway.

[209] Sandu, A., Daescu, D. N. and Carmichael, G. R. (2003). *Direct and adjoint sensitivity analysis of chemical kinetic systems with KKP: I. Theory and software tools.* Atmospheric Environment, 37: 5083-5096.

[210] Sandu, A., Daescu, D. N., Carmichael, G. R. and Chai, T. (2005). *Adjoint sensitivity analysis of regional air quality models.* Journal of Computational Physics, 2005; to appear.

[211] Sandu, A., Potra, F. A., Carmichael, G. R. and Damian, V. (1996). *Efficient implementation of fully implicit methods for atmospheric chemical kinetics.* J. Comp. Phys., 129: 101-110.

[212] Sandu, A., Verwer, J. G., van Loon, M., Carmichael, G. R., Potra, F. A., Dabdub, D. and Seinfeld, J. H. (1997). *Benchmarking stiff ODE systems for atmospheric chemistry problems: I. Implicit versus Explicit.* Atmospheric Environment, 31: 3151-3166.

[213] Sandu, A., Verwer, J. G., Bloom, J. G., Spee, E. J. and Carmichael, G. R. (1997). *Benchmarking stiff ODE systems for atmospheric chemistry problems: II. Rosenbrock solvers.* Atmospheric Environment, 31: 3459-3472.

[214] San Jose, R. Rodriguez, M. A., Cortes, E. and Gonzalez, R. M. (1999). *EMMA model: an advanced operational mesoscale air quality model for urban and regional environments.* Environmental Management and Health, 10: 258-266.

[215] Schell, B., Ackermann, I. J., Hass, H., Binkowski, F. S. and Ebel, A. (2001). *Modelling the formation of secondary organic aerosol within a comprehensive air quality model system.* Journal of Geophysical Research, 106: 28275-28293.

[216] Segerlind, L. J. (1976). *Applied Finite Element Analysis.* Wiley, New York.

[217] Shampine, L. F. (1994). *Numerical solution of ordinary differential equations.* Chapman and Hall, New York.

[218] Shampine, L. F. and Gordon, M. K. (1975). *Computer solution of ordinary differential equations: The initial value problem.* Freeman, San Francisco.

[219] Sherman, A. H. (1978). *Algorithms for sparse Gaussian elimination with partial pivoting.* ACM Trans. Math. Software, 4: 330-338.

[220] Shampine, L. F., Reichelt, M. W. and Kierzenka, J. A. (1999). *Solving Index-1 DAEs in MATLAB and Simulink.* SIAM Rev., 41: 538-552.

[221] Shieh, D. S., Chang, Y. and Carmichael, G. R. (1988). *The evaluation of numerical techniques for solution of stiff ordinary differential equations arising from chemical kinetic problems.* Environ. Software, 3: 28-38.

[222] Simpson, D. (1992). *Long-period modelling of photochemical oxidants in Europe. Model calculations for July 1985.* Atmospheric Environment, 26A: 1609-1634.

[223] Simpson, D. (1993). *Photochemical model calculations over Europe for two extended summer periods: 1985 and 1989. Model results and comparisons with observations.* Atmospheric Environment, 27A: 921-943.

[224] Simpson, D., Andersson-Sköld, Y. and Jenkin, M. E. (1993). *Updating the chemical scheme for the EMEP MCD-W oxidant model: current status.* EMEP/MSC-W Report 2/93. Meteorological Synthesizing Centre-West, Norwegian Meteorological Institute, P. O. Box 43-Blindern, N0313 Oslo 3, Norway.

[225] Simpson, D., Guenther, A., Hewitt, C. N. and Steinbrecher, R. (1995). *Biogenic emissions in Europe: I. Estimates and uncertainties.* Journal of Geophysical Research, 100: 22875-22890.

[226] Simpson, D., Tuovinen, J.-P., Emberson, L. D. and Ashmore, M. R. (2003). *Characteristics of an ozone deposition module II. densitivity analysis.* Water, Air and Soil Pollution, 143: 123-137.

[227] Skelboe, S. and Zlatev, Z. (1997). *Exploiting the natural partitioning in the numerical solution of ODE systems arising in atmospheric chemistry.* In: Numerical Analysis and Its Applications (Vulkov, L., Wasniewski, J. and Yalamov, P., eds.), pp. 458-465. Springer, Berlin.

[228] SMP/E Internet Library (2005). http://www-03.ibm.com/servers/eserver/zseries/zos/smpe/smpepubs.html.

[229] Solin, P. *Partial differential equations and the finite element method.* Wiley, New York, 2005.

[230] Sonneveld, P. (1989). *CGS, a fast Lanczos-type solver for nonsymmetric linear systems.* SIAM J. Sci. Statist. Comput., 10: 36-52.

[231] Spaulding, M. L. (1988). *A state-of-the-art review of oil spill trajectory and fate modelling.* Oil and Chemical Pollution, 4: 39-55.

[232] Stewart, G. W. (1973). *Introduction to matrix computations.* Academic Press, New York.

[233] Stockwell, W. R., Kirchner, F. and Kuhn, M. (1997). *A new mechanism for regional atmospheric chemistry modelling.* Journal of Geophysical Research, 102: 25847-25879.

[234] Stockwell, W. R., Middleton, P., Chang, J. S. and Tang, X. (1990). *A second generation regional acid deposition model chemical mechanism for regional air quality modeling.* Journal of Geophysical Research, 95: 16343-16367.

[235] Strang, G. (1968). *On the construction and comparison of difference schemes.* SIAM J. Numer. Anal., 5: 505-517.

[236] Strang, G. and Fix, G. J. (1973). *An analysis of the finite element method.* Prentice-Hall, Englewood Cliffs, New Jersey.

[237] Strikwerda, J. C. *Finite difference schemes and partial differential equations.* Second Edition, SIAM, Philadelphia, 2004.

[238] Su, N. (2004). *Generalization of various hydrological and environmental transport models using the Fokker-Planck equation.* Environmental Modelling and Software, 19: 345-356.

[239] Su, N., Liu, F. and Anh, V. (2003). *Tides as phase-modulated waves inducing periodic groundwater flow in coastal aquifers overlaying a sloping impervious base.* Environmental Modelling and Software, 18: 937-942.

[240] Swart, J. and Blom, J. (1996). *Experience with sparse matrix solvers in parallel ODE software.* Comp. Math. Appl., 31: 43-55.

[241] Swarztrauber, P. N. (1982). *Vectorizing FFT's.* In: Parallel Computations (Rodrigue, G., ed.). Academic Press, London - New York.

[242] Syrakov, D. (1997). *On the TRAP advection scheme - descriptions, tests and applications.* In: Regional modelling of Air Pollution in Europe (Geernaert, G, Walløe Hansen, A. and Zlatev, Z., eds.), pp. 141-152. Proceedings of the First REMAPE Workshop, Copenhagen September 1996, National Environmental Research Institute, Roskilde, Denmark.

[243] Syrakov, D. and Galperin, M. (1999). *On some flux-type advection schemes for dispersion modelling application.* In: Large-Scale Computations in Air Pollution Modelling (Zlatev, Z. Brandt, J. Builtjes, P. J. H., Carmichael, G., Dimov, I. Dongarra, J., van Dop, H., Georgiev, K. Hass, H., and San Jose, R., eds.), pp. 303-312. NATO Science Series, 2. Environmental Security - Vol. 57. Kluwer Academic Publishers, Dordrecht-Boston-London.

[244] Syrakov, D. and Galperin, M. (2000) *On some explicit advection schemes for dispersion modelling applications.* International Journal of Environment and Pollution, 14: 267-277.

[245] Talagrand, O. and Courtier, Ph. (1987). *Variational assimilation of meteorological observations with the adjoint vorticity equation. I: Theory.* Quarterly Journal of the Royal Meteorological Society, 113: 1311-1328.

[246] Temperton, C. (1979). *Fast Fourier transforms on Cray-1.* Report No. 21. European Centre for Medium Range Weather Forecasts, Reading, England.

[247] Tewarson, R. P. (1973). *Sparse matrices.* Academic Press, New York.

[248] Tikhonov, A. N. and Samarski, A. A. (1977). *Equations of the Mathematical Physics.* Nauka, Moscow.

[249] Tkalich, P. (2005). *A CFD solution of oil spil problems.* Environmental Modelling & Software, to appear (available at www.sciencedirect.com).

[250] Tomlin, A. S., Ghorai, S., Hart, G. and Berzins, M. (1999). *3D adaptive unstructured meshes for air pollution modelling.* Environmental Management and Health, 10: 267-274.

[251] Tuovinen, J.-P. (2000). *Assessing vegetation exposure to ozone: properties of the AOT40 index and modifications by deposition modelling.* Environmental Pollution, 109: 361-372.

[252] Trefethen, L. N. and Bau III, D. (1997). *Numerical Linear Algebra.* SIAM, Philadelphia.

[253] Umgiesser, G., Canu, D. M., Solidoro, C. and Ambrose, R. (2003). *A finite element ecological model: a first application to the Venice Lagoon.* Environmental Modelling & Software, 18: 131-145.

[254] UNECE (1999). *Protocol to the 1979 Convention on Long-Range Air Pollution to abate acidification, eutrophication and ground level ozone.* EB.AIR/1999/1, Gothenburg, Sweden, 15. October 1999.

[255] van der Vorst, H. A. (1992). *BI-CGSTAB: A fast and smoothly converging variant of BI-CG for the solution of nonsymmetric linear systems*, SIAM J. Sci. Statist. Comput., 13: 631-644.

[256] van der Vorst, H. A. (2003). *Iterative Krylov Methods for Large Linear Systems.* Cambridge University Press, Cambridge.

[257] Venkatram, A., Karamchandani, P. K. and Misra, P. K. (1988). *Testing a comprehensive acid deposition model.* Atmospheric Environment, 22: 737-747.

[258] Verwer, J. G., Blom, J. G, van Loon, M. and Spee, E. J. (1996). *A comparison of stiff ODE solvers for atmospheric chemistry problems.* Atmospheric Environment, 30: 49-58.

[259] Verwer, J. G. and van Loon, M. (1994). *An evaluation of explicit pseudo-steady state approximation for stiff ODE systems from chemical kinetics.* J. Comput. Phys., 113: 347-352.

[260] Verwer, J. G. and Simpson, D. (1995). *Explicit methods for stiff ODE's from atmospheric chemistry.* Appl. Numer. Math., 18: 413-430.

[261] Vestreng, V. and Støren, E. (2000). *Analysis of UNECE/EMEP emission data.* EMEP MSC-W Note 1/00. Meteorological Synthesizing Centre - West, Norwegian Meteorological Institute, P. O. Box 43 - Blindern, N-0313 Oslo 3, Norway.

[262] Vinsome, P. (1976). *ORTHOMIN, an iterative method for solving sparse sets of simultaneous linear equations.* Proc. Fourth Sympos. on Reservoir Simulation, Society of Petr. Eng. of AIME.

[263] Vuik, C. and van der Vorst, H. A. (1992). *A comparison of some GMRES-like methods.* Lin. Alg. Appl., 160: 131-160.

[264] Wang, K. Y., Lary, D. J., Shallcross, D. E., Hall, S. M. and Pyle, J. A. (2001). *A review on the use of the adjoint method in four-dimensional atmospheric-chemistry data assimilation.* Quarterly Journal of the Royal Meteorological Society, 127B: 2181-2204.

[265] WEB-site for OPEN MP tools, (1999). http://www.openmp.org

[266] Wilkinson, J. H. (1963). *Rounding errors in algebraic processes.* Notes in Applied Sciences, No. 32, HMSO, London.

[267] Wilkinson, J. H. (1965). *The algebraic eigenvalue problem.* Oxford University Press, Oxford-London.

[268] Willoughby, R. A. (1970). *Sparse matrix algorithms and their relation to problem classes and computer architecture.* In: Large Sparse Sets of Linear Equations (Reid, J. K., ed.), pp. 255-277, Academic Press, London-New York.

[269] Yanenko, N. N. (1962). *On convergence of the splitting method for heat equation with variable coefficients.* Journal of Computational Mathematics and Mathematical Physics, 2(5): 933-937 (in Russian).

[270] Yang, U. M. and K. Gallivan, K. (1999). *A family of preconditioned iterative solvers for sparse linear systems.* Applied Numerical Mathematics, 30: 155-173.

[271] Zienkiewicz, O. C. (1971). *The finite element method in engineering science.* McGraw-Hill, London.

[272] Zlatev, Z. (1980). *On some pivotal strategies in Gaussian elimination by sparse technique.* SIAM Journal on Numerical Analysis, 17: 18-30.

[273] Zlatev, Z. (1981). *Modified diagonally implicit Runge-Kutta methods.* SIAM Journal on Scientific and Statistical Computing, 2: 321-334.

[274] Zlatev, Z. (1982). *Use of iterative refinement in the solution of sparse linear systems.* SIAM Journal on Numerical Analysis, 19: 381-399.

[275] Zlatev, Z. (1983). *Consistency and convergence of general multistep variable stepsize variable formula methods.* Computing, 31: 47-67.

[276] Zlatev, Z. (1984). *Application of predictor-corrector schemes with several correctors in solving air pollution problems.* BIT, 24: 700-715.

[277] Zlatev, Z. (1985). *Mathematical model for studying the sulphur pollution in Europe.* Journal of Computational and Applied Mathematics, 12: 651-666.

[278] Zlatev, Z. (1989). *Treatment of some mathematical models describing long-range transport of air pollutants on vector processors.* Parallel Computing, 6: 87-98.

[279] Zlatev, Z. (1989). *Advances in the theory of variable stepsize variable formula methods for ordinary differential equations.* Applied Mathematics and Computation, 31: 209-249.

[280] Zlatev, Z. (1990) *Running large air pollution models on high speed computers.* Math. Comput. Modelling, 14: 737-740.

[281] Zlatev, Z. (1991). *Computational methods for general sparse matrices.* Kluwer Academic Publishers, Dordrecht-Boston-London.

[282] Zlatev, Z. (1995). *Computer treatment of large air pollution models.* Kluwer Academic Publishers, Dordrecht-Boston-London.

[283] Zlatev, Z. (2001). *Partitioning ODE systems with an application to air pollution models.* Computers and Mathematics with Applications, 42: 817-832.

[284] Zlatev, Z. and Ambelas Skjøth, C. (1999) *The WEB-site of the Danish Eulerian Model developed at the National Environmental Research Institute (NERI).* http://www.dmu.dk/AtmosphericEnvironment/DEM.

[285] Zlatev, Z., Bergström, R., Brandt, J., Hongisto, M., Jonson, J. E. Langner, J. and Sofiev, M. (2001). *Studying sensitivity of air pollution levels caused by variations of different key parameters.* TemaNord 2001:569. Nordic Council of Ministers, Copenhagen.

[286] Zlatev, Z. and Berkowicz, R. (1988). *Numerical treatment of large-scale air pollution models.* Computers and Mathematics with Applications, 15: 93-109.

[287] Zlatev, Z., Berkowicz, R. and Prahm, L. P. (1983). *Testing subroutines solving advection-diffusion equations in atmospheric environments.* Comput. Fluids, 11: 12-38.

[288] Zlatev, Z., Berkowicz, R. and Prahm, L. P. (1983). *Three dimensional advection-diffusion modelling for regional scale.* Atmospheric Environment, 17: 491-499.

[289] Zlatev, Z., Berkowicz, R. and Prahm, L. P. (1983). *Stability restrictions on time stepsize for numerical integration of first-order partial differential equations.* Journal of Computational Physics, 51: 1-27.

[290] Zlatev, Z., Berkowicz, R. and Prahm, L. P. (1984). *Implementation of a variable stepsize variable formula method in the time-integration part of a code for long-range transport of air pollutants.* Journal of Computational Physics, 55: 279-301.

[291] Zlatev, Z. Brandt, J. Builtjes, P. J. H., Carmichael, G., Dimov, I. Dongarra, J., van Dop, H., Georgiev, K. Hass, H. and San Jose, R. (1999). *Large Scale Computations in Air Pollution Modelling.* Kluwer Academic Publishers, Dordrecht-Boston-London.

[292] Zlatev, Z., Christensen, J. and Eliassen, A. (1993). *Studying high ozone concentrations by using the Danish Eulerian Model.* Atmospheric Environment, 27A: 845-865.

[293] Zlatev, Z., Christensen, J. and Hov, Ø. (1992). *An Eulerian air pollution model for Europe with nonlinear chemistry.* Journal of Atmospheric Chemistry, 15: 1-37.

[294] Zlatev, Z., Dimov, I. and Georgiev, K. (1994). *Studying long-range transport of air pollutants.* Computational Science and Engineering, 1(3): 45-52.

[295] Zlatev, Z., Dimov, I. and Georgiev, K. (1996). *Three-dimensional version of the Danish Eulerian Model.* Zeitschrift für Angewandte Mathematik und Mechanik, 76(S4): 473-476.

[296] Zlatev, Z., Dimov, I., Ostromsky, Tz., Geernaert, G., Tzvetanov, I. and Bastrup-Birk, A. (2001). *Calculating losses of crops in Denmark caused by high ozone levels.* Environmental Modeling and Assessment, 6: 35-55.

[297] Zlatev, Z., Fenger, J. and Mortensen, L. (1996). *Relationships between emission sources and excess ozone concentrations.* Computers and Mathematics with Applications, 32(11): 101-123.

[298] Zlatev, Z., Geernaert, G. and Skov, H. (1999). *A study of ozone critical levels in Denmark.* EUROSAP Newsletter 36: 1-9.

[299] Zlatev, Z. and Syrakov, D. (2004). *A fine resolution modelling study of pollution levels in Bulgaria. Part 1: SO_x and NO_x pollution.* International Journal of Environment and Pollution, 22 (No. 1-2): 186-202.

[300] Zlatev, Z. and Syrakov, D. (2004). *A fine resolution modelling study of pollution levels in Bulgaria. Part 2: High ozone levels.* International Journal of Environment and Pollution, 22 (No. 1-2): 203-222.

Symbol Table

- $(x, y, z) \in \Omega \subset \mathbf{R}^3$ - point in $\mathbf{R}^3$
- $\mathbf{R}^3$ - 3-dimensional Euclidean space
- Ω - domains in the Euclidean space
- $t \in [0, T]$ - time
- Δt or τ - time-stepsize
- N_x, N_y, N_z - numbers of grid-points along the coordinate axes in the space domain
- $\| f \|$ - norm of f
- $A \in \mathbf{R}^{m \times n}$ - $m \times n$ matrix
- a_{ij} - element in the ith row and jth column of the matrix A
- x^T - transposed vector
- A^T - transposed matrix
- Λ - diagonal matrix containing the eigenvalues of $P^{-1}H$
- $g_n^{[i]}$ - approximation to the vector g at time-step n and at iteration i
- $J_k^{[0]}$ - Jacobian matrix of function f calculated at the Newton iterative procedure at time step k
- $\mathbf{F} = \{F_1, F_2, \ldots, F_m\}$ - predictor-corrector (PC) method (set of predictor-corrector schemes F_j)
- F_j, (where $j \in M$, $M = \{1, 2, \ldots, m\}$) - predictor-corrector (PC) scheme $PEC_1EC_2 \ldots C_{q_j}E$ in which P is some predictor formula, E denotes an evaluation of the right-hand-side of the equation and C_r, $r = 1, 2, \ldots, q_j$, are correctors

- $G_K^* = \{t_k | t_0 = a, t_k = t_{k-1} + \triangle t_k, \triangle t_k > 0, k = 1(1)K, t_K = b\}$ - non-equidistant grid
- $c_i, i = 1, \ldots, q$ - concentration of ith pollutant
- $N_s = q$ - number of pollutants
- u, v, w - wind velocities (projections of the wind field on axes x, y and z, correspondingly)
- K_x, K_y, K_z - diffusion coefficients (projections of the diffusion field on axes x, y and z, correspondingly)
- E_s - function describing the emission sources in the space domain
- κ_{1s}, κ_{2s} - deposition coefficients
- $Q_s(c_1, c_2, \ldots, c_q)$ - non-linear functions describing the chemical reactions used in the model
- $AOTc = \sum_{i=1}^{N} \max(c_i - c)$ - **A**ccumulated exposition **O**ver **t**hreshold of c $ppb \times hours$

Author Index

Subject Index

D

E